Andreas Meister
Jens Struckmeier

Hyperbolic
Partial Differential Equations

W0043839

Andreas Meister
Jens Struckmeier

Hyperbolic Partial Differential Equations

Theory, Numerics and Applications

vieweg

Hochschuldozent Dr. Andreas Meister
Medizinische Universität zu Lübeck
Institut für Mathematik
Wallstraße 40
D-23560 Lübeck, Germany

Professor Dr. Jens Struckmeier
Universität Hamburg
Fachbereich Mathematik
Bundesstraße 55
D-20146 Hamburg, Germany

meister@math.mu-luebeck.de
struckmeier@math.uni-hamburg.de

Die Deutsche Bibliothek – CIP-Cataloguing-in-Publication-Data
A catalogue record for this publication is available from
Die Deutsche Bibliothek.

First edition, March 2002

Vieweg is a company in the specialist publishing group BertelsmannSpringer.
www.vieweg.de

Cover design: Ulrike Weigel, www.CorporateDesignGroup.de

Printed on acid-free paper

ISBN-13:978-3-322-80229-3   e-ISBN-13:978-3-322-80227-9
DOI: 10.1007/978-3-322-80227-9

Preface

The following chapters summarize lectures given in March 2001 during the summerschool on **Hyperbolic Partial Differential Equations** which took place at the Technical University of Hamburg–Harburg in Germany. This type of meeting is originally funded by the Volkswagenstiftung in Hannover (Germany) with the aim to bring together well–known leading experts from special mathematical, physical and engineering fields of interest with PhD–students, members of Scientific Research Institutes as well as people from Industry, in order to learn and discuss modern theoretical and numerical developments.

Hyperbolic partial differential equations play an important role in various applications from natural sciences and engineering. Starting from the classical Euler equations in fluid dynamics, several other hyperbolic equations arise in traffic flow problems, acoustics, radiation transfer, crystal growth etc. The main interest is concerned with nonlinear hyperbolic problems and the special structures, which are characteristic for solutions of these equations, like shock and rarefaction waves as well as entropy solutions. As a consequence, even numerical schemes for hyperbolic equations differ significantly from methods for elliptic and parabolic equations: the transport of information runs along the characteristic curves of a hyperbolic equation and consequently the direction of transport is of constitutive importance. This property leads to the construction of upwind schemes and the theory of Riemann solvers. Both concepts are combined with explicit or implicit time stepping techniques whereby the chosen order of accuracy usually depends on the expected dynamic of the underlying solution.

Numerical schemes for hyperbolic equations should always satisfy two contrary goals:

- smooth solutions should be approximated with a high spatial accuracy and

- discontinuities should be resolved as sharp as possible.

These two opposite purposes might even be found for scalar conservation laws in one space dimension, like the inviscid Burgers equation. Hence, even on an elementary level, it is possible to introduce to non–experts crucial concepts like stability, monotonicity or the TVD–property (**Total Variation Dimishing**). A quite nice concept is the classification of numerical schemes only using a numerical flux function, which approximates the flux function defined by the given hyperbolic equation.

In the current book, all concepts mentioned above are discussed. Starting from basic ideas on conservation principles to model various problems from natural and engineering science, the lecturers develop the basic concepts from numerical analysis up to topics which are actually treated by researchers working on numerical schemes for hyperbolic equations.

Exactly this concept the editors had in mind when planning the summerschool. The aim was to build a bridge from the elementary modeling of various phenomena using scalar conservation laws up to the newest developments, like multi–scale analysis for low–Mach number flows and the combination of computational fluid dynamics with accoustics. In order to be more concrete, the five chapters of this book contain contributions from (in alphabetical order)

- Michael Junk on Hyperbolic conservation laws and industrial applications

- Claus–Dieter Munz on Computational fluid dynamics and aeroacoustics for low mach number

- Milovan Perić, Ismet Demirdžić and Samir Muzaferija on Pressure–correction methods for all flow speeds

- Giovanni Russo on Central schemes and systems of balance laws

- Thomas Sonar on Methods on unstructured grids including WENO and ENO recovery techniques

The first chapter covers the theoretical aspects of conservation laws. Starting from initial and boundary value problems for scalar equations, the concept of weak solutions and entropy up to the theory of systems of equations are introduced. Additionally, the first chapter contains a lot of physical and technical examples modeled by conservation laws. The second chapter is devoted to the description of higher order central schemes on rectangular grids using CWENO (Central-Weighted Essentially Non-Oscillatory) reconstruction techniques. Beside the scalar case also systems of conservation laws are considered and the treatment of stiff source terms is discussed. The objective of the third chapter is the development of finite volume approximations of hyperbolic conservation laws on unstructured grids. Thereby, primary as well as secondary grid methods are presented and different time stepping schemes are explained. Special attention is laid on ENO- and WENO-recovery techniques and grid adaptivity. The main principles of various pressure correction methods like SIMPLE, SIMPLEC and PISO are outlined in chapter 4. Therein, a particular finite volume scheme for the discretization of the Navier-Stokes equations is described. The applicability of the method for the simulation of flow fields at all speeds is demonstrated by means of different laminar and turbulent test cases. Numerical schemes for low Mach number flows based on an asymptotic single as well as multiple scale analysis are considered in the concluding fifth chapter. Thereby, the compressible pressure correction approach and the multiple pressure variable method are outlined. Furthermore, a comprehensive description concerning the coupling of methods for fluid calculation with schemes for sound propagation due to acoustic waves is presented.

First of all, the editors would like to thank the Volkswagen-Stiftung in Hannover, which financially supported the summerschool in the context of the DMV–GAMM–GI Fachgruppe Numerische Software. Next this book would be impossible without the contributions of the lecturers of the school. Hence, we would like to thank Michael Junk (University of Kaiserslautern), Claus–Dieter Munz (University of Stuttgart), Milovan Perić (Technical University of Hamburg-Harburg), Giovanni Russo (University of Catania, Italy) and Thomas Sonar (Technical University of Braunschweig) for their impressive talks as well as their fascinating and well-written contributions. Moreover, we are also grateful to Wolfgang Mackens, Professor at the Technical University of Hamburg–Harburg for initiating a large number of summerschools, for giving us the possibility to realize this course and for his collaboration and help during the organization and execution. Finally, we would like to thank Frau Schmickler–Hirzebruch from Vieweg publishers, who was the responsible person to edit the book.

Hamburg, November 2001   Andreas Meister and Jens Struckmeier.

Contents

1 Hyperbolic Conservation Laws and Industrial Applications

1.1 Transport theorem and balance laws

The starting point of many mathematical models in science and technology are balance considerations of the physical quantities mass, momentum, and energy. As examples we will present the Euler and Navier-Stokes equations, the Saint-Venant equations, and kinetic equations like the Vlasov or Boltzmann equations.

In other cases, one is interested in the number distribution of certain objects (cars in traffic models, individuals in population models, or silver halide crystals in the production of photographic emulsions), and again, balance considerations are the key for setting up a suitable model.

1.1.1 The flux vector

An important concept in the derivation of balance laws is the *flux vector* $q(t,x)$. Let us consider a situation in which mass is transported. This process is completely described by the mass flux $q(t,x)$ which provides the amount of mass m that is transported through a surface element ΔS with normal n at a point x during a time interval $[t, t + \Delta t]$

$$m = n \cdot q(t,x)\Delta S \Delta t.$$

Note that the flux is maximal for n parallel to q (see Fig. 1.1), $m = 0$ if q is tangential to the surface, and $m < 0$ if the direction of transport is opposite to n.

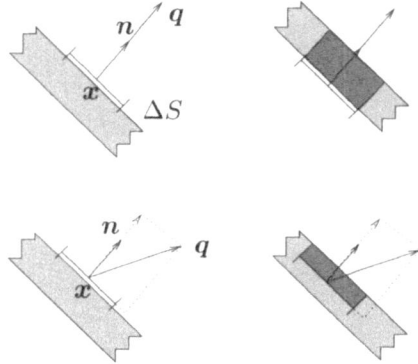

Figure 1.1 Flux through a surface element ΔS with different directions of q

Let us now consider a volume A which contains the mass $M_t(A)$ at time t. The increase in mass $\Delta M_t(A) = M_{t+\Delta t}(A) - M_t(A)$ in a small tie interval of length Δt can be due to a flux through ∂A or due to a mass production (or consumption) within A.

Assuming that we know q, the total flux through the surface is obtained by integrating contributions $\Delta t(-n) \cdot q\,dS$ along the surface (the negative sign accounts for the fact that n is the outward normal field so that the mass in A increases if q points in direction $-n$ – see Fig. 1.2).

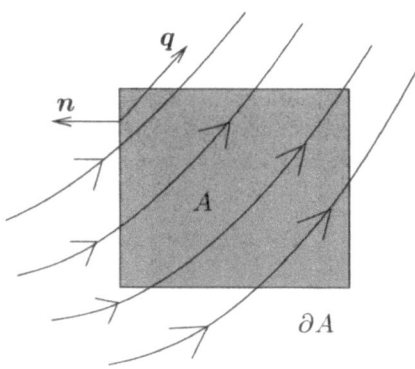

Figure 1.2   Mass increase due to flux through the surface

If $Q(t,x)$ is the rate of mass production per volume (which is negative if mass disappears), we obtain the balance relation

$$\Delta M_t(A) = \Delta t \int_{\partial A} (-n) \cdot q\,dS + \Delta t \int_A Q\,dx$$

or in the limit $\Delta t \to 0$

$$\frac{dM_t(A)}{dt} = -\int_{\partial A} n \cdot q\,dS. + \int_A Q\,dx.$$

If the measure $A \mapsto M_t(A)$ has a density u with respect to Lebesgue measure, we obtain the integral form of the balance law

$$\frac{d}{dt} \int_A u\,dx + \int_{\partial A} n \cdot q\,dS - \int_A Q\,dx = 0$$

which holds for all (sufficiently regular) volumes A. If, in addition, q is smooth, we can apply the divergence theorem to transform the surface integral into a volume integral. Dividing by the volume $|A|$ of A and assuming that u is differentiable with respect to t, we find

$$\frac{1}{|A|} \int_A \frac{\partial u}{\partial t} + \operatorname{div} q - Q\,dx = 0.$$

Hence, for a sequence $A_{n,x}$ of volumes which shrinks to a point x for $n \to \infty$, the mean value theorem implies

$$\frac{\partial u}{\partial t} + \operatorname{div}\boldsymbol{q} - Q = 0. \tag{1.1}$$

The balance law (1.1) is called *conservation law*, if $Q \equiv 0$, i.e. if mass is neither produced nor destroyed.

Introducing the space–time variable $\boldsymbol{y} = (t,\boldsymbol{x})$, we can rewrite (1.1) in a simple divergence form

$$\operatorname{div}_{\boldsymbol{y}}\mathcal{A} = Q, \qquad \mathcal{A} = (u,\boldsymbol{q}) \tag{1.2}$$

which will be very useful later.

Up to know, we have assumed that we know both the flux vector $\boldsymbol{q}$ and the source term Q. In a specific application, the modeling therefore consists in deriving expressions for $\boldsymbol{q}$ and Q. This modeling step typically requires a lot of insight into the problem because the structure of $\boldsymbol{q}$ and Q will not be given by general rules. However, if the flux of the balanced quantity is partly caused by a spatial motion, the *transport theorem* helps to specify $\boldsymbol{q}$.

1.1.2 The transport theorem

We describe a flow in the space $\mathbb{R}^d$ by the flow map $\Pi : \mathbb{R}^+ \times \mathbb{R}^d \to \mathbb{R}^d$ which assigns to a point $\boldsymbol{x} \in \mathbb{R}^d$ its location $\Pi(t,\boldsymbol{x}) = \Pi_t(\boldsymbol{x})$ after following the flow for a time $t \in \mathbb{R}^+$ (see Fig. 1.3).

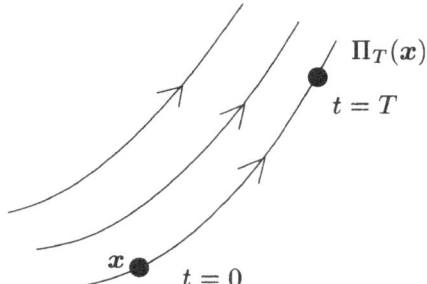

Figure 1.3 Action of the flow map Π

The flow velocity $\boldsymbol{a}$ associated to this movement is given by the partial time derivative of Π

$$\boldsymbol{a}(t,\Pi_t(\boldsymbol{x})) = \frac{\partial \Pi_t(\boldsymbol{x})}{\partial t}.$$

In many practical cases, the modeling step yields an expression for the flow velocity $\boldsymbol{a}$ and $\Pi_t(\boldsymbol{x}) = \boldsymbol{z}(t)$ is then defined as solution of the ordinary differential equation $\dot{\boldsymbol{z}}(t) = \boldsymbol{a}(t,\boldsymbol{z}(t))$, $\boldsymbol{z}(0) = \boldsymbol{x}$.

In analogy to our considerations in the previous section, we want to set up a balance relation but now for volumes $\Pi_t(A)$ which move with the flow. An expression for the rate of change of $M_t(\Pi_t(A))$ is then given by the transport theorem:

Theorem 1   *Let* $\Pi \in C^1(\mathbb{R}^+ \times \mathbb{R}^d)$, *and* $\Pi_t : \mathbb{R}^d \to \mathbb{R}^d$ *be injective with* $\det(\nabla \Pi_t) > 0$ *for every* $t \in \mathbb{R}^+$. *Assume further that*

$$M_t(A) = \int_A u(t,\boldsymbol{x}) \, d\boldsymbol{x}$$

is a measure with density $u \in C^1(\mathbb{R}^+ \times \mathbb{R}^d)$ *and that* A *is an open bounded set with a* C^1 *-boundary. Then*

$$\frac{d}{dt} M_t(\Pi_t(A)) = \int_{\Pi_t(A)} \frac{\partial u}{\partial t} + \mathrm{div}(u\boldsymbol{a}) \, d\boldsymbol{y} \tag{1.3}$$

where $\boldsymbol{a}(t, \Pi_t(\boldsymbol{x})) = \frac{\partial \Pi_t(\boldsymbol{x})}{\partial t}$ *is the flow velocity.*

Proof: We only present the main idea of the proof and refer to [2] for details. First, we introduce the change of variables $\boldsymbol{y} = \Pi_t(\boldsymbol{x})$ so that

$$M_t(\Pi_t(A)) = \int_A u(t, \Pi_t(\boldsymbol{x})) \det D_t(\boldsymbol{x}) \, d\boldsymbol{x}$$

where $D_t(\boldsymbol{x})$ has been introduced as abbreviation for $\nabla \Pi_t(\boldsymbol{x})$. Pulling the time derivative under the integral and observing the definition of the flow velocity, we obtain with product and chain rule

$$\frac{d}{dt} M_t(\Pi_t(A)) = \int_A \left(\frac{\partial u}{\partial t} + \boldsymbol{a}\nabla u \right) \det D_t + u \frac{\partial}{\partial t} \det D_t \, d\boldsymbol{x} \tag{1.4}$$

For every fixed $\boldsymbol{x} \in A$, we now consider $D_t = D_t(\boldsymbol{x})$ as an invertible matrix which smoothly depends on a parameter $t \in \mathbb{R}^+$. Using the definition of the determinant, one can show that in such a case

$$\frac{d}{dt} \det D_t = \mathrm{tr}(\dot{D}_t D_t^{-1}) \det D_t \tag{1.5}$$

where tr denotes the trace of a matrix. Using now the special form of $D_t = \nabla \Pi_t$ it turns out that

$$\mathrm{tr}(\dot{D}_t D_t^{-1}) = \mathrm{div}\boldsymbol{a}. \tag{1.6}$$

The result follows by inserting (1.5) and (1.6) in (1.4) and transforming the variables back with $\boldsymbol{x} = \Pi_t^{-1}(\boldsymbol{y})$. ∎

With the expression (1.3) for the rate of change of mass in a transported volume, we now set up a balance equation. If Q describes again the rate of mass production per volume and $\tilde{q}$ is the mass flux *relative to the movement*, then the balance law reads

$$\frac{d}{dt} M_t(\Pi_t(A)) = \int_{\partial \Pi_t(A)} (-\boldsymbol{n}) \cdot \tilde{q} \, dS + \int_{\Pi_t(A)} Q \, d\boldsymbol{x}.$$

Using the transport theorem and letting A shrink to a point, we arrive at the differential form of the balance law

$$\frac{\partial u}{\partial t} + \operatorname{div}(u\boldsymbol{a} + \tilde{\boldsymbol{q}}) = Q. \tag{1.7}$$

In particular, if the quantity u under consideration is conserved $(Q = 0)$ and if there are no mechanisms leading to a flux apart from the flow Π, the conservation law is fully determined by the flow velocity

$$\frac{\partial u}{\partial t} + \operatorname{div}(u\boldsymbol{a}) = 0.$$

In the following sections, we see several examples which are based on these assumptions.

1.1.3 Continuum hypothesis

Very often, the assumption that $M_t(A)$ has a density $u(t,\boldsymbol{x})$ with respect to Lebesgue measure is violated. To give a concrete example, let us consider a model of a large population with a typical size of $\bar{N} = 10^8$ individuals. We assume that $N_t(A)$ is the number of individuals with age $x \in A \subset \mathbb{R}^+$. Obviously, N_t is a discrete measure. In fact, N_t can be written as sum of Dirac measures

$$N_t = \sum_{i \in \mathbb{N}} \omega_i(t)\delta_{x_i(t)}$$

where $x_i(t)$ is the age of individual i at time t and $\omega_i(t) \in \{0,1\}$ indicates whether the individual is alive $(\omega_i(t) = 1)$, or dead $(\omega_i(t) = 0)$. Since the Dirac measure has no density with respect to the Lebesgue measure, the same holds for N_t and we cannot proceed as in section 1.1.1 and 1.1.2. However, taking into account that the total population is quite large, it makes sense to approximate N_t by a continuous measure. To illustrate the idea, we consider the cumulative distribution function $x \mapsto N_t([0,x])$ which is a piecewise constant function with jumps at the points $x_i(t)$ and typical jump height one (see fig. 1.4(a)).

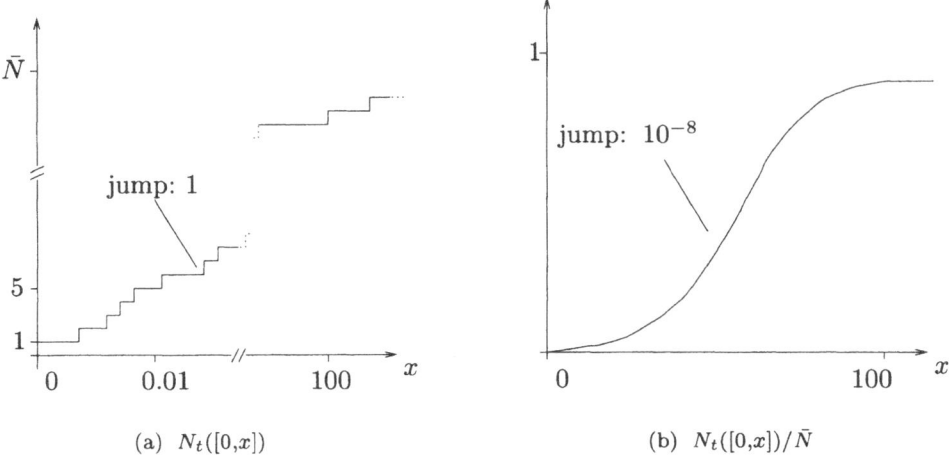

(a) $N_t([0,x])$ (b) $N_t([0,x])/\bar{N}$

Figure 1.4 Unscaled and scaled cumulative age distribution function

The scaled measure $N_t/\bar{N}$, on the other hand, leads to a distribution function with only very small jumps $\bar{N}^{-1}$ and for this function one can postulate the *continuum hypothesis:*

$N_t/\bar{N}$ can be approximated by a measure M_t with (smooth) density $u(t,x)$.

Note that the approximation M_t is a good substitute for $N_t/\bar{N}$ only on sets with relatively large N_t measure. This should be kept in mind when interpreting results derived with u in terms of the underlying population.

1.1.4 Examples of balance laws

Population model

Applying the continuum hypothesis of the previous section, we introduce

$$M_t(I) = \int_I u(t,x)\, dx$$

as the number of individuals with age $x \in I \subset \mathbb{R}^+$. The only reason for a flow in the "age-space" $\mathbb{R}^+$ is the aging process: an individual with age x at time 0 has age $x + t$ at time t. This gives rise to $\Pi_t(x) = x + t$ and the associated flow speed

$$a(t,x) = \frac{\partial \Pi_t}{\partial t} = 1.$$

The lack of any further flux mechanism implies $\tilde{q} = 0$ in the general balance law (1.7). Since individuals can "disappear" at any age with a certain age dependent death rate $\mu(t,x)$, we obtain the negative production rate $Q(t,x) = -\mu(t,x)u(t,x)$, giving rise to the linear balance law

$$\frac{\partial u}{\partial t} + \frac{\partial u}{\partial x} = -\mu u \tag{1.8}$$

The domain of definition of (1.8) is the (t,x) quarter plane $\Omega = \mathbb{R}^+ \times \mathbb{R}^+$. Natural conditions at the boundaries are $u(0,x) = u^0(x)$ for the condition at $t = 0$ (for example, the situation after a census) and

$$u(t,0) = \int_0^\infty \lambda(t,x)u(t,x)\, dx$$

models the creation of individuals with age $x = 0$ based on an age dependent fertility rate $\lambda(t,x) \geq 0$. This population model is known as Von Foerster model (see [11]).

Traffic model

Applying the continuum hypothesis (heavy traffic), we introduce

$$M_t(I) = \int_I u(t,x)\, dx$$

as the number of vehicles in an interval I of the single lane one-way street $\mathbb{R}$. The only reason for a transport along the road is the movement of the vehicles with velocity $a(t,x)$. In a simple model, we assume that $a(t,x)$ only depends on the local traffic density, i.e. $a(t,x) = V(u(t,x))$ where V is determined from measurements and typically has a shape as indicated in fig. 1.5.

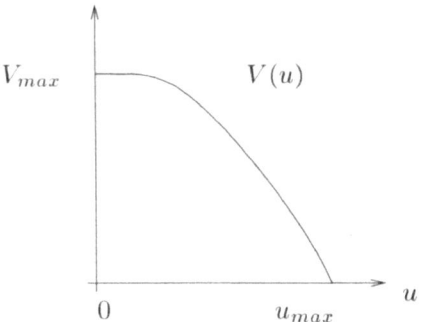

Figure 1.5 Typical dependence of vehicle velocity on traffic density

As additional simplification, we assume that the road has no entrance or exit which yields $Q = 0$. Altogether, we obtain the nonlinear conservation law

$$\frac{\partial u}{\partial t} + \frac{\partial}{\partial x}(uV(u)) = 0 \tag{1.9}$$

with domain of definition $\Omega = \mathbb{R}^+ \times \mathbb{R}$. More complicated models are used to predict traffic situations and to run traffic control systems as shown in fig. 1.6.

Figure 1.6 Traffic control systems are based on mathematical traffic models

While the assumption of a density dependent vehicle velocity is quite intuitive, it can lead to a behavior which is not observed in real traffic flow. The reason is that drivers choose their speed not only based on the local density but also on the rate of change of this density. For example, if the density increases in front, the velocity is already reduced. This behavior gives rise to a more realistic flux

$$q = uV(u) - \alpha\frac{\partial u}{\partial x}, \quad \alpha > 0$$

The corresponding balance equation

$$\frac{\partial u}{\partial t} + \frac{\partial}{\partial x}(uV(u)) = \alpha\frac{\partial^2 u}{\partial x^2} \tag{1.10}$$

is of parabolic type and formally reduces to the hyperbolic model (1.9) for $\alpha \to 0$. Later, when we discuss the concept of weak solutions, the refined parabolic model will be useful to single out the correct (i.e. realistic) weak solution of (1.9).

Before going over to the next example, let us briefly focus on the concept of *linearization*: if the vehicle density only slightly deviates from a constant density $\bar{u}(t,x) = c$, one can approximate the nonlinear model (1.9) by a simpler, linear equation. Formally, the linearization process consists of the following steps: first, the equation is written in terms of a function F depending on u and derivatives of u

$$0 = \frac{\partial u}{\partial t} + q'(u)\frac{\partial u}{\partial x} = F\left(u, \frac{\partial u}{\partial t}, \frac{\partial u}{\partial x}\right)$$

with $F(w_0,w_1,w_2) = w_1 + q'(w_0)w_2$ and $w \in \mathbb{R}^3$. Then, the idea is to replace the nonlinear function F by its linearization

$$F(w) \approx F(\bar{w}) + (w - \bar{w}) \cdot \nabla_w F(\bar{w})$$

Substituting w and $\bar{w}$ by

$$w = (u, \frac{\partial u}{\partial t}, \frac{\partial u}{\partial x}), \qquad \bar{w} = (\bar{u}, \frac{\partial\bar{u}}{\partial t}, \frac{\partial\bar{u}}{\partial x})$$

and using the fact that $\bar{u}$ is a specific solution of (1.9) so that $F(\bar{w}) = 0$, we arrive at the linear equation $(w - \bar{w}) \cdot \nabla_w F(\bar{w}) = 0$. In the case under consideration, we have $\bar{w} = (c,0,0)$, $\nabla_w F(\bar{w}) = (0,1,q'(c))$, and hence

$$\frac{\partial u}{\partial t} + q'(c)\frac{\partial u}{\partial x} = 0 \tag{1.11}$$

as linearization of (1.9).

Kinetic equations

We consider a large number of microscopic particles which behave according to the laws of classical mechanics. In particular, the movement of a single particle in the force field $F(x)$ is determined by Newton's law which we write in the form

$$\dot{x} = v, \qquad \dot{v} = F(x).$$

The distribution of the particles in physical space as well as in velocity space is given by the measure

$$M_t(A \times B) = \int_A \int_B f(t,x,v)\, dv dx$$

which describes the mass of particles in $A \subset \mathbb{R}^3$ having velocities in $B \subset \mathbb{R}^3$. Introducing the phase space variable $\boldsymbol{y} = (\boldsymbol{x}, \boldsymbol{v})$, the flow velocity which describes the movement of the particles in phase space is given by

$$\boldsymbol{a}(\boldsymbol{y}) = (\boldsymbol{v}, \boldsymbol{F}(\boldsymbol{x})), \qquad \boldsymbol{y} = (\boldsymbol{x}, \boldsymbol{v}).$$

If there is no additional flux but a source or sink term Q, we obtain the balance law

$$\frac{\partial f}{\partial t} + \operatorname{div}(f\boldsymbol{a}) = Q. \tag{1.12}$$

Since $\operatorname{div}_{\boldsymbol{y}}(f\boldsymbol{a}) = \boldsymbol{a} \cdot \nabla_{\boldsymbol{y}} f + f \operatorname{div}_{\boldsymbol{y}} \boldsymbol{a}$ and $\operatorname{div}_{\boldsymbol{y}} \boldsymbol{a} = \operatorname{div}_{\boldsymbol{x}} \boldsymbol{v} + \operatorname{div}_{\boldsymbol{v}} \boldsymbol{F}(\boldsymbol{x}) = 0$, we can rewrite (1.12) in the form (with $Q = 0$)

$$\frac{\partial f}{\partial t} + \boldsymbol{v} \cdot \nabla_{\boldsymbol{x}} f + \boldsymbol{F}(\boldsymbol{x}) \cdot \nabla_{\boldsymbol{v}} f = 0 \tag{1.13}$$

which is called *Vlasov equation*. Typically, the force field $\boldsymbol{F}$ is generated by the particles themselves (electrical or gravitational field) which gives rise to an additional equation

$$\boldsymbol{F} = -\nabla_{\boldsymbol{x}} \phi, \quad \Delta\phi = \pm \int_{\mathbb{R}^3} f \, d\boldsymbol{v}$$

Important industrial applications of the Vlasov equation arise in the design of systems which use plasma (like plasma engines in space vehicles).

If the particle interactions result from short range forces, it is useful to avoid a detailed description of the force field $\boldsymbol{F}$ and model the interactions with a source term Q. To illustrate this approach, let us consider the case in which two particles with velocities $\boldsymbol{v}'$ and $\boldsymbol{v}'_*$ collide in a point $\boldsymbol{x}$ which means that the velocities instantaneously change into outgoing velocities $\boldsymbol{v}$ and $\boldsymbol{v}_*$ (see fig. 1.7).

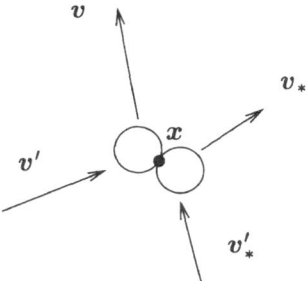

Figure 1.7 Collision of particles

In other words, the particles characterized by $(\boldsymbol{x}, \boldsymbol{v}')$ and $(\boldsymbol{x}, \boldsymbol{v}'_*)$ disappear, and the pair $(\boldsymbol{x}, \boldsymbol{v})$ and $(\boldsymbol{x}, \boldsymbol{v}_*)$ suddenly appears. Without going further into details, we mention that the corresponding source term has the form

$$Q(f) = \int_{\mathbb{R}^3} \int_{S_+^2} B(\boldsymbol{v}, \boldsymbol{v}_*, \boldsymbol{\xi})(f' f'_* - f f_*) \, d\boldsymbol{\xi} \, d\boldsymbol{v}_* \tag{1.14}$$

where f',f'_*,f_* are equal to f evaluated at the velocities v',v'_*,v_*, where v',v'_* depend on v,v_* and the vector $\xi \in S^2_+$ (for details see [1]). Using (1.14) together with the flow velocity $a(,v) = (v,0)$, we obtain the balance law

$$\frac{\partial f}{\partial t} + v \cdot \nabla_x f = Q(f)$$

which is the *Boltzmann equation*. Important applications of this equation are rarefied gas flows which appear, for example, during reentry of space vehicles.

Euler equation

The Euler equation is the paradigmatic case of a hyperbolic system of balance laws. Since it describes the flow of gases, it obviously has many technical applications.

The balanced quantities are mass, momentum, and energy with corresponding densities $\rho,\rho u,\rho E$. The field $u = \rho u/\rho$ is the flow velocity of the gas. If the flow with velocity u is the only reason for a mass transport (no mass diffusion), and if there are no mass sources, we immediately deduce the balance law for ρ

$$\frac{\partial \rho}{\partial t} + \operatorname{div}(\rho u) = 0 \tag{1.15}$$

The momentum balance is derived with Newton's law which relates the rate of change of momentum of a control volume to forces acting on the volume and its surface. Introducing the stress tensor σ, the force on a surface element dS of the control volume with normal n is $\sigma n \, dS$. In particular, the ith component of the surface force is given by $\sigma_i \cdot n$, where $\sigma_i = (\sigma_{i,1},\sigma_{i,2},\sigma_{i,3})$, so that $-\sigma_i$ acts as additional flux in the balance law for the ith component of ρu. Volume forces which act on every point inside the control volume are incorporated as source term. For example, gravitational forces give rise to the source term ρg where g is the gravitational acceleration. Altogether, the balance of forces yields

$$\frac{\partial \rho u_i}{\partial t} + \operatorname{div}(\rho u_i u) = \operatorname{div}\sigma_i + \rho g_i, \qquad i = 1,2,3. \tag{1.16}$$

Since the forces acting on the control volume provide or require energy, each force term in the momentum equation leads to a corresponding term in the energy equation. Additional contributions arise from a possible heat flux q and a heat source Q. The energy balance equation has the form

$$\frac{\partial \rho E}{\partial t} + \operatorname{div}(\rho E u) = \operatorname{div}(\sigma u) + \rho g \cdot u - \operatorname{div}q + Q \tag{1.17}$$

Since equations (1.15), (1.16),(1.17) contain more unknowns than equations, the system is not closed and additional closure relations are required. For the Euler equation of an ideal gas, one assumes that the surface forces are only due to pressure $\sigma_{ij} = -p\delta_{ij}$, where pressure is related to temperature $p = \rho RT$, and temperature to energy via $E = |u|^2/2 + RT/(\gamma - 1)$ (R is the universal gas constant and γ is the ratio of heat capacities at constant pressure and constant volume). Heat flux and heat production are typically neglected $q = 0, Q = 0$.

As in the case of the traffic model, the hyperbolic system is only a first model. A more refined approach is given by the Navier-Stokes-Fourier closure, where also friction forces are taken into account and where heat flux is caused by a temperature gradient

$$\sigma_{ij} = -p\delta_{ij} + \mu \left(\frac{\partial u_i}{\partial x_j} + \frac{\partial u_j}{\partial x_i} \right), \qquad q = -\lambda \nabla T.$$

The parameters μ and λ are viscosity and heat conduction coefficients respectively. Equations (1.15), (1.16),(1.17) with these closure relations are called *compressible Navier-Stokes equations*.

Apart from density, pressure, and temperature, another important thermodynamical variable is the entropy ρS

$$\rho S = -\rho \ln \left(\frac{T}{\rho^{\gamma-1}} \right)$$

which is a convex function in terms of the variables $(\rho, \rho u, \rho E)$ of the Euler system. It turns out that for smooth solutions of the Euler system, ρS automatically satisfies the conservation law

$$\frac{\partial \rho S}{\partial t} + \mathrm{div}(\rho S u) = 0 \tag{1.18}$$

which means that entropy is just transported with the velocity u. For more general thermodynamical processes (e.g. piecewise smooth solutions of the Euler system with jumps in the variables), one physically expects that the entropy can also decrease (note that we have chosen the negative logarithm in the definition of S), so that a physically more general entropy equation is given by

$$\frac{\partial \rho S}{\partial t} + \mathrm{div}(\rho S u) \leq 0. \tag{1.19}$$

Starting with the Navier-Stokes-Fourier closure, one can, in fact, derive (1.19) by letting μ, λ tend to zero in the equation for the entropy. It turns out that the sign condition in (1.19) is very important to single out the physically correct solution of the Euler system once the solution is not smooth. Another observation is that relations similar to (1.19) are obtained when ρS is replaced by a more general expression of the form

$$\eta_\phi = \rho \phi \left(\frac{T}{\rho^{\gamma-1}} \right)$$

where $\phi : \mathrm{I\!R}^+ \to \mathrm{I\!R}$ satisfies $\phi' \leq 0$, $\phi'' \geq 0$. Starting again from the Navier-Stokes-closure, we obtain an evolution equation for the quantity η_ϕ which depends on the unknowns $\rho, \rho u$, an ρE. Letting μ, λ tend to zero in this equation, one ends up with

$$\frac{\partial \eta_\phi}{\partial t} + \mathrm{div}(\eta_\phi u) \leq 0 \tag{1.20}$$

which are additional conditions on the physically correct solution of the Euler system (which is the limit system of the Navier-Stokes equation for $\mu \searrow 0$). Although η_ϕ is, strictly speaking, not an entropy, one nevertheless calls η_ϕ a (Lax-) entropy and (1.20) an *entropy condition*. The notion of entropy plays an important role in the theory of hyperbolic conservation laws.

Restricting to the case of smooth solutions of the Euler system and replacing the variable E by the entropy S, equation (1.17) (with the Euler closure and $\boldsymbol{g} = 0$) turns into (1.18) (remember that smooth solutions of the Euler system satisfy (1.18) automatically). Assuming moreover that S is initially constant, equation (1.18) in connection with (1.15) implies that S stays constant for all times. Hence, the Euler system reduces to the mass and momentum equation in this *isentropic case* and $p = \rho RT$ implies that the pressure is a function of density and the constant entropy (via T). More precisely, the isentropic Euler system has the form

$$\frac{\partial \rho}{\partial t} + \operatorname{div}(\rho \boldsymbol{u}) = 0,$$

$$\frac{\partial \rho u_i}{\partial t} + \operatorname{div}(\rho u_i \boldsymbol{u}) + \frac{\partial p}{\partial x_i} = 0, \qquad p(\rho) = C\rho^\gamma.$$

Especially in applications where weak acoustic fields are to be calculated, it is reasonable to linearize the isentropic Euler system around the specific solution $\bar{\rho}(t,\boldsymbol{x}) = \rho_0$, $\bar{\boldsymbol{u}}(t,\boldsymbol{x}) = 0$. The linearization process works as in the case of the traffic flow model and we will explain it only briefly for the mass conservation equation (here we use Einstein's summation convention for repeated indices)

$$0 = \frac{\partial \rho}{\partial t} + u_j \frac{\partial \rho}{\partial x_j} + \rho \frac{\partial u_j}{\partial x_j} = F(\rho,\boldsymbol{u},\partial_t\rho,\nabla\rho,\operatorname{div}\boldsymbol{u})$$

where $F(w_0,\boldsymbol{w}_1,w_2,\boldsymbol{w}_3,w_4) = w_2 + \boldsymbol{w}_1 \cdot \boldsymbol{w}_3 + w_0 w_4$ with $w_0,w_2,w_4 \in \mathbb{R}$ and $\boldsymbol{w}_1,\boldsymbol{w}_3 \in \mathbb{R}^3$. Replacing f in this equation by its linearization at the point

$$\bar{\boldsymbol{w}} = (\bar{\rho},\bar{\boldsymbol{u}},\partial_t\bar{\rho},\nabla\bar{\rho},\operatorname{div}\bar{\boldsymbol{u}}) = (\rho_0,0,0,0,0)$$

we obtain the linear problem

$$\frac{\partial \rho}{\partial t} + \rho_0 \operatorname{div}\boldsymbol{u} = 0 \tag{1.21}$$

Similarly, the momentum conservation equation turns into a linear equation for the velocity

$$\frac{\partial \boldsymbol{u}}{\partial t} + \frac{p'(\rho_0)}{\rho_0}\nabla\rho = 0 \tag{1.22}$$

If the density ρ in (1.21) and (1.22) is replaced by a scaled, linear pressure $p = (p(\rho_0) + p'(\rho_0)\rho)/(p'(\rho_0)\rho_0)$, one obtains the equations of acoustics

$$\frac{\partial p}{\partial t} + \operatorname{div}\boldsymbol{u} = 0, \qquad \frac{\partial \boldsymbol{u}}{\partial t} + p'(\rho_0)\nabla p = 0 \tag{1.23}$$

which will be discussed in detail at the end of section 1.2.

1.2   Linear initial and boundary value problems

1.2.1   The method of characteristics

Recalling the general form of a scalar balance law, we have

$$\frac{\partial u}{\partial t} + \mathrm{div}_{\boldsymbol{x}}\boldsymbol{q} = Q,$$

or with the combined variable $\boldsymbol{y} = (t,\boldsymbol{x})$ and the combined flux $\mathcal{A}(\boldsymbol{y},v) = (v,\boldsymbol{q}(\boldsymbol{y},v))$

$$\mathrm{div}_{\boldsymbol{y}}\mathcal{A}(\boldsymbol{y},u(\boldsymbol{y})) = Q(\boldsymbol{y},u(\boldsymbol{y})). \tag{1.24}$$

Equation (1.24) is linear if $\mathcal{A}$ and Q are (affine-) linear in u, i.e.

$$\mathcal{A}(\boldsymbol{y},v) = \boldsymbol{a}(\boldsymbol{y})v, \qquad Q(\boldsymbol{y},v) = c(\boldsymbol{y}) + d(\boldsymbol{y})v.$$

Denoting $b(\boldsymbol{y}) = -\mathrm{div}_{\boldsymbol{y}}\boldsymbol{a}(\boldsymbol{y}) + d(\boldsymbol{y})$, we obtain

$$\boldsymbol{a}(\boldsymbol{y}) \cdot \nabla_{\boldsymbol{y}} u(\boldsymbol{y}) = b(\boldsymbol{y})u(\boldsymbol{y}) + c(\boldsymbol{y}), \qquad \boldsymbol{y} \in \Omega. \tag{1.25}$$

We will consider such problems in connection with boundary values

$$u(\boldsymbol{y}) = \phi(\boldsymbol{y}) \quad \text{for } \boldsymbol{y} \in \Gamma \subset \partial\Omega. \tag{1.26}$$

We remark that $\boldsymbol{a} \cdot \nabla$ in (1.25) is a *directional derivative* in direction $\boldsymbol{a}$ so that $\boldsymbol{a} \cdot \nabla u = bu + c$ describes the rate of change of u along integral curves of the field $\boldsymbol{a}$ (i.e. solution curves of the ODE $\boldsymbol{z}'(s) = \boldsymbol{a}(\boldsymbol{z}(s))$ – see fig. 1.8).

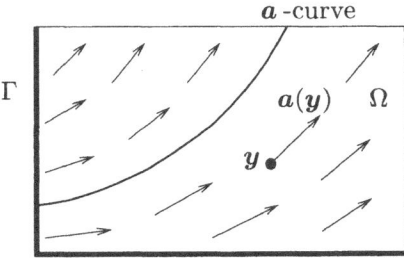

Figure 1.8 Integral curve of the vector field $\boldsymbol{a}$

More precisely, if $\boldsymbol{z}(s)$ is the solution of the ordinary differential equation $\boldsymbol{z}'(s) = \boldsymbol{a}(\boldsymbol{z}(s))$, we find

$$\frac{d}{ds}u(\boldsymbol{z}(s)) = \nabla u(\boldsymbol{z}(s)) \cdot \boldsymbol{a}(\boldsymbol{z}(s)) = b(\boldsymbol{z}(s))u(\boldsymbol{z}(s)) + c(\boldsymbol{z}(s)).$$

This simple observation is the basis for a solution method which is called the *method of characteristics*: In order to find a solution of equation (1.25) together with the boundary condition (1.26) in a point $\boldsymbol{y} \in \Omega$, we first have to find an $\boldsymbol{a}$-curve which connects $\boldsymbol{y}$ with a point $\bar{\boldsymbol{y}} \in \Gamma$ (fig. 1.9).

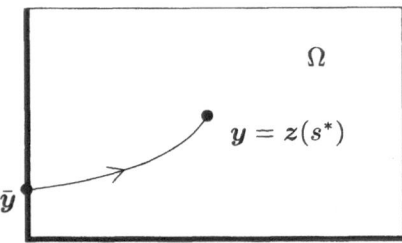

Figure 1.9 Integral curve connecting y with a boundary point $\bar{y}$

In other words, we look for a solution $z(s)$ of $z'(s) = a(z(s))$ with the property $z(0) = \bar{y}$ and $z(s^*) = y$ (the curve z is called *characteristic ground curve*). Having found this curve, we use our knowledge about both the value $v(0) = \phi(\bar{y})$ of the solution at the starting point, and the rate of change $v'(s) = b(z(s))v(s) + c(z(s))$ of the solution along a-curves to compute the value $u(y) = v(s^*)$ (fig. 1.10).

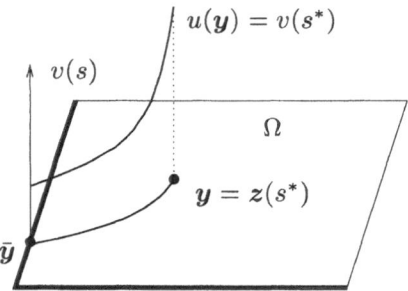

Figure 1.10 Values of the solution along the characteristic curve

Written in a compact form, the method of characteristics consists of three steps:

1) determine the solution $z_{\bar{y}}, v_{\bar{y}}$ of the *characteristic system*

$$z' = a(z), \qquad\qquad z(0) = \bar{y}$$
$$v' = b(z)v + c(z), \qquad\qquad v(0) = \phi(\bar{y})$$

2) find s^* and $\bar{y} \in \Gamma$ such that $z_{\bar{y}}(s^*) = y$

3) set $u(y) = v_{\bar{y}}(s^*)$.

We now consider several applications of this method.

Advection equation

The most simple application of the characteristic method is given by the advection equation (1.11) which we have found by linearizing the traffic flow model around the constant traffic density c

$$\frac{\partial u}{\partial t} + q'(c)\frac{\partial u}{\partial x} = 0, \quad u(0,x) = u^0(x).$$

Translating this equation into our general setup, we have $y = (t,x)$, $\Omega = \mathbb{R}^+ \times \mathbb{R}$, $\Gamma = \partial\Omega$, $\phi(y) = u^0(x)$, and $a(y) = (1,q'(c))$, $b = c = 0$. This leads to the characteristic system

$$z' = (1,q'(c)), \qquad\qquad\qquad z(0) = (0,\bar{x})$$
$$v' = 0, \qquad\qquad\qquad\qquad v(0) = u^0(\bar{x}).$$

Since the field a is independent of y, the characteristic ground curves are straight lines (see fig. 1.11) and since $v' = 0$, the solution is constant along these lines.

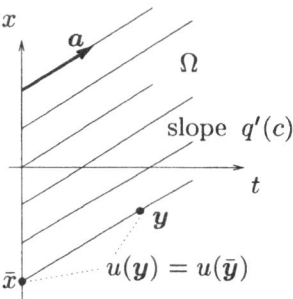

Figure 1.11 Characteristic ground curves for the advection equation

Formally, we have

1) the solution of the characteristic system

$$z_{\bar{y}}(s) = (s,\bar{x} + sq'(c)), \quad v_{\bar{y}}(s) = u^0(\bar{x})$$

2) for $y = (t,x) \in \Omega$ we have with $s^* = t$ and $\bar{x} = x - tq'(c)$ that $z_{\bar{y}}(s^*) = y$

3) the solution has the form

$$u(t,x) = v_{\bar{y}}(s^*) = u^0(x - tq'(c)).$$

Geometrically, the initial value u^0 is simply advected with velocity $q'(c)$ (see fig. 1.12).

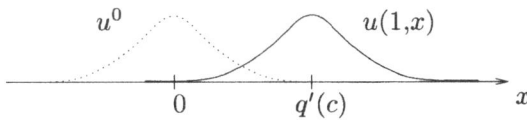

Figure 1.12 Solution of the advection equation at time $t = 1$

Note that the speed of transport $q'(c)$ is different from the car speed $V(c)$ since $q(u) = uV(u)$ so that $q'(c) = V(c) + cV'(c)$. Since V is typically a decreasing function (see fig. 1.5), $q'(c)$ will be less than $V(c)$ and can even be negative. This result coincides with the every-day observation that disturbances in the traffic density travel backwards along the road.

Stationary radiative transport

In radiative transport, the balanced quantity is the energy of radiation where

$$M(A \times B) = \int_A \int_B I(\boldsymbol{x}, \boldsymbol{\xi}) \, d\boldsymbol{\xi} \, d\boldsymbol{x}$$

is the energy in the domain $A \subset \mathbb{R}^d$ with ray directions $\boldsymbol{\xi} \in S^{d-1}$. The density $I(\boldsymbol{x}, \boldsymbol{\xi})$ is called intensity. The rate of change of the intensity along rays is influenced by absorption (parameter $\sigma_a \geq 0$), scattering (e.g. isotropic with parameter $\sigma_s \geq 0$), and possible sources $\tilde{Q}$

$$\boldsymbol{\xi} \cdot \nabla I = -\sigma_a I - \sigma_s \left(I - \frac{1}{4\pi} \int_{S^2} I \, d\boldsymbol{\eta} \right) + \tilde{Q}.$$

To focus on the difficulties in the characteristic method, we assume $\sigma_s = 0$ and $\tilde{Q} = 0$. Then, with the combined variable $\boldsymbol{y} = (\boldsymbol{x}, \boldsymbol{\xi})$, the characteristic system is based on $\boldsymbol{a}(\boldsymbol{y}) = (\boldsymbol{\xi}, 0)$ and $b(\boldsymbol{y}) = -\sigma_a$ and has the simple solution

$$\boldsymbol{z}_{\bar{\boldsymbol{y}}}(s) = (\bar{\boldsymbol{x}} + s\bar{\boldsymbol{\xi}}, \bar{\boldsymbol{\xi}}), \quad v_{\bar{\boldsymbol{y}}}(s) = I(\bar{\boldsymbol{x}}, \bar{\boldsymbol{\xi}}) e^{-\sigma_a s}$$

where $\bar{\boldsymbol{y}} = (\bar{\boldsymbol{x}}, \bar{\boldsymbol{\xi}})$ is a point on $\Gamma \subset \partial\Omega$. The more difficult task in the method of characteristics is given by the inversion problem: given some point $\boldsymbol{y} \in \Omega$, find the corresponding point $\bar{\boldsymbol{y}} = (\bar{\boldsymbol{x}}, \bar{\boldsymbol{\xi}})$ on the boundary. This means that intersection points of straight lines (the characteristic ground curves) with, in general, curved boundaries have to be detected which can be quite complicated (see fig. 1.13).

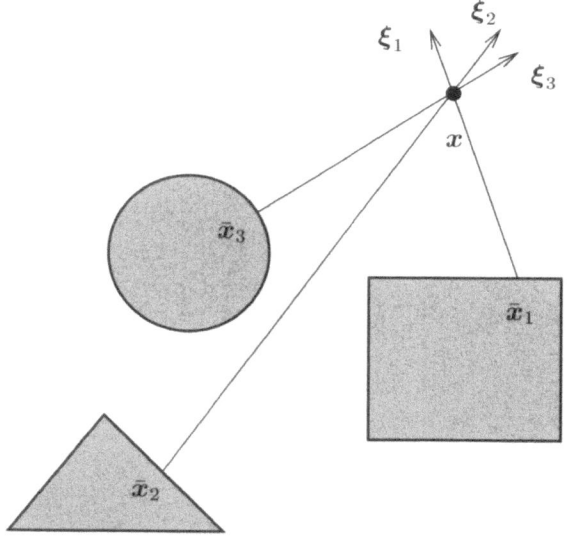

Figure 1.13 Intersection of rays with general boundaries in $\boldsymbol{x}$-space

To illustrate the phenomenon of non-smooth solutions, let us consider the following example: as spatial domain $\Omega^s \subset \mathbb{R}^2$, we take the annulus between the two circles $\Gamma_0^s = S^1$ and $\Gamma_1^s = RS^1$ with $R > 1$ (see fig. 1.14) which gives rise to the $\boldsymbol{y}$-domain $\Omega = \Omega^s \times S^1$.

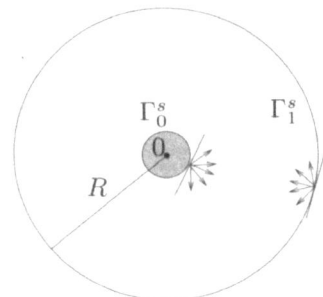

Figure 1.14 Spatial geometry of the radiation problem

The boundary on which we want to prescribe values consists of two parts

$$\Gamma_0 = \{(\boldsymbol{x},\boldsymbol{\xi}) : \boldsymbol{x} \in \Gamma_0^s, \boldsymbol{\xi} \in S^1, \boldsymbol{x} \cdot \boldsymbol{\xi} > 0\},$$
$$\Gamma_1 = \{(\boldsymbol{x},\boldsymbol{\xi}) : \boldsymbol{x} \in \Gamma_1^s, \boldsymbol{\xi} \in S^1, \boldsymbol{x} \cdot \boldsymbol{\xi} < 0\}.$$

On Γ_0 we assume Lambert radiation

$$I(\boldsymbol{x},\boldsymbol{\xi}) = I_0 \boldsymbol{x} \cdot \boldsymbol{\xi}, \quad (\boldsymbol{x},\boldsymbol{\xi}) \in \Gamma_0, \quad I_0 > 0$$

and on Γ_1 the (rather artificial) case of constant radiation in all directions

$$I(\boldsymbol{x},\boldsymbol{\xi}) = I_1, \quad (\boldsymbol{x},\boldsymbol{\xi}) \in \Gamma_1, \quad I_1 \geq 0.$$

Taking as parameters $R = 20$, $\sigma_a = 1/10$, $Q = 0$, $\sigma_s = 0$, $I_0 = 1$, and $I_1 = 0$, we calculate the solution at three different points in the domain (see fig. 1.15(a)).

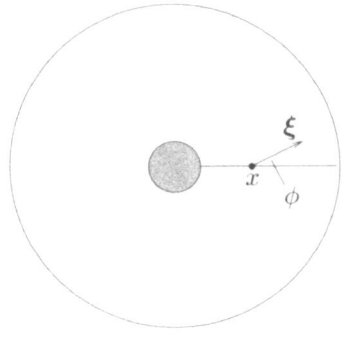

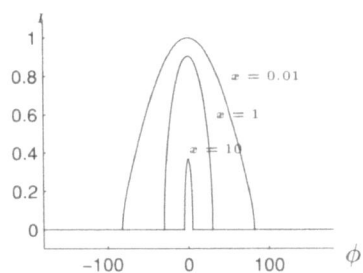

(a) Position x and angle ϕ (b) Intensity at different x

Figure 1.15 Solution of the radiation problem for background intensity $I_1 = 0$

For fixed $\boldsymbol{x} = (1 + x,0)$, the solution still depends on $\boldsymbol{\xi}$ which we express in terms of the angle ϕ (measured in degree in fig. 1.15(b)). Note that for increasing distance from the radiating surface Γ_0^s, the angle decreases in which radiation is received (the circle appears smaller in larger distance). Also, the maximal intensity decreases because of absorption. We also remark that the solution is not differentiable for all $\boldsymbol{\xi}$ because of the obvious jump in the slopes of the solution. This situation is even more pronounced, if we choose $I_1 > 0$ (see fig. 1.16).

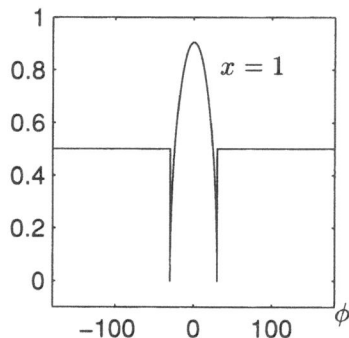

Figure 1.16 Solution of the radiation problem for background intensity $I_1 = 1/2$

Then, the solution is not even continuous despite the fact that all given data and boundaries are perfectly smooth. The reason for this behavior is explained schematically in fig. 1.17: we assume that along the curve Γ the very smooth data ϕ are prescribed and that the solution is constant along characteristic ground curves. Now consider the situation that one of the $\boldsymbol{a}$-curves starts in $\boldsymbol{y}^*$ and touches the boundary again in $\bar{\boldsymbol{y}}$ (i.e. $\boldsymbol{n}(\bar{\boldsymbol{y}}) \cdot \boldsymbol{a}(\bar{\boldsymbol{y}}) = 0$).

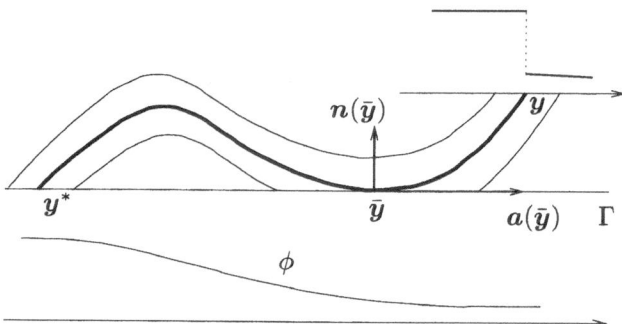

Figure 1.17 Discontinuous solution because of characteristic points on Γ

If we plot the solution in a neighborhood of $\boldsymbol{y} \in \Omega$ we see that curves which arrive left of the one which touches the boundary carry large values of ϕ while those on the right carry small values. Consequently, the solution has a jump in $\boldsymbol{y}$. This behavior is obviously caused by the geometric property of Γ to be parallel to $\boldsymbol{a}$ in the point $\bar{\boldsymbol{y}}$ (Γ is called *characteristic* in $\bar{\boldsymbol{y}}$). Note that prescribing data on surfaces which are characteristic at some points leads to solvability problems: on the one hand, the value of the solution in $\bar{y}$ is given by $\phi(\bar{y})$, on the other hand, it is given by $\phi(y^*)$ because u is constant along characteristics in our example. This leads to a contradiction since $\phi(\bar{y}) \neq \phi(y^*)$ (the problem is actually shared by all points between y^* and $\bar{y}$). We remark that the jump in the radiation problem (fig. 1.16) occurs exactly for those ray directions which belong to rays starting in x and being tangential to the small circle Γ_0^s. Note that such jumps are extremely important for the human visual orientation because they allow to distinguish between objects like the circle Γ_0^s and the background Γ_1^s.

Vlasov-equation with constant force

A straight forward application of the method of characteristics yields solutions of the initial value problem for the Vlasov equation (1.13) with constant force field

$$\frac{\partial f}{\partial t} + v \cdot \nabla_x f + g \cdot \nabla_v f = 0, \quad f(0,x,v) = f^0(x,v)$$

Setting $y = (t,x,v)$, $\Omega = \mathbb{R}^+ \times \mathbb{R}^3 \times \mathbb{R}^3$, $\Gamma = \partial\Omega$, we find the characteristic system with $a(y) = (1,v,g)$, $b = c = 0$. Since the boundary Γ has normal $n = (-1,0,0)$, it is nowhere-characteristic and arbitrary boundary data f^0 can be described along Γ. The solution of the characteristic system has the form

$$z(s) = (s, \bar{x} + s\bar{v} + s^2 g/2, \bar{v} + sg).$$

Solving $y = (t,x,v) = z_{\bar{y}}(s^*)$ for s^* and $\bar{y}$, we eventually obtain

$$f(t,x,v) = f^0(x - tv + t^2 g/2, v - tg).$$

Crystal precipitation

In the production of photographic paper or films, silver halide crystals (i.e. silver bromide, iodide, or chloride) of a certain size distribution are required (between 0.5 and 7 μm). We present a simple model for the prediction of crystal size distributions (taken from [4]).

Starting with a well stirred solution with salt concentration $c(t)$ at time t, we are interested in the number $M_t(I)$ of crystals with typical diameter $l \in I \subset \mathbb{R}^+$

$$M_t(I) = \int_I u(t,L)\, dL.$$

To describe the "transport" in the space of typical diameters, we assume that a crystal of size L at time t changes its size according to

$$\frac{dL}{dt} = G(t,L)$$

where we take the model of *Ostwald ripening*

$$G(t,L) = \begin{cases} k_g(c(t) - \exp(\alpha/L))^g & L \geq \alpha/\log c(t) \\ -k_d(\exp(\alpha/L) - c(t))^d & L < \alpha/\log c(t) \end{cases}$$

where $\alpha, k_g, k_d > 0$, $1 \le g, d \le 2$. For simplicity, we assume that there are no sources $(Q = 0)$ and we obtain the conservation law

$$\frac{\partial u}{\partial t} + \frac{\partial}{\partial L}(uG) = 0, \quad u(0,L) = u^0(L) > 0 \tag{1.27}$$

Note, however, that if crystals grow, they take salt atoms out of the solution. The mass used within the crystals (compared to the initial mass) is given by

$$\rho k_v \int_0^\infty L^3 (u(t,L) - u^0(L)) \, dL$$

where k_v is a parameter which relates the third power of the typical diameter to the volume of the crystal. Obviously, the consumed mass changes the salt concentration which satisfies

$$c(t) = c_0 - \rho k_v \int_0^\infty L^3 (u(t,L) - u^0(L)) \, dL \tag{1.28}$$

so that $c(t)$ depends on the solution u. Note that c enters in the formula for the flow velocity G so that the coupled system (1.27) and (1.28) is actually non-linear. A typical solution technique for such a problem is based on a fixed point formulation: we first assume that $c(t)$ is a *given* function and solve the (linear) problem (1.27) (for example, with the method of characteristics). Having obtained the solution u depending on the given function c, we define the map

$$P(c) = c_0 - \rho k_v \int_0^\infty L^3 (u(t,L) - u^0(L)) \, dL.$$

A solution of the non-linear problem is then given by the fixed point $P(c) = c$ and the corresponding solution of (1.27). A typical characteristic field for equation (1.27) is shown in fig. 1.18 where the concentration has been chosen as $c(t) = 2 \exp(-t)$, $g = d = 1.5$, $k_g = k_d = 1.0$, $\alpha = 0.1$. One can distinguish two parts in the (t,L)-space. Above the thick curve, the crystals are growing and in the lower part crystals disappear.

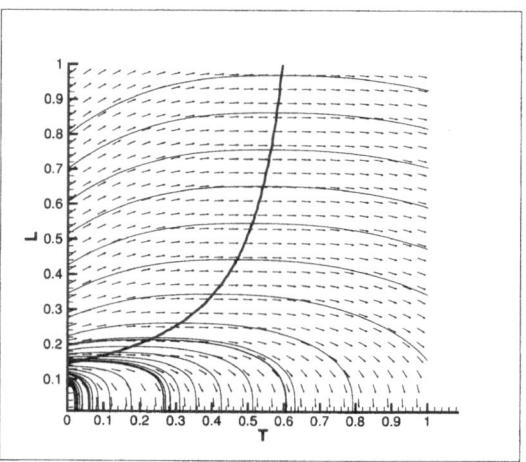

Figure 1.18 Characteristic ground curves for the crystal growth equation

1.2.2 Linear systems

We consider a system of conservation laws for the variables $\boldsymbol{U} = (U_1, \ldots, U_m)$

$$\frac{\partial U_i}{\partial t} + \mathrm{div}_{\boldsymbol{x}} \boldsymbol{q}_i(\boldsymbol{U}) = 0, \quad i = 1, \ldots, m.$$

The system is called linear if the fluxes $\boldsymbol{q}_i = (q_i^{(1)}, \ldots, q_i^{(d)})$ are linear in $\boldsymbol{U}$, i.e.

$$q_i^{(k)}(\boldsymbol{U}) = \sum_{j=1}^{m} a_{ij}^{(k)} U_j.$$

Introducing the matrices $A^{(k)} = (a_{ij}^{(k)})$, we can write the system in the form

$$\frac{\partial \boldsymbol{U}}{\partial t} + \sum_{k=1}^{d} A^{(k)} \frac{\partial \boldsymbol{U}}{\partial x_k} = 0. \tag{1.29}$$

A system of the form (1.29) is hyperbolic if all eigenvalues of the matrix $\sum_{k=1}^{d} \xi_k A^{(k)}$ are real for all $\boldsymbol{\xi} \in \mathbb{R}^d$. Considering the acoustics equations (1.23)

$$\frac{\partial p}{\partial t} + \mathrm{div} \boldsymbol{u} = 0, \quad \frac{\partial u_i}{\partial t} + \mathrm{div}(c^2 p \boldsymbol{e}_i) = 0 \tag{1.30}$$

with $c^2 = P'(\rho_0)$ and $\boldsymbol{e}_i$ being the ith unit vector, we can write (1.30) in the form (1.29) with

$$A^{(1)} = \begin{pmatrix} 0 & 1 & 0 & 0 \\ c^2 & 0 & 0 & 0 \\ 0 & 0 & 0 & 0 \\ 0 & 0 & 0 & 0 \end{pmatrix}, \quad A^{(2)} = \begin{pmatrix} 0 & 0 & 1 & 0 \\ 0 & 0 & 0 & 0 \\ c^2 & 0 & 0 & 0 \\ 0 & 0 & 0 & 0 \end{pmatrix}, \quad A^{(3)} = \begin{pmatrix} 0 & 0 & 0 & 1 \\ 0 & 0 & 0 & 0 \\ 0 & 0 & 0 & 0 \\ c^2 & 0 & 0 & 0 \end{pmatrix}$$

To check whether (1.30) is hyperbolic, we could try to calculate the eigenvalues of $\sum_{k=1}^{d} \xi_k A^{(k)}$ directly. However, one can use an important structure of (1.30) to show hyperbolicity.

Symmetric hyperbolic systems

If we multiply the acoustics equations written in general form (1.29) with the symmetric, positive definite (spd) matrix

$$B^{(0)} = \begin{pmatrix} c^2 & 0 & 0 & 0 \\ 0 & 1 & 0 & 0 \\ 0 & 0 & 1 & 0 \\ 0 & 0 & 0 & 1 \end{pmatrix}$$

from the left, we observe that the matrices $B^{(k)} = B^{(0)} A^{(k)}$ are symmetric

$$B^{(1)} = \begin{pmatrix} 0 & c^2 & 0 & 0 \\ c^2 & 0 & 0 & 0 \\ 0 & 0 & 0 & 0 \\ 0 & 0 & 0 & 0 \end{pmatrix}, \quad B^{(2)} = \begin{pmatrix} 0 & 0 & c^2 & 0 \\ 0 & 0 & 0 & 0 \\ c^2 & 0 & 0 & 0 \\ 0 & 0 & 0 & 0 \end{pmatrix}, \quad B^{(3)} = \begin{pmatrix} 0 & 0 & 0 & c^2 \\ 0 & 0 & 0 & 0 \\ 0 & 0 & 0 & 0 \\ c^2 & 0 & 0 & 0 \end{pmatrix}.$$

Hence, also linear combinations $\sum_{k=1}^{d} \xi_k B^{(k)} = B^{(0)} \sum_{k=1}^{d} \xi_k A^{(k)}$ are symmetric and from this property we can immediately conclude that $A = \sum_{k=1}^{d} \xi_k A^{(k)}$ has only real eigenvalues. To see this, we introduce a scalar product on $\mathbb{C}^m$

$$\left\langle B^{(0)} x, \bar{y} \right\rangle = \sum_{i,j} B_{ij}^{(0)} x_j \bar{y}_i$$

where the bar denotes complex conjugation. If $e \in \mathbb{C}^m$ is an eigenvector of A with eigenvalue $\lambda \in \mathbb{C}$ and the normalization $\langle B^{(0)} e, \bar{e} \rangle = 1$, we find

$$\lambda - \bar{\lambda} = \left\langle B^{(0)} \lambda e, \bar{e} \right\rangle - \left\langle B^{(0)} e, \bar{\lambda} e \right\rangle.$$

Using the symmetry of $B^{(0)}$, we conclude further

$$\lambda - \bar{\lambda} = \left\langle B^{(0)} A e, \bar{e} \right\rangle - \left\langle e, B^{(0)} A \bar{e} \right\rangle$$

and with the symmetry of the product $B^{(0)} A$ it follows

$$\lambda - \bar{\lambda} = \left\langle B^{(0)} A e, \bar{e} \right\rangle - \left\langle B^{(0)} A e, \bar{e} \right\rangle = 0,$$

showing that λ is, in fact, real and that the acoustics equations are hyperbolic.

In general, one calls a system (1.29) *symmetric hyperbolic* if there exists a symmetric positive definite matrix $B^{(0)}$ such that all matrices $B^{(0)} A^{(k)}$ are symmetric. For systems having this structure, a complete existence and uniqueness theory is available, even in the case where the matrices depend on space and time (see [3]). A key feature which is used in the existence and uniqueness proofs are energy estimates which are obtained through an additional conservation law for the quantity

$$\eta(U) = \frac{1}{2} \left\langle B^{(0)} U, U \right\rangle$$

which is a convex function of U. Indeed, if U is a solution of the symmetric hyperbolic system (1.29), we find with the symmetry assumptions

$$\frac{\partial \eta(U)}{\partial t} = \left\langle B^{(0)} U, \frac{\partial U}{\partial t} \right\rangle = - \left\langle B^{(0)} U, \sum_{k=1}^{d} A^{(k)} \frac{\partial U}{\partial x_k} \right\rangle$$

$$= - \left\langle U, \sum_{k=1}^{d} B^{(0)} A^{(k)} \frac{\partial U}{\partial x_k} \right\rangle = - \sum_{k=1}^{d} \left\langle B^{(0)} A^{(k)} U, \frac{\partial U}{\partial x_k} \right\rangle = - \sum_{k=1}^{d} \frac{\partial \phi_k(U)}{\partial x_k}$$

where $\phi_k(U) = \frac{1}{2} \left\langle B^{(0)} A^{(k)} U, U \right\rangle$. In other words, the solution of a linear symmetric hyperbolic problem automatically satisfies the additional conservation law

$$\frac{\partial \eta(\boldsymbol{U})}{\partial t} + \mathrm{div}\phi(\boldsymbol{U}) = 0$$

with a strictly convex function η (note the similarity to the entropy concept for the Euler equation – in fact, if a hyperbolic system of conservation laws admits a strictly convex entropy, one can show that its linearizations are symmetric hyperbolic systems). In the case of system (1.30), the additional conserved quantity and its flux are

$$\eta(p,\boldsymbol{u}) = \frac{1}{2}c^2 p^2 + \frac{1}{2}|\boldsymbol{u}|^2, \qquad \phi(p,\boldsymbol{u}) = c^2 p\boldsymbol{u}.$$

A problem with a moving boundary

In the following, we want to consider the linear equations of acoustics in a cylindrical tube with a moving piston on the left and a fixed wall on the right (see fig. 1.19).

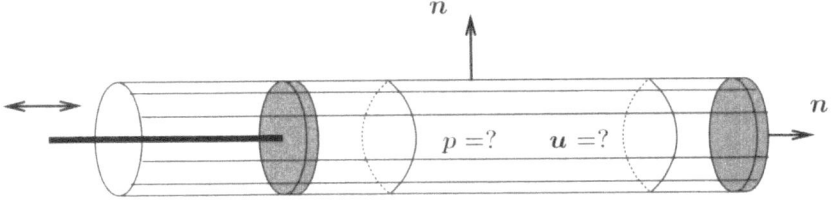

Figure 1.19   Model problem for the linear equations of acoustics

The piston movement will be prescribed as function $s(t)$ and we assume that all walls are impenetrable. If $\boldsymbol{n}$ is the outer normal field of this geometry, the boundary conditions lead to the equations $\boldsymbol{u} \cdot \boldsymbol{n} = -\dot{s}(t)$ on the piston and $\boldsymbol{u} \cdot \boldsymbol{n} = 0$ on the remaining walls. As initial condition we choose the constant state $p = p_0$, $\boldsymbol{u} = 0$. Because of axial symmetry, we look for a solution with the structure

$$p(x,y,z) = \hat{p}(x), \quad \boldsymbol{u}(x,y,z) = (\hat{u}(x),0,0), \quad x \in (s(t),1).$$

Plugging these expressions into the system (1.30), we obtain the one-dimensional problem

$$\frac{\partial \hat{p}}{\partial t} + \frac{\partial \hat{u}}{\partial x} = 0, \quad \frac{\partial \hat{u}}{\partial t} + c^2\frac{\partial \hat{p}}{\partial x} = 0$$
$$\hat{u}(0,x) = 0, \quad \hat{p}(0,x) = 1, \quad x \in (s(0),1)$$
$$\hat{u}(t,s(t)) = \dot{s}(t), \quad \hat{u}(t,1) = 0, \quad t > 0.$$

Note that, in case of uniqueness, a solution of the reduced system gives rise to the correct solution in three space dimensions.

Before we continue with the solution of the model problem, let us consider the general case of linear hyperbolic systems in a single space dimension

$$\frac{\partial \boldsymbol{U}}{\partial t} + A\frac{\partial \boldsymbol{U}}{\partial x} = 0, \quad (t,x) \in \Omega \qquad (1.31)$$

where $U = (U_1, \ldots, U_m)$. Assuming that we have a basis of eigenvectors of A

$$A r_i = \lambda_i r_i, \qquad i = 1, \ldots, m \tag{1.32}$$

we set up the matrices $R = (r_1 \ldots r_m)$ and $\Lambda = \mathrm{diag}(\lambda_1, \ldots, \lambda_m)$ such that (1.32) can be written as

$$R^{-1} A R = \Lambda.$$

If U is a solution of (1.31) one can show that the characteristic variables $\alpha = R^{-1} U$ solve a much simpler problem

$$\frac{\partial \alpha}{\partial t} = R^{-1} \frac{\partial U}{\partial t} = -R^{-1} A \frac{\partial U}{\partial x} = -R^{-1} A R \frac{\partial \alpha}{\partial x} = -\Lambda \frac{\partial \alpha}{\partial x}$$

which are actually m decoupled advection equations

$$\frac{\partial \alpha_i}{\partial t} + \lambda_i \frac{\partial \alpha_i}{\partial x} = 0, \qquad i = 1, \ldots, m.$$

In particular, the pure initial value problem

$$\frac{\partial U}{\partial t} + A \frac{\partial U}{\partial x} = 0, \quad U(0,x) = U^0(x)$$

has the solution $U(t,x) = R\alpha(t,x)$ where

$$\alpha_i(t,x) = \alpha_i^0(x - t\lambda_i), \qquad \alpha^0 = R^{-1} U^0.$$

For more general boundary value problems we will still use the fact that α_i is constant along the characteristics with slope λ_i (see fig. 1.20).

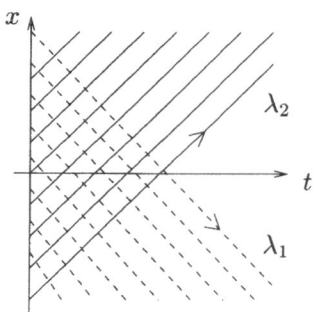

Figure 1.20   Characteristic ground curves for a system of two linear equations

Writing the reduced acoustic model in matrix vector form (and dropping the hat superscript), we get

$$\frac{\partial}{\partial t} \begin{pmatrix} p \\ u \end{pmatrix} + \begin{pmatrix} 0 & 1 \\ c^2 & 0 \end{pmatrix} \frac{\partial}{\partial x} \begin{pmatrix} p \\ u \end{pmatrix} = \begin{pmatrix} 0 \\ 0 \end{pmatrix} \tag{1.33}$$

The eigenvalues of the system matrix are $\lambda_1 = -c$, $\lambda_2 = +c$ with eigenvectors $r_1 = (1, -c)$, $r_2 = (1,c)$. This gives rise to the characteristic variables

$$\begin{pmatrix} \alpha_1 \\ \alpha_2 \end{pmatrix} = \begin{pmatrix} (p - u/c)/2 \\ (p + u/c)/2 \end{pmatrix}, \qquad \begin{pmatrix} p \\ u \end{pmatrix} = \begin{pmatrix} \alpha_2 + \alpha_1 \\ c(\alpha_2 - \alpha_1) \end{pmatrix} \qquad (1.34)$$

The domain on which we want to solve (1.33) has the form as shown in fig. 1.21, if we assume that the piston follows a piecewise linear movement with speed $\pm 2A$.

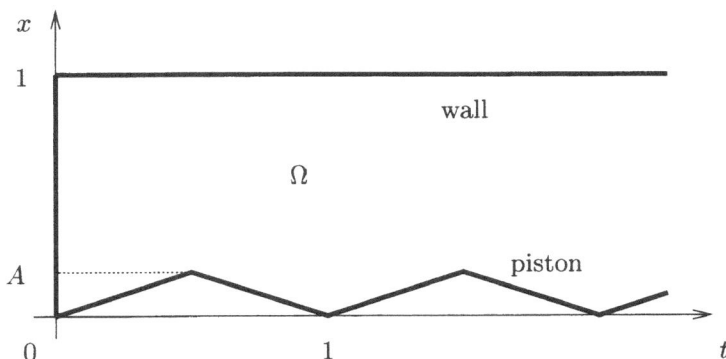

Figure 1.21 Geometry of the space time domain for the moving piston problem

Since $\lambda_1 < 0$, the characteristic curves of the first family run in negative x direction which means that α_1 is either determined by the initial values or by its values at the right wall (fig. 1.22).

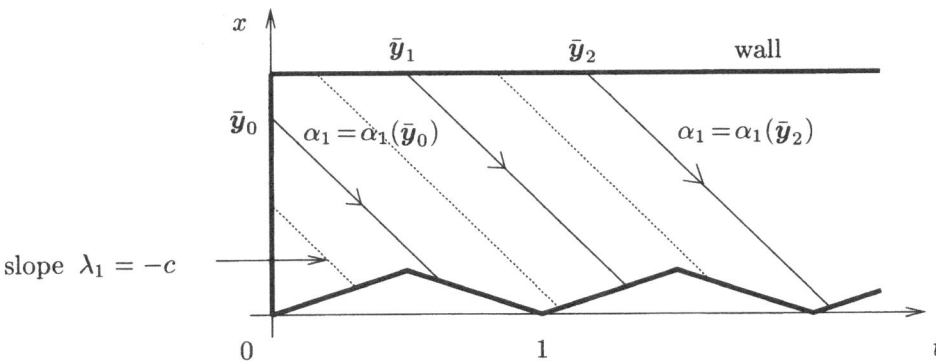

Figure 1.22 Lines along which α_1 is constant

Conversely, the characteristic curves of the second family run in positive x direction and α_2 is either determined by its value at the initial surface $t = 0$ or those at the moving boundary (see fig. 1.23).

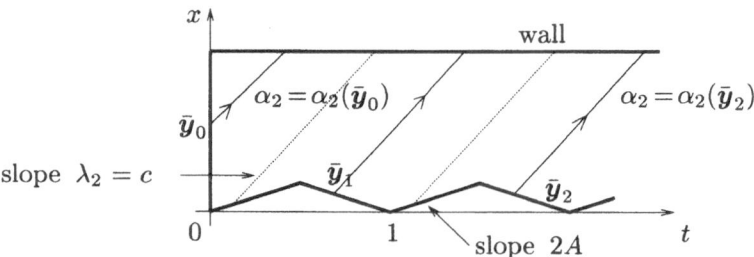

Figure 1.23 Lines along which α_2 is constant

At points $(\bar{t},\bar{x})$ of the left and the right x-boundary we have a condition of the form $u(\bar{t},\bar{x}) = v$ where $v = 0$ for $\bar{x} = 1$ and $v = \pm 2A$ for $\bar{x} = s(\bar{t})$. Using the transformation (1.34), this condition translates into $c(\alpha_2 - \alpha_1) = v$. Since the value of one of the variables α_1, α_2 is always available with the method of characteristics, the value of the other can be computed with the boundary data. Hence, the provided information suffices to solve the problem. As example, consider the situation depicted in fig. 1.24. To find the solution $p(\boldsymbol{y}), u(\boldsymbol{y})$, we need both α_1 and α_2 in the point $\boldsymbol{y} \in \Omega$.

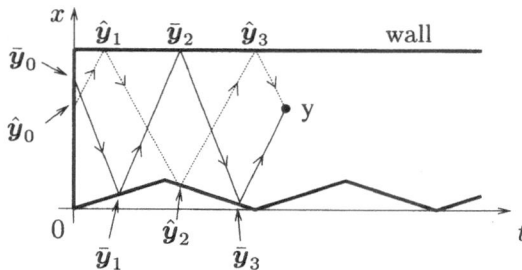

Figure 1.24 Solving the initial- boundary value problem with the method of characteristics

The value of α_1 is given by its value at $\hat{\boldsymbol{y}}_3$ which, according to the boundary condition is the same as $\alpha_2(\hat{\boldsymbol{y}}_3)$. To obtain $\alpha_2(\hat{\boldsymbol{y}}_3)$, we trace back the λ_2-characteristic to the piston where we find $\alpha_2(\hat{\boldsymbol{y}}_3) = \alpha_2(\hat{\boldsymbol{y}}_2) = \alpha_1(\hat{\boldsymbol{y}}_2) - 2A/c$. Now, the missing value is $\alpha_1(\hat{\boldsymbol{y}}_2)$ which is equal to $\alpha_1(\hat{\boldsymbol{y}}_1)$ and finally we arrive at the initial surface $\alpha_1(\hat{\boldsymbol{y}}_1) = \alpha_2(\hat{\boldsymbol{y}}_1) = \alpha_2(\hat{\boldsymbol{y}}_0) = p_0/2$. For the value at $\boldsymbol{y}$ this implies $\alpha_1(\boldsymbol{y}) = p_0/2 - 2A/c$. The same procedure for α_2 yields $\alpha_2(\boldsymbol{y}) = p_0/2$ so that, after transforming α_1, α_2 into p, u

$$p(\boldsymbol{y}) = p_0 - \frac{2A}{c}, \qquad u(\boldsymbol{y}) = 2A.$$

Note that the switching between α_1 and α_2 at the boundaries corresponds to a wave reflection at the piston or wall respectively. We also remark that solvability problems arise if we prescribe, for example, velocity *and* pressure at the fixed wall. Since α_2 is already determined by the initial values or the values at the piston, it is, in general, not possible to enforce the values at the fixed wall. More generally, if we have a system of the form (1.31) in a domain Ω, we can prescribe in a point $\bar{\boldsymbol{y}} \in \partial\Omega$ only those characteristic variables α_i for which

$$\sigma \begin{pmatrix} 1 \\ \lambda_i \end{pmatrix} \cdot n(\bar{y}) < 0$$

where one has to choose $\sigma = 1$ for initial and $\sigma = -1$ for end-value problems. Geometrically, this means that values can only be prescribed for incoming characteristic curves (respectively outgoing characteristic curves in the case of end-value problems – see fig. 1.25).

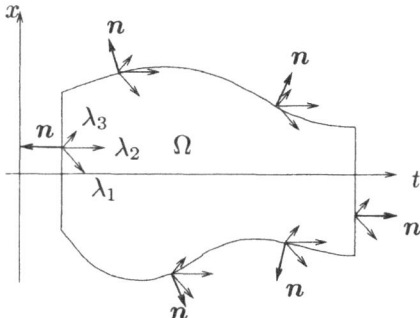

Figure 1.25 The number of incoming (outgoing) characteristic curves determines the number of boundary conditions

1.3 Weak solutions and entropy

Similar to the case of linear scalar balance laws, we rewrite the equation

$$\frac{\partial u}{\partial t} + \mathrm{div}_x q = Q$$

in divergence form using $y = (t,x)$ and $A(y,v) = (v,q(y,v))$

$$\mathrm{div}_y A(y,u(y)) = Q(y,u(y)).$$

Setting $a(y,v) = \frac{\partial}{\partial v} A(y,v)$ and $b(y,v) = -\mathrm{div}_y A(y,v) + Q(y,v)$, we then obtain the quasilinear form

$$a(y,u(y)) \cdot \nabla_y u(y) = b(y,u(y)), \qquad y \in \Omega$$

which we consider in connection with boundary conditions $u(y) = \phi(y)$ for $y \in \Gamma \subset \Omega$. The difference to the linear case is now that the form of the characteristic ground curve depends also on the values of the solution along the curve, giving rise to the coupled characteristic system

$$\begin{aligned} z' &= a(z,v), & z(0) &= \bar{y} \\ v' &= b(z,v), & v(0) &= \phi(\bar{y}). \end{aligned}$$

This coupling leads to the phenomenon that characteristic ground curves can cross in the nonlinear problem which is the breakdown of the characteristic method. The situation is illustrated in fig. 1.26 for the case $b = 0$ (i.e. u is constant along a curves).

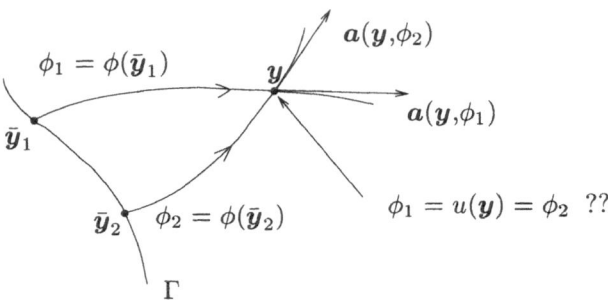

Figure 1.26 Characteristic ground curves can cross in the nonlinear case

If two characteristic ground curves cross in a point $y \in \Omega$, it is impossible to trace back the value of u since ϕ_1 and ϕ_2 are generally different. To show that crossing characteristics appear naturally, we consider again the traffic flow model

$$\frac{\partial u}{\partial t} + \frac{\partial}{\partial x} q(u) = 0, \quad u(0,x) = u^0(x), \quad x \in \mathbb{R} \tag{1.35}$$

where $q(u) = u V(u)$ and $V(u)$ is the vehicle velocity. In order to completely specify the model, we adopt the simple choice shown in fig. 1.27(a), which at least gives qualitatively reasonable results.

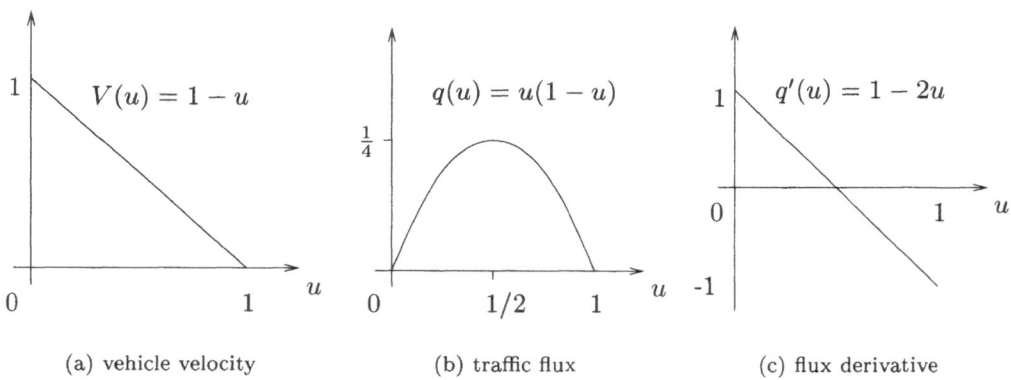

(a) vehicle velocity   (b) traffic flux   (c) flux derivative

Figure 1.27 Simple model for the density dependent vehicle velocity

Written in quasilinear form, equation (1.35) has the form $a(u) \cdot \nabla u = 0$ with $a(u) = (1, q'(u))$. The corresponding characteristic system

$$z' = (1, q'(v)), \qquad\qquad z(0) = (0, \bar{x})$$
$$v' = 0, \qquad\qquad v(0) = u^0(\bar{x})$$

has the solution $v(s) = u^0(\bar{x})$, $z(s) = (s, \bar{x} + s q'(u^0(\bar{x})))$ and for the initial value u^0 shown in fig. 1.28(a), the characteristic ground curves are straight lines with different slopes which really cross in a certain subdomain of Ω (see fig. 1.28(b)).

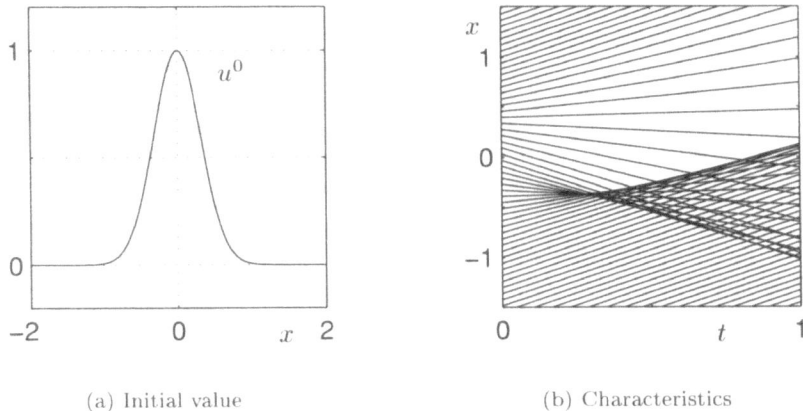

(a) Initial value   (b) Characteristics

Figure 1.28 A typical initial value leading to crossing of characteristic ground curves

Solving (1.35) with the methods of characteristics therefore leads, after some time, to a multi-valued solution (see fig. 1.29(b)) which is unphysical because the traffic density along the road should be a single valued function.

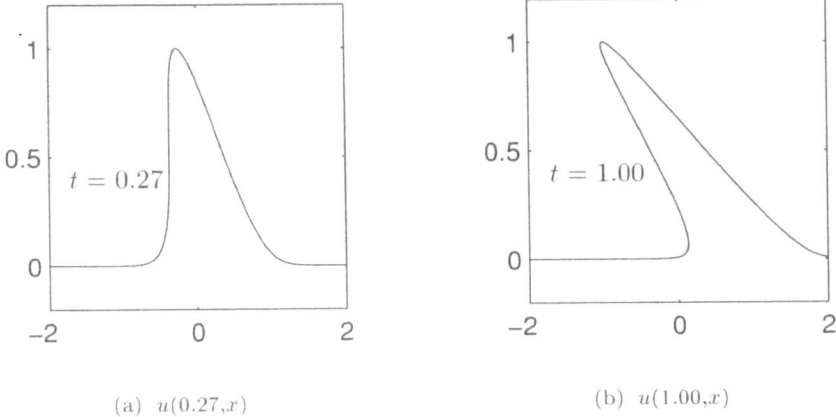

(a) $u(0.27,x)$ (b) $u(1.00,x)$

Figure 1.29 Solution of the method of characteristics for the initial value in fig. 1.28(a)

We will now show that the unexpected behavior of the model is related to the assumption that the vehicle speed is only a function of the traffic density. If we consider the acceleration felt by a vehicle at position $x(t)$ which is moving with speed $\dot{x}(t) = V(u(t,x(t)))$, we find

$$\ddot{x} = V'(u)\left(\frac{\partial u}{\partial t} + \dot{x}\frac{\partial u}{\partial x}\right) = (-1)\left(-q'(u)\frac{\partial u}{\partial x} + V(u)\frac{\partial u}{\partial x}\right) = -u\frac{\partial u}{\partial x} = -\frac{\partial}{\partial x}\left(\frac{1}{2}u^2\right)$$

Just before the failure of the characteristic method, it turns out that the density u exhibits a strong gradient (see fig. 1.29(a)) which even becomes infinite. In view of the expression for the acceleration, this means that the drivers have to break dramatically at the location of this strong increase which is due to the fact that they do not control their speed in view of the change in density ahead. On the other hand, the anticipation of the drivers is built into the refined model

$$\frac{\partial u_\alpha}{\partial t} + \frac{\partial}{\partial x}q(u_\alpha) = \alpha\frac{\partial^2 u_\alpha}{\partial x^2}, \quad u_\alpha(0,x) = u^0(x), \quad x \in \mathbb{R} \tag{1.36}$$

and it turns out that solutions of (1.36) do not show any unphysical behavior. Again, there is a formation of a steep gradient but it never exceeds a certain maximal slope (in fig. 1.30, the slope appears to be infinite only because of plotting accuracy).

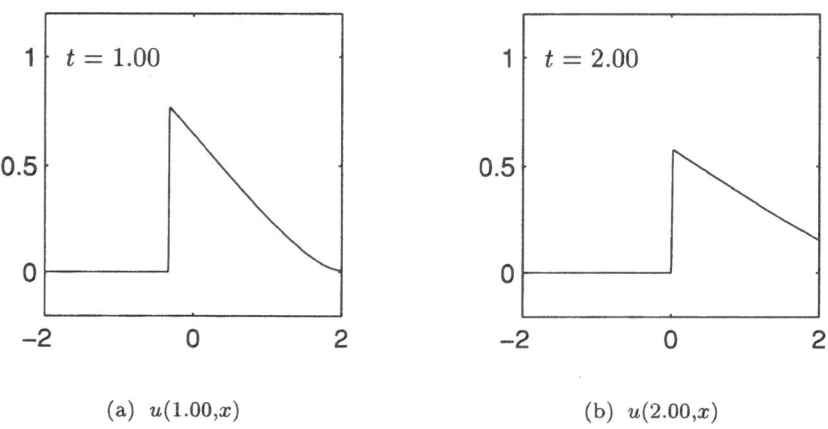

(a) $u(1.00,x)$ (b) $u(2.00,x)$

Figure 1.30 Solution of the model (1.36) with anticipating drivers and very small $\alpha > 0$

1.3.1 Weak solutions

Since we have introduced $\alpha > 0$ only as a small parameter to model the tendency that drivers react also on spatial variations in density, the question arises whether one can get rid of this parameter in the limit $\alpha \to 0$ while keeping the basic structure of the solution. Obviously, this limiting procedure cannot be carried out in the strong sense because $\|\partial_x q(u_\alpha)\|_{\mathbb{L}^\infty} \to \infty$ for $\alpha \to 0$. As remedy, we transform the equation (1.36) into a weak form in which the limit can be carried out safely. To do this, we first mention a quite general result about the divergence in the distributional sense.

Theorem 2   *Let $\Omega_1, \Omega_2 \subset \mathbb{R}^N$ be open, $\mathbb{R}^N = \bar{\Omega}_1 \cup \bar{\Omega}_2$, and let $\Gamma = \partial\Omega_1 = \partial\Omega_2$. Assume $w : \mathbb{R}^N \to \mathbb{R}^N$ is a function whose restrictions $w^{(i)} = w|_{\Omega_i}$ are smooth such that the divergence theorem can be applied to $\Omega_i, w^{(i)}$ (see fig. 1.31). Then, in the distributional sense, we have*

$$\mathrm{div}\,w = \mathrm{div}^{(p)}w + [w] \cdot n\delta_\Gamma \tag{1.37}$$

where $\mathrm{div}^{(p)}$ *is the pointwise divergence defined in each subset* Ω_i

$$\mathrm{div}^{(p)}\boldsymbol{w}(\boldsymbol{y}) = \begin{cases} \mathrm{div}\boldsymbol{w}^{(i)}(\boldsymbol{y}) & \boldsymbol{y} \in \Omega_i \\ 0 & else. \end{cases}$$

The Distribution δ_Γ *is defined by*

$$\langle \delta_\Gamma, \psi \rangle = \int_\Gamma \psi(\boldsymbol{y}) \, dS(\boldsymbol{y}), \qquad \forall \psi \in \mathscr{D}(\mathrm{IR}^N)$$

and the jump $[\boldsymbol{w}]$ *of* $\boldsymbol{w}$ *in the boundary points is given by*

$$[\boldsymbol{w}](\boldsymbol{y}) = \lim_{s \searrow 0} \left(\boldsymbol{w}(\boldsymbol{y} + s\boldsymbol{n}(\boldsymbol{y})) - \boldsymbol{w}(\boldsymbol{y} - s\boldsymbol{n}(\boldsymbol{y})) \right) \quad \boldsymbol{y} \in \Gamma.$$

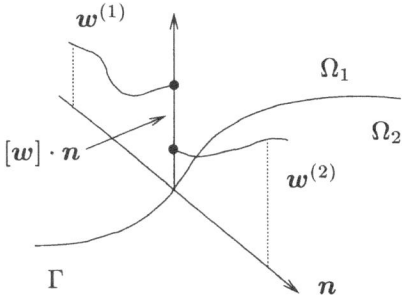

Figure 1.31 A piecewise smooth function with jump on the surface Γ

Proof: The proof relies on the divergence theorem. Taking any $\psi \in \mathscr{D}(\mathrm{IR}^N)$, we have

$$\langle \mathrm{div}\boldsymbol{w}, \psi \rangle = -\sum_{j=1}^{N} \langle w_j, \partial_{x_j}\psi \rangle = -\sum_{k=1}^{2}\sum_{j=1}^{N} \int_{\Omega_k} w_j^{(k)} \partial_{x_j}\psi \, d\boldsymbol{y}$$

$$= -\sum_{k=1}^{2} \int_{\Omega_k} \mathrm{div}(\psi\boldsymbol{w}^{(k)}) - \psi\mathrm{div}\boldsymbol{w}^{(k)} \, d\boldsymbol{y}$$

$$= \sum_{k=1}^{2} \left(-\int_{\partial\Omega_k} \psi\boldsymbol{w}^{(k)} \cdot \boldsymbol{n} \, dS(\boldsymbol{y}) + \int_{\Omega_k} \psi\mathrm{div}\boldsymbol{w}^{(k)} \, d\boldsymbol{y} \right)$$

$$= \int_\Gamma \psi(\boldsymbol{w}^{(2)} - \boldsymbol{w}^{(1)}) \cdot \boldsymbol{n} \, dS(\boldsymbol{y}) + \int_{\mathrm{IR}^N} \psi\mathrm{div}^{(p)}\boldsymbol{w} \, d\boldsymbol{y}$$

$$= \left\langle [\boldsymbol{w}] \cdot \boldsymbol{n}\delta_\Gamma + \mathrm{div}^{(p)}\boldsymbol{w}, \psi \right\rangle$$

∎

Using the divergence relation (1.37) of the above theorem, we can formulate (1.36) in the distributional sense. First, we extend the solution $u_\alpha(t,x)$ which is defined on $\Omega_2 = \{(t,x) : t > 0, x \in \mathrm{IR}\}$ to the complement $\Omega_1 = \{(t,x) : t < 0, x \in \mathrm{IR}\}$ by setting it equal to zero

$$u_\alpha(t,x) = \begin{cases} 0 & t < 0 \\ u_\alpha(t,x) & t > 0 \end{cases}$$

Using the space-time flux $\mathcal{A}(u_\alpha) = (u_\alpha, q(u_\alpha))$ we then conclude immediately

$$\mathrm{div}_y^{(p)} \mathcal{A}(u_\alpha) = \alpha \frac{\partial^2 u_\alpha}{\partial x^2}, \qquad y \in \Omega_i, \quad i = 1,2.$$

Moreover, on Γ_0 we have a jump in the normal component of $\mathcal{A}$ which is just given by the initial value

$$[\mathcal{A}(u_\alpha)] \cdot \boldsymbol{n} = u^0 \quad \text{on } \Gamma_0.$$

Using (1.37) we thus get in the sense of distributions

$$\mathrm{div}_{(t,x)} \mathcal{A}(u_\alpha) = \alpha \frac{\partial^2 u_\alpha}{\partial x^2} + u^0 \delta_{\Gamma_0} \qquad \text{in } \mathscr{D}'(\mathrm{I\!R}^2)$$

which means that for all $\psi \in \mathscr{D}(\mathrm{I\!R}^2)$

$$-\int_{\mathrm{I\!R}^2} \mathcal{A}(u_\alpha) \cdot \nabla \psi \, dy = \alpha \int_{\mathrm{I\!R}^2} u_\alpha \frac{\partial^2 \psi}{\partial x^2} \, dy + \int_{\Gamma_0} u^0 \psi \, dS(y)$$

In this weak formulation of the problem, the limit $\alpha \to 0$ can be carried out under the mild assumption that $\|u_\alpha\|_{\mathrm{L}^\infty}$ is uniformly bounded and that $u_\alpha \to u$ almost everywhere. The limit of the solutions u_α then satisfies the condition

$$-\int_{\mathrm{I\!R}^2} \mathcal{A}(u) \cdot \nabla \psi \, dy = \int_{\Gamma_0} u^0 \psi \, dS(y) \qquad \forall \psi \in \mathscr{D}(\mathrm{I\!R}^2)$$

or, in the distributional sense, $\mathrm{div}_{(t,x)} \mathcal{A}(u) = u^0 \delta_{\Gamma_0}$. This observation gives rise to the definition of an extended solution concept for the hyperbolic problem

$$\frac{\partial u}{\partial t} + \frac{\partial}{\partial x} q(u) = 0, \qquad u(0,x) = u^0(x). \tag{1.38}$$

Definition 3   Let $u^0 \in \mathrm{L}^\infty(\mathrm{I\!R})$, $u \in \mathrm{L}^\infty_{loc}(\mathrm{I\!R}^2)$, $u(t,x) = 0$ for $t < 0$, $\Gamma_0 = \{0\} \times \mathrm{I\!R}$. *Then, the function* u *is called weak solution of* (1.38), *provided*

$$\mathrm{div}_{(t,x)} \mathcal{A}(u) = u^0 \delta_{\Gamma_0} \quad \text{in } \mathscr{D}'(\mathrm{I\!R}^2)$$

We remark that classical solutions are automatically weak solutions which is an easy consequence of the divergence relation (1.37). In fact, if u is a classical solution of (1.38) then $\mathrm{div}^{(p)} \mathcal{A}(u) = 0$ and thus $\mathrm{div}\mathcal{A}(u) = u^0 \delta_{\Gamma_0}$ in $\mathscr{D}'$ which is the condition on u to be a weak solution. More generally, one can show with the same argument that a piecewise classical solution is a weak solution if and only if $[\mathcal{A}(u)] \cdot \boldsymbol{n} = 0$ on the jump line Γ_s (see fig. 1.32).

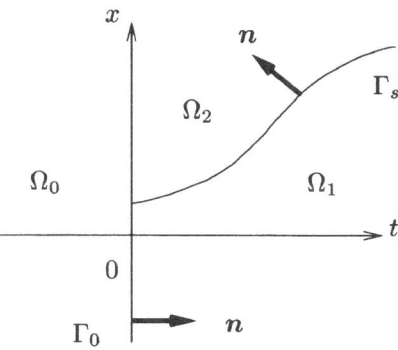

Figure 1.32 A piecewise classical solution is a weak solution if $[\mathcal{A}(u)] \cdot \boldsymbol{n} = 0$ on Γ_s

The proof of the statement again relies on the divergence relation (but now for the case of at least three sets Ω_i). Since the function u is a piecewise classical solution if and only if $\mathrm{div}^{(p)}\mathcal{A}(u) = 0$ in each subset, the relation

$$\mathrm{div}_{(t,x)}\mathcal{A}(u) = \mathrm{div}^{(p)}\mathcal{A}(u) + [\mathcal{A}(u)] \cdot \boldsymbol{n}\delta_{\Gamma_s} + u^0\delta_{\Gamma_0} \quad \text{in } \mathscr{D}'(\mathrm{I\!R}^2)$$

yields the condition $[\mathcal{A}(u)] \cdot \boldsymbol{n} = 0$ on Γ_s on u to be a weak solution. The jump condition $[\mathcal{A}(u)] \cdot \boldsymbol{n} = 0$ is called *Rankine-Hugoniot condition* and can be reformulated if we parameterize Γ_s according to

$$\Gamma_s = \{(t,s(t)) : t \in I\}, \quad I = (\bar{t},T).$$

Since the tangential vector on Γ_s is given by $\boldsymbol{\tau} = n_x(1, -n_t/n_x)$ (see fig. 1.33), the slope of s is given by $\dot{s} = -n_t/n_x$.

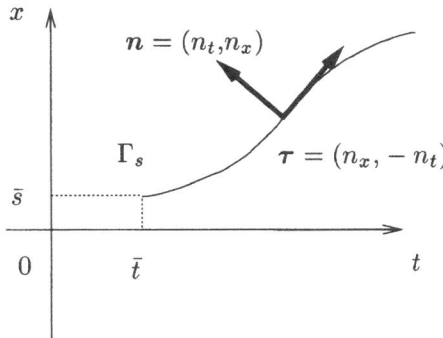

Figure 1.33 Normal and tangential vector at a point on the jump line

Hence $0 = [\mathcal{A}(u)] \cdot \boldsymbol{n} = [u]n_t + [q(u)]n_x$ implies the following condition on the so called shock curve s

$$\dot{s}(t) = \frac{[q(u)]}{[u]}(t,s(t)), \quad s(\bar{t}) = \bar{s}.$$

This ordinary differential equation can be explicitly solved for our traffic flow model with $V(u) = 1 - u$, if we use a piecewise linear initial value as shown in fig. 1.34(a).

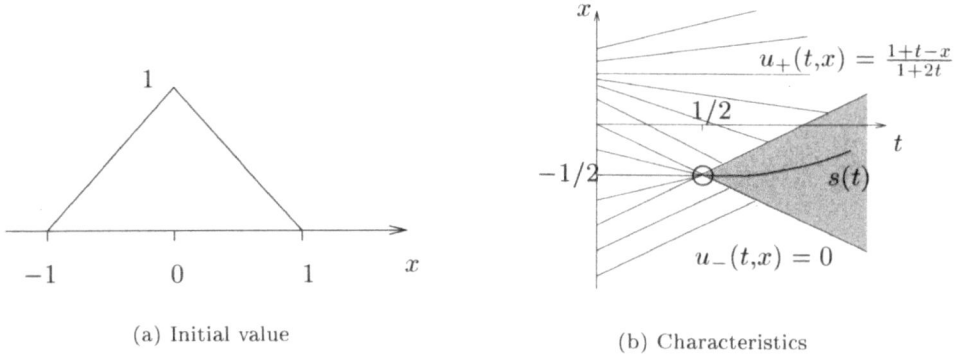

(a) Initial value   (b) Characteristics

Figure 1.34 Shock solution for a piecewise linear initial condition

One easily finds out that the characteristic ground curves intersect for the first time in $\bar{s} = -1/2$ at $\bar{t} = 1/2$. In the conical shaped domain emanating from $(\bar{t},\bar{s})$ in fig. 1.34(b), one finds two possible classical solutions with the method of characteristics which are

$$u_-(t,x) = 0, \qquad u_+(t,x) = \frac{1 + t - x}{1 + 2t}.$$

This leads to the ordinary differential equation

$$\dot{s}(t) = \frac{q(u_+) - q(u_-)}{u_+ - u_-}(t,s(t)) = \frac{t + s(t)}{1 + 2t}, \qquad s(1/2) = -1/2$$

which has the solution $s(t) = 1 + t - \sqrt{2 + 4t}$ for $t > 1/2$.

While the concept of weak solutions allows to cope with shock solutions, additional problems arise if uniqueness of weak solutions is considered. To give an example, we consider the initial value shown in fig. 1.35(a).

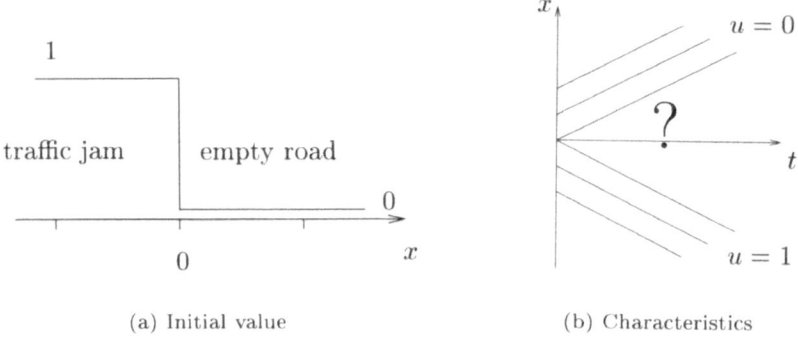

(a) Initial value   (b) Characteristics

Figure 1.35 Characteristic ground curves leave a gap in (t,x) space

Since $q'(u) = 1 - 2u \in \{-1,1\}$ for $u \in \{0,1\}$, the characteristic ground curves leave a conical gap emanating from $(t,x) = (0,0)$ (fig. 1.35(b)). One method to fill this gap is to set $u(t,x) = 0$ for $x > 0$ and $u(t,x) = 1$ for $x < 0$. The jump surface $\Gamma_s = \{(t,0) : t > 0\}$ is then a valid shock surface because $[q(u)]/[u] = 0$ on Γ_s. Practically, the solution constructed in this way is stationary which means that no car in the traffic jam starts moving although the street is completely empty for $x > 0$ (see fig. 1.36(a)).

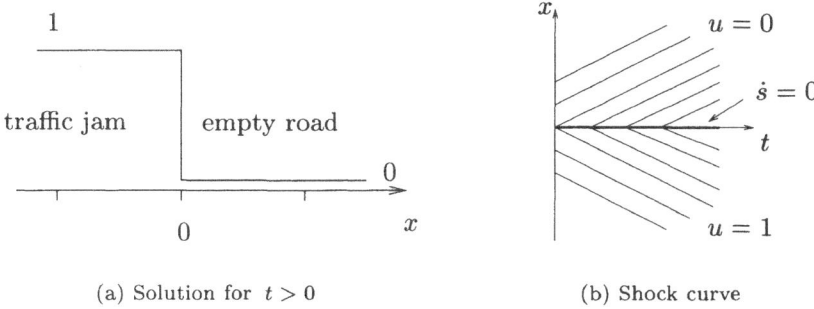

(a) Solution for $t > 0$ (b) Shock curve

Figure 1.36 Stationary shock solution for a piecewise constant initial condition

This very unrealistic behavior again seems to be the result of non-anticipating drivers. A different weak solution of the same problem, on the other hand, is much more reasonable. In fact, we can find a continuous weak solution of the problem by trying the ansatz $u(t,x) = \tilde{u}(x/t)$. Plugging this form into (1.35), we find the condition

$$\tilde{u}'(\xi)(q'(\tilde{u}(\xi)) - \xi) = 0, \qquad \xi = \frac{x}{t}$$

which is obviously satisfied for all ξ if

$$\tilde{u}(\xi) = (q')^{-1}(\xi) = \frac{1-\xi}{2}.$$

Since $\tilde{u}(1) = 0$ and $\tilde{u}(-1) = 1$, we therefore obtain a piecewise classical solution of the initial value problem by setting

$$u(t,x) = \begin{cases} 1 & x/t < -1 \\ (1-\xi)/2 & -1 \leq x/t \leq 1 \\ 0 & x/t > 1 \end{cases}$$

Note that u is a weak solution because it is piecewise classical and satisfies the jump condition $[\mathcal{A}(u)] \cdot n = 0$ across the lines $x/t = \pm 1$, in fact, because of continuity we even have $[\mathcal{A}(u)] = 0$. This type of solution (see fig. 1.37(a)) is called *rarefaction wave*. It is characterized by the fact that it appears as limit of the solutions u_α of (1.36) for $\alpha \to 0$.

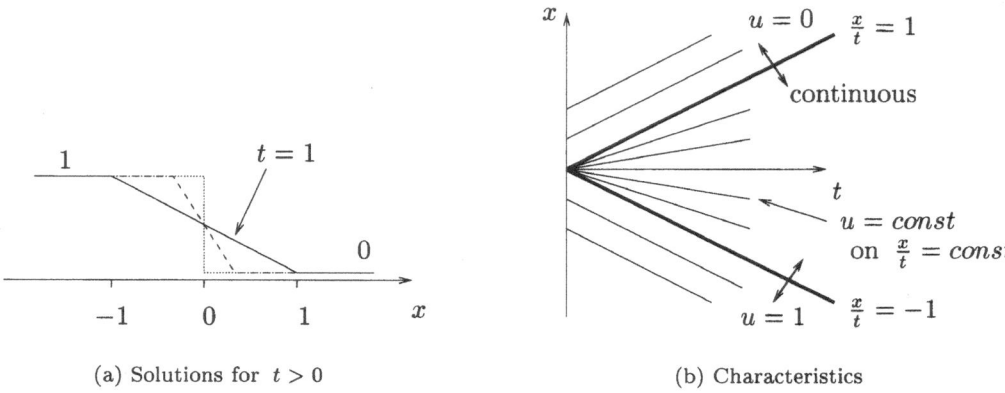

(a) Solutions for $t > 0$ (b) Characteristics

Figure 1.37   Rarefaction solution for a piecewise constant initial condition

1.3.2   Entropy condition

Since we accepted the model of the anticipating driver as a guiding line for the reduced model, it is natural to adopt the rarefaction solution and to discard the unrealistic shock solution. Unfortunately, the criterion to accept only the unique weak solution of (1.38) which is the limit of u_α for $\alpha \to 0$ is difficult to handle in practice. A more useful approach is given by the *entropy condition* which we will describe in the following. Let $\eta \in C^1(\mathbb{R})$ be a convex function (an *entropy*) and let ϕ be a primitive of $\eta' q'$ (an *entropy flux*). The pair $\mathcal{E}(u) = (\eta(u), \phi(u))$ is then called entropy-entropy flux pair and satisfies the relation

$$\mathcal{E}'(u) = \eta'(u)(1, q'(u)) = \eta'(u)\mathcal{A}'(u)$$

In particular, any classical solution of the hyperbolic conservation law $\mathrm{div}_{\boldsymbol{y}}\mathcal{A}(u) = 0$ satisfies automatically

$$\mathrm{div}_{\boldsymbol{y}}\mathcal{E}(u) = \mathcal{E}'(u) \cdot \nabla_{\boldsymbol{y}}u = \eta'(u)\mathcal{A}'(u) \cdot \nabla_{\boldsymbol{y}}u = \eta'(u)\mathrm{div}_{\boldsymbol{y}}\mathcal{A}(u) = 0 \qquad (1.39)$$

which is an additional conservation law

$$\frac{\partial \eta(u)}{\partial t} + \frac{\partial \phi(u)}{\partial x} = 0.$$

Similarly, one obtains for the extended model (1.36)

$$\mathrm{div}_{\boldsymbol{y}}\mathcal{E}(u_\alpha) = \eta'(u_\alpha)\mathrm{div}_{\boldsymbol{y}}\mathcal{A}(u_\alpha) = \alpha\eta'(u_\alpha)\frac{\partial^2 u_\alpha}{\partial x^2}.$$

Since η is convex, we also have

$$\frac{\partial^2 \eta(u_\alpha)}{\partial x^2} = \underbrace{\eta''(u_\alpha)}_{\geq 0}\underbrace{\left(\frac{\partial u_\alpha}{\partial x}\right)^2}_{>0} + \eta'(u_\alpha)\frac{\partial^2 u_\alpha}{\partial x^2}$$

so that altogether

$$\operatorname{div}\mathcal{E}(u_\alpha) \leq \alpha \frac{\partial^2 \eta(u_\alpha)}{\partial x^2}.$$

Extending again u_α according to

$$u_\alpha(t,x) = \begin{cases} 0 & t < 0 \\ u_\alpha(t,x) & t > 0 \end{cases}$$

we obtain in the distributional sense

$$\operatorname{div}_{\boldsymbol{y}}\mathcal{E}(u_\alpha) \leq \alpha \frac{\partial^2 \eta(u_\alpha)}{\partial x^2} + \eta(u^0)\delta_{\Gamma_0} \quad \text{in } \mathscr{D}'(\mathbb{R}^2)$$

which means more explicitly

$$-\int_{\mathbb{R}^2} \mathcal{E}(u_\alpha) \cdot \nabla_{\boldsymbol{y}}\psi \, d\boldsymbol{y} \leq \alpha \int_{\mathbb{R}^2} \eta(u_\alpha) \frac{\partial^2 \psi}{\partial x^2} \, d\boldsymbol{y} + \int_{\Gamma_0} \eta(u^0)\psi \, dS(\boldsymbol{y})$$

for all $\psi \in \mathscr{D}(\mathbb{R}^2)$ with $\psi \geq 0$. If $\|u_\alpha\|_{\mathbb{L}^\infty}$ is uniformly bounded and $u_\alpha \to u$ almost everywhere, we find that the limit u is a weak solution of (1.38) which satisfies for every convex entropy-entropy flux pair $\mathcal{E}$ the additional relation

$$\operatorname{div}_{\boldsymbol{y}}\mathcal{E}(u) \leq \eta(u^0)\delta_{\Gamma_0} \quad \text{in } \mathscr{D}'(\mathbb{R}^2)$$

It turns out that this *entropy condition* really singles out a unique weak solution among all possible weak solutions which is then, of course, the limit of the refined model (1.36) for $\alpha \to 0$. More precisely, one has the general result (for a proof see [8])

Theorem 4   *Let $u^0 \in \mathbb{L}^\infty(\mathbb{R})$ and $q \in C^1(\mathbb{R})$. Then there exists exactly one weak solution of*

$$\frac{\partial u}{\partial t} + \frac{\partial}{\partial x}q(u) = 0, \quad u(0,x) = u^0(x), \quad x \in \mathbb{R},$$

which satisfies $\operatorname{div}_{(t,x)}\mathcal{E}(u) \leq \eta(u^0)\delta_{t=0}$ in $\mathscr{D}'(\mathbb{R}^2)$ for all convex entropy-entropy flux pairs $\mathcal{E}$.

For the particular case of piecewise classical solutions, the entropy condition can be cast in a form similar to the Rankine-Hugoniot condition. More precisely, a piecewise classical solution satisfies the entropy condition if for all convex entropy-entropy flux pairs the jump condition $[\mathcal{E}(u)] \cdot \boldsymbol{n} \leq 0$ is satisfied on the shock curves Γ_s. The proof again relies on the divergence relation (1.37)

$$\operatorname{div}_{\boldsymbol{y}}\mathcal{E}(u) = \operatorname{div}^{(p)}\mathcal{E}(u) + [\mathcal{E}(u)] \cdot \boldsymbol{n}\delta_{\Gamma_s} + \eta(u^0)\delta_{\Gamma_0}$$

and on the fact that for piecewise classical solutions $\operatorname{div}^{(p)}\mathcal{E}(u) = 0$ according to (1.39). An additional consideration allows to restrict the entropy requirement to special entropy-entropy flux pairs of the form

$$\mathcal{E}_k(u) = \big(|u - k|, \operatorname{sign}(u - k)(q(u) - q(k))\big), \quad k \in \mathbb{R}$$

(for details see [5]). Using the pairs $\mathcal{E}_k$ and the condition $[\mathcal{E}_k(u)] \cdot n \leq 0$, elementary considerations show that a piecewise classical solution with jump *upwards* (see fig. 1.38(a)) satisfies the entropy condition if and only if the flux function $q(u)$ lies *above* the chord between $q(u_-)$ and $q(u_+)$ (fig. 1.38(b)).

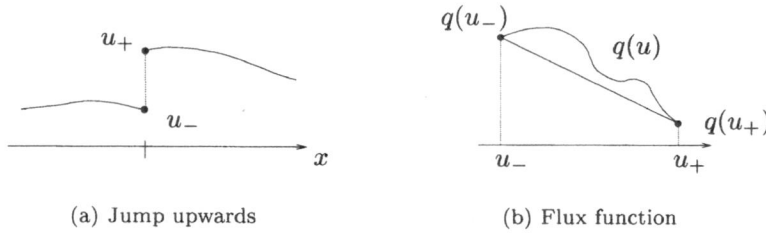

(a) Jump upwards   (b) Flux function

Figure 1.38 Entropy condition for jumps upward

Conversely, a weak piecewise classical solution with jump *downward* (see fig. 1.39(a)) satisfies the entropy condition exactly when the flux function $q(u)$ lies *below* the chord between $q(u_+)$ and $q(u_-)$ (fig. 1.39(b)).

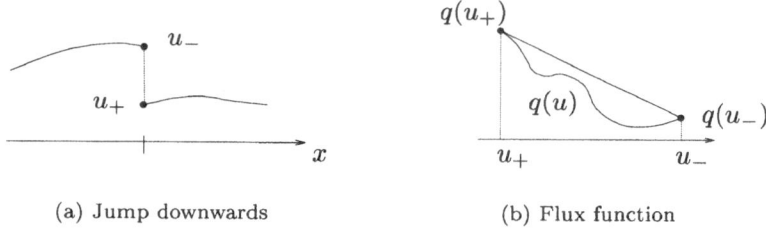

(a) Jump downwards   (b) Flux function

Figure 1.39 Entropy condition for jumps downward

In the case of traffic flow, the flux function $q(u) = u(1-u)$, $u \in [0,1]$ is concave so that only jumps upwards can be entropy consistent. Equivalently, one can cast this condition in the form $q'(u_-) \geq \dot{s} \geq q'(u_+)$ which also holds in the case of convex fluxes and which has the nice geometrical interpretation of characteristic curves running into the shock (see fig. 1.40).

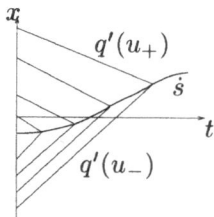

Figure 1.40 Characteristic curves running into the shock curve

1.3.3   Ion etching in semiconductor production

We consider an industrial application of a hyperbolic conservation law with a flux function which is neither convex nor concave. The example arises in connection with ion etching where a material (e.g. the semiconductor material GaAs) is etched at its surface by bombardment with a beam of ions (see fig. 1.41(a)).

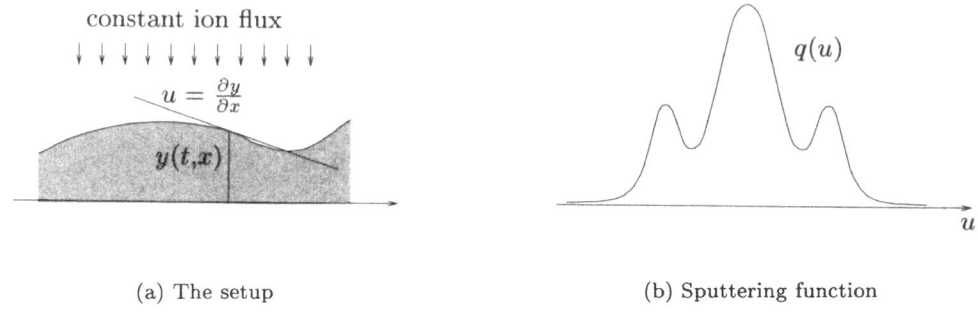

constant ion flux

$u = \frac{\partial y}{\partial x}$

$y(t,x)$

$q(u)$

u

(a) The setup   (b) Sputtering function

Figure 1.41   Etching of a material surface with a beam of ions

The removal of semiconductor material is proportional to the slope $u(t,x) = \frac{\partial y}{\partial x}$ of the surface $y(t,x)$ which leads to the evolution

$$\frac{\partial y}{\partial t} = -q\left(\frac{\partial y}{\partial x}\right) \tag{1.40}$$

The function q may have shapes as shown in fig. 1.41(b) where multiple local extrema can be caused by the crystalline structure of the semiconductor material. While (1.40) is a Hamilton-Jacobi equation for y, it can be transformed into a conservation law by going over to the x-derivative $u(t,x)$ of $y(t,x)$ which satisfies

$$\frac{\partial u}{\partial t} + \frac{\partial}{\partial x}q(u) = 0, \quad u(0,x) = y_0'(x) = u^0(x). \tag{1.41}$$

In the following, we will consider the Riemann problem consisting of equation (1.41) with initial value of jump type

$$u^0(x) = \begin{cases} u_- & x < 0 \\ u_+ & x > 0 \end{cases}.$$

In terms of the semiconductor surface, upward jumps $u_+ > u_-$ correspond to configurations as in fig. 1.42(a) and downward jumps $u_+ < u_-$ to situations as in fig. 1.42(b). Before going into details of the ion etching example, we mention a general property of the unique entropy solution u of a Riemann problem. Setting $u_\beta(t,x) = u(\beta t,\beta x)$ for some $\beta > 0$, it is easily checked that u_β is also an entropy solution which implies that $u = u_\beta$ (because of uniqueness).

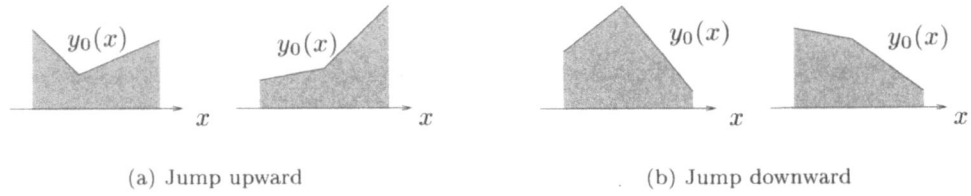

(a) Jump upward   (b) Jump downward

Figure 1.42 Geometric meaning of Riemann initial data for the slope of the surface

In particular, for some fixed $(\bar{t},\bar{x})$ we find

$$u(\bar{t},\bar{x}) = u_{1/\bar{t}}(\bar{t},\bar{x}) = u(1,\bar{x}/\bar{t}).$$

This result shows that the solution of the Riemann problem depends only on the quotient x/t (self similarity). Hence, it suffices to consider the solution at $t = 1$.

Let us now consider a jump upwards as initial value for equation (1.41) with flux function q shown in fig. 1.43(b).

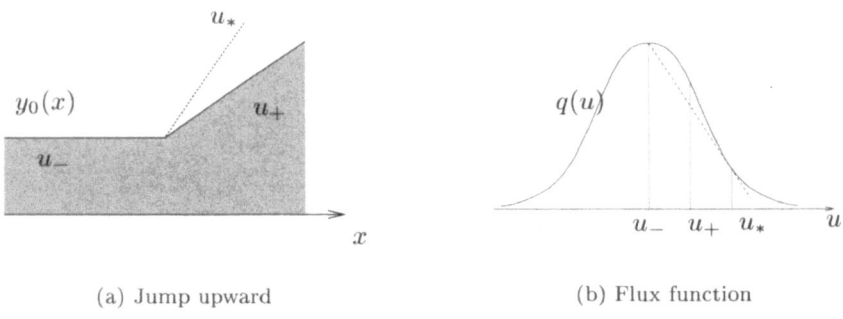

(a) Jump upward   (b) Flux function

Figure 1.43 Riemann initial value for the ion etching equation

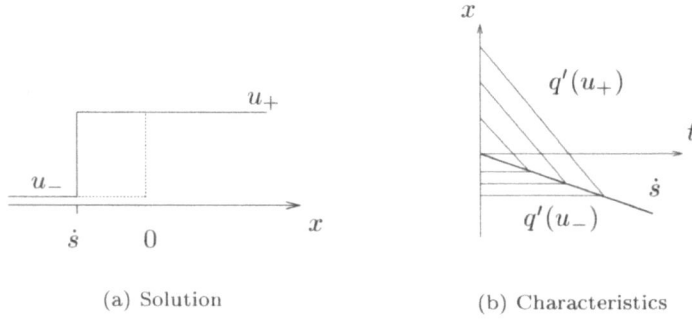

(a) Solution   (b) Characteristics

Figure 1.44 Entropy solution for the initial value in fig. 1.43(a)

We first assume that the value u_- is zero and u_+ is chosen such that the chord between $q(u_-)$ and $q(u_+)$ is below the flux function $q(u)$ on the interval $[u_-,u_+]$ (the maximal u_+ satisfying this property is called u_*). According to the entropy condition, the correct entropy solution is then given by a shock which runs into negative x direction because $[q(u)]/[u] < 0$ (see fig. 1.44(a)). The effect of the moving jump discontinuity on the slope of the surface is obtained by integration

$$y(t,x) = y(t,\bar{x}) + \int_{\bar{x}}^{x} u(t,\sigma)\, d\sigma \tag{1.42}$$

where the height of the surface at the fixed point $\bar{x}$ is obtained from relation (1.40) $\frac{\partial y}{\partial t}(t,\bar{x}) = -q(u(t,\bar{x}))$. In particular, if $\bar{x}$ is chosen positive then $u(t,\bar{x}) = u_+$ since the shock is moving to the left. Hence

$$y(t,\bar{x}) = y_0(\bar{x}) - tq(u_+).$$

Together with (1.42) we conclude that the surface is, on the one hand, moving downwards because of removal of material and, on the other hand, the kink is moving to the left (see fig. 1.45).

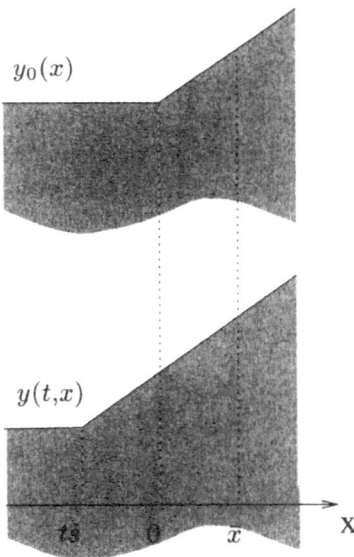

Figure 1.45 Evolution of the material surface with a kink traveling to the left

In our second example, we assume that the jump upward is stronger with $u_+ > u_*$ (see fig. 1.46(a)). Now, a single shock is no longer an entropy solution of the problem because $q(u)$ is located above and below the chord between $q(u_-)$ and $q(u_+)$. However, we can construct a weak piecewise classical solution by combining a shock wave with a rarefaction wave. In order for the shock wave to be entropy consistent, we choose again the right state $\bar{u} \leq u_*$.

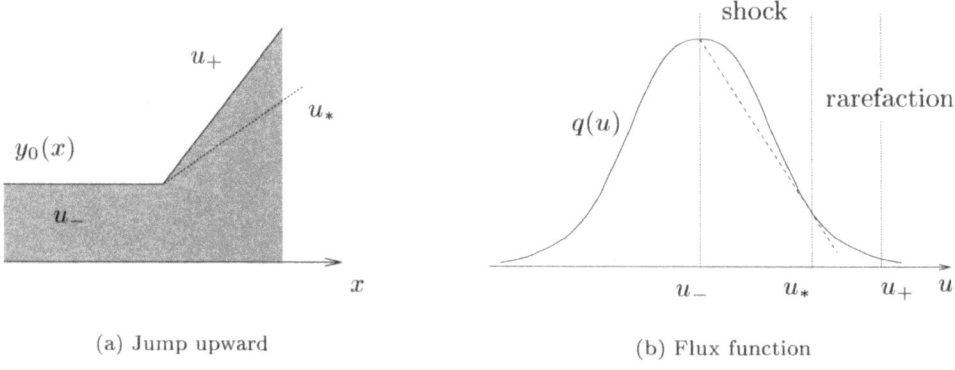

(a) Jump upward   (b) Flux function

Figure 1.46 Riemann initial value with strong jump for the ion etching equation

Then we try to continue the solution with a rarefaction wave between the values $\bar{u}$ and u_+ which requires q' to be invertible on $[\bar{u}, u_+]$. Since the combined solution must be a weak solution, we have the requirement that the right value $(q')^{-1}(\dot{s})$ generated by the rarefaction wave at the shock location $x/t = \dot{s}$ is equal to $\bar{u}$ since otherwise the shock speed $\dot{s} = (q(\bar{u}) - q(u_-))/(\bar{u} - u_-)$ is not correct. This leads to the requirement that $(q(\bar{u}) - q(u_-))/(\bar{u} - u_-) = q'(\bar{u})$, i.e. the chord between $q(u_-)$ and $q(\bar{u})$ must be tangent to the flux function in $\bar{u}$. The only point which satisfies this requirement is actually u_* giving rise to the entropy solution shown in fig. 1.47(a).

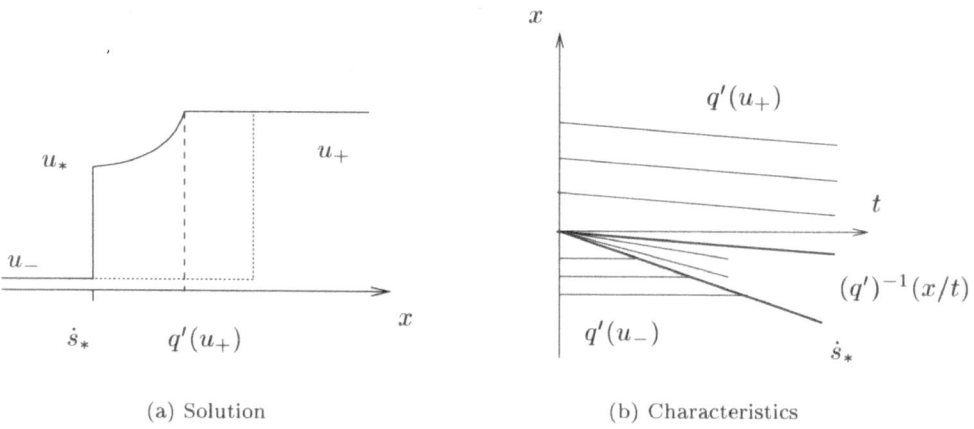

(a) Solution   (b) Characteristics

Figure 1.47 Entropy solution for the initial value in fig. 1.46(a)

For a downward jump the situation is reversed: if $u_- = 0$ and $u_+ > -u_*$, the solution simply consists of a rarefaction wave (fig. 1.48(a)) and if $u_+ < -u_*$ it is a combination of a shock and a rarefaction wave (fig. 1.48(b)).

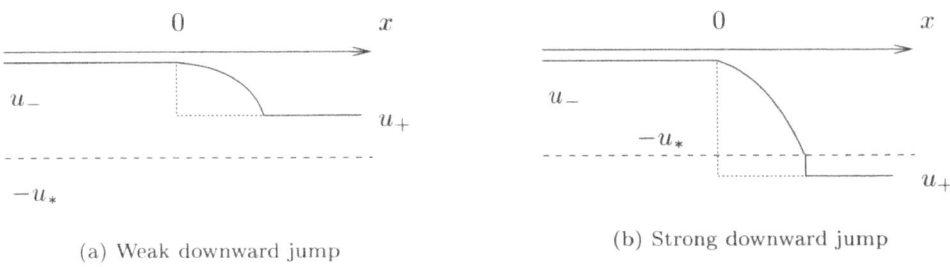

(a) Weak downward jump

(b) Strong downward jump

Figure 1.48 Entropy solutions for initial values with downward jump in the slopes

Finally, for a more complicated material with a flux having several extrema, situations like the one depicted in fig. 1.49 can appear. For the configuration of the initial surface shown in fig. 1.50(a), the entropy solution now consists of a jump, followed by a (very small) rarefaction wave which is again followed by a stationary jump and then, symmetrically, again a small rarefaction wave and a final jump to u_+ (see fig. 1.50(b)).

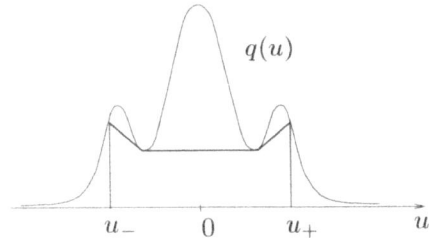

Figure 1.49 Flux function modeling a more complicated material behavior

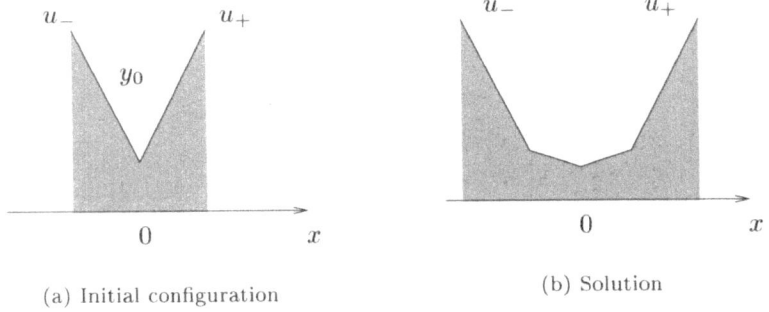

(a) Initial configuration

(b) Solution

Figure 1.50 Entropy solution in the case of complicated material behavior

For further details concerning the ion beam etching (also in the case of compound materials which leads to free boundary value problems), we refer to [4].

1.4   Systems of conservation laws

1.4.1   Characteristic surfaces

We begin with some general comments on non-linear systems of balance laws for the variables $\boldsymbol{U} = (U_1, \ldots, U_m) \in \mathcal{S} \subset \mathbb{R}^m$. Starting in the form

$$\operatorname{div}_{\boldsymbol{y}} \mathcal{A}_k(\boldsymbol{U}) = Q_k(\boldsymbol{U}), \quad k = 1, \ldots, m \tag{1.43}$$

with generalized fluxes $\mathcal{A}_k = (\mathcal{A}_k^{(1)}, \ldots, \mathcal{A}_k^{(N)}) \in C^1(\mathcal{S}, \mathbb{R}^N)$, we rewrite the problem in quasi-linear form

$$\sum_{j=1}^m (\boldsymbol{a}_{kj}(\boldsymbol{U}) \cdot \nabla) U_j = Q_k, \quad k = 1, \ldots, m \tag{1.44}$$

where the vectors $\boldsymbol{a}_{kj} = (a_{kj}^{(1)}, \ldots, a_{kj}^{(N)})$ are defined by $a_{kj}^{(i)} = \frac{\partial \mathcal{A}_k^{(i)}}{\partial U_j}$. We recall that the system is hyperbolic if the matrix $A(\boldsymbol{\xi}, \boldsymbol{U}) = (\boldsymbol{\xi} \cdot \boldsymbol{a}_{kj}(\boldsymbol{U}))$ has only real eigenvalues for all $\boldsymbol{\xi} \in \mathbb{R}^N$ and all $\boldsymbol{U} \in \mathcal{S}$. Let us now consider (1.44) together with boundary conditions on a hypersurface $\Gamma \subset \mathbb{R}^N$. If we want to determine $\boldsymbol{U}(\boldsymbol{y})$ at some point $\boldsymbol{y} \notin \Gamma$ starting from known values $\boldsymbol{\phi} \in \Gamma$, we need information about the rate of change of $\boldsymbol{U}$ in a direction normal to Γ. This information should be provided by the system (1.44) which, in fact, consists of linear combinations of directional derivatives $\boldsymbol{a}_{kj} \cdot \nabla$. In particular, we can hope to get information about the rate of change of $\boldsymbol{U}$ in normal direction if the directions $\boldsymbol{a}_{kj}$ have suitable normal components. Splitting the derivative according to

$$\boldsymbol{a}_{kj} \cdot \nabla = (\boldsymbol{n} \cdot \boldsymbol{a}_{kj})(\boldsymbol{n} \cdot \nabla) + (\boldsymbol{a}_{kj} - (\boldsymbol{n} \cdot \boldsymbol{a}_{kj})\boldsymbol{n}) \cdot \nabla$$

in a normal and a tangential derivative and combining the source term with the tangential derivative along Γ

$$b_k(\boldsymbol{U}|_\Gamma) = Q_k(\boldsymbol{U}|_\Gamma) - \sum_{j=1}^m (\boldsymbol{a}_{kj} - (\boldsymbol{n} \cdot \boldsymbol{a}_{kj})\boldsymbol{n}) \cdot \nabla U_j|_\Gamma,$$

we reformulate (1.44) on Γ as

$$A(\boldsymbol{n}, \boldsymbol{U}|_\Gamma)(\boldsymbol{n} \cdot \nabla)\boldsymbol{U} = \boldsymbol{b}(\boldsymbol{U}|_\Gamma) \tag{1.45}$$

where $A_{kj}(\boldsymbol{\xi}, \boldsymbol{U}) = (\boldsymbol{\xi} \cdot \boldsymbol{a}_{kj}(\boldsymbol{U}))$. Prescribing $\boldsymbol{U}|_\Gamma = \boldsymbol{\phi}$, we see that all normal derivatives $(\boldsymbol{n} \cdot \nabla)\boldsymbol{U}$ are obtained form the partial differential equation provided $A(\boldsymbol{n}, \boldsymbol{\phi})$ is invertible. We therefore distinguish two cases:

1) If $\det A(\boldsymbol{n}(\boldsymbol{y}), \boldsymbol{\phi}(\boldsymbol{y})) \neq 0$ for all $\boldsymbol{y} \in \Gamma$ (i.e. Γ is *nowhere-characteristic*), the differential equation provides all normal derivatives

$$(\boldsymbol{n} \cdot \nabla)\boldsymbol{U} = A^{-1}(\boldsymbol{n}, \boldsymbol{\phi})\boldsymbol{b}(\boldsymbol{\phi})$$

so that $\boldsymbol{U}$ is determined in a neighborhood of Γ by the given values $\boldsymbol{\phi}$ and the partial differential equation. Hence, nowhere-characteristic surfaces are suitable for prescribing arbitrary data $\boldsymbol{\phi}$.

2) The situation is very much different in points $y \in \Gamma$ where Γ is *characteristic*, i.e. $\det A(n(y),\phi(y)) = 0$. In this case the system (1.45) yields compatibility conditions on the data ϕ. In fact, let $0 \neq r$ be orthogonal to the image of the matrix $A(n(y),\phi(y))$. Then,

$$0 = r \cdot A(n(y),\phi(y))(n \cdot \nabla)U = r \cdot b(\phi),$$

giving rise to the condition $r \cdot b(\phi) = 0$ which puts restrictions on ϕ. Consequently, a surface Γ having characteristic points is not suited for prescribing general data ϕ. However, if all points of Γ are characteristic (i.e. if Γ is a characteristic surface), we have additional knowledge about the behavior of the solution along Γ. Exactly this property has been used in the method of characteristics where y was a two-dimensional variable and the hypersurface Γ is one-dimensional (called characteristic ground curve). In this case, b is just the derivative along Γ if there are no sources $Q = 0$, so that $r \cdot b(\phi) = 0$ implies that the solution must be constant along characteristic ground curves.

1.4.2 The Saint-Venant system

As example for a hyperbolic system, we will now consider the Saint-Venant (or shallow water) equations. These equations have a wide range of applications, for example in the prediction of water levels along rivers. A good prediction of the water height is necessary to load ships maximally under the constraint that they can reach their destination without the problem of ground contacts.

Figure 1.51 For optimal loading of ships, a prediction of the water height along the river is mandatory

The shallow water system is also used in the prediction of pollutant transport in rivers and for estimating the effects of breaking dams.

Considering, for example, the river Rhine we find typical dimensions

extension	typical size
length	1000 km
width	100 m
depth	5 m

which indicates that a one-dimensional model should be adequate, mapping a river segment of length L to the interval $[0,L]$. Then, the Saint-Venant equations are derived by a balancing consideration for water mass and momentum (for details and references we refer to [9]). Assuming constant density, the water mass contained in a slice of width dx at position x is $dm = \rho F(t,x)dx$ where F is the river cross section at position x and time t (see fig. 1.52).

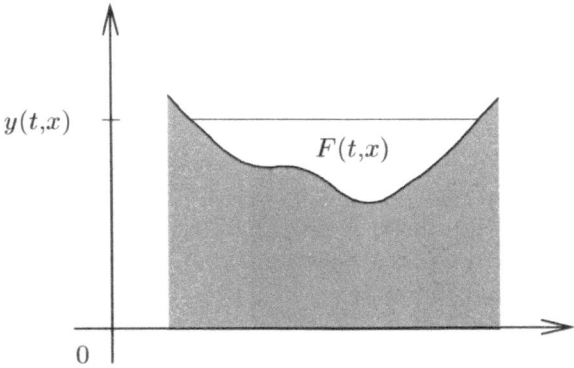

Figure 1.52 River cross section at position $x \in [0,L]$

Similarly, the mass flux is given by $\rho D(t,x)dt$ where $D = vF$ and $v(t,x)$ is an averaged flow velocity at position x. Taking into account sources d caused by influx from the sides, we obtain

$$\frac{\partial F}{\partial t} + \frac{\partial D}{\partial x} = d$$

The momentum balance leads to an equation of the form

$$\frac{\partial D}{\partial t} + \frac{\partial}{\partial x}(vD) = -gF\left(\frac{\partial y}{\partial x} + S_f\right) + v'd$$

where g is the gravitational constant and y is the water height over sea level which is related to F by some (measured) function $y = H(x,F)$. The term S_f is an empirical friction model taking into account momentum loss due to bottom friction. A typical form is $S_f = \alpha|v|v/R^{4/3}$ where $R = F/U$ and U is the wetted perimeter. Finally, v' is the velocity of the lateral inflows in the direction of the main flow. Writing the equations in quasi-linear form, we find

$$\frac{\partial F}{\partial t} + v\frac{\partial F}{\partial x} + F\frac{\partial v}{\partial x} = d$$

$$\frac{\partial v}{\partial t} + v\frac{\partial v}{\partial x} + g\frac{\partial H}{\partial F}\frac{\partial F}{\partial x} = -gF\left(\frac{\partial H}{\partial x} + S_f\right) + (v' - v)d$$

(1.46)

To investigate the hyperbolicity, we set up the matrix A

$$A(\boldsymbol{\xi},F,v) = \begin{pmatrix} \xi_1 + \xi_2 v & \xi_2 F \\ \xi_2 g\frac{\partial H}{\partial F} & \xi_1 + \xi_2 v \end{pmatrix}$$

which has eigenvalues $\lambda_{1/2} = \xi_1 + \xi_2 v \mp \sqrt{\xi_2^2 F g \frac{\partial H}{\partial F}}$. Obviously, the eigenvalues are real because at a fixed position x, larger cross section F implies a larger water height $H(x,F)$ giving rise to a positive derivative $\frac{\partial H}{\partial F} > 0$. Hence, the Saint-Venant equations are of hyperbolic type. Considering (1.46) on the domain $\Omega = \mathbb{R}^+ \times (0,L)$ we have to prescribe boundary values. For the surface $\Gamma_0 = \{0\} \times (0,L)$ we find the normal vector $\boldsymbol{n} = (-1,0)$ so that

$$A(\boldsymbol{n},F,v) = \begin{pmatrix} -1 & 0 \\ 0 & -1 \end{pmatrix}$$

always a nonzero determinant. Consequently, Γ_0 is nowhere-characteristic and arbitrary initial data can be prescribed. Conversely, let us now determine the characteristic surfaces Γ for which $\det A(\boldsymbol{n},F,v) = 0$. Using the parameterization

$$\Gamma = \{(t,\gamma(t)) : t > 0\}$$

we obtain with $\boldsymbol{n} = \alpha(\dot{\gamma}(t), -1)$, $\alpha = (1 + \dot{\gamma}^2)^{-1/2}$ the condition

$$0 = \det\begin{pmatrix} \dot{\gamma} - v & -F \\ -g\frac{\partial H}{\partial F} & \dot{\gamma} - v \end{pmatrix} = \det(\dot{\gamma}\,\mathrm{id} - A((0,1),F,v))$$

which is satisfied if and only if $\dot{\gamma}$ is an eigenvalue of $A((0,1),F,v)$. Hence, there exist two families of characteristic curves (see fig. 1.53) which are defined by

$$\dot{\gamma} = \lambda_1(\gamma) \qquad\qquad \lambda_1 = v - \sqrt{Fg\frac{\partial H}{\partial F}}$$

$$\dot{\gamma} = \lambda_2(\gamma) \qquad\qquad \lambda_2 = v + \sqrt{Fg\frac{\partial H}{\partial F}}$$

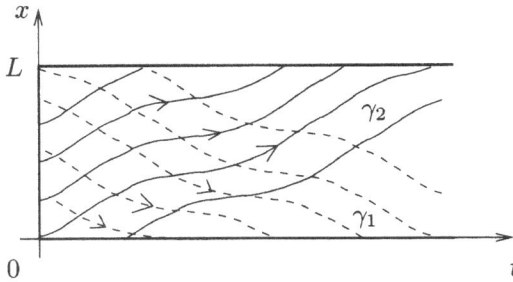

Figure 1.53 Two families of characteristic curves for the shallow water system

The orientation of the two families of curves at the boundaries $\mathbb{R}^+ \times \{0\}$ and $\mathbb{R}^+ \times \{L\}$ are important to determine the number of boundary conditions which have to be supplied. In the case $|v| < \sqrt{Fg\frac{\partial H}{\partial F}}$ (subcritical flow) one has to prescribe one condition at each boundary (for example the water level at the upper and lower end of the observed river segment). The supercritical case $|v| > \sqrt{Fg\frac{\partial H}{\partial F}}$ is characterized by the fact that two boundary conditions (level and flux) have to be prescribed at one and no boundary condition at the other boundary (see fig. 1.54).

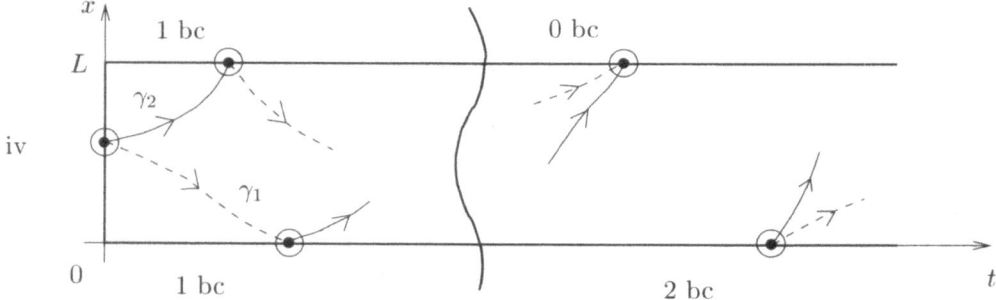

Figure 1.54 Number of boundary conditions (bc) for sub- and supercritical flow

In the following we are going to consider the Riemann problem for a simplified version of the shallow water system. First, we assume that the river has no slope and always a rectangular cross section. As variable for the height we introduce $h(t,x)$ so that $F(t,x) = bh(t,x)$ (see fig. 1.55).

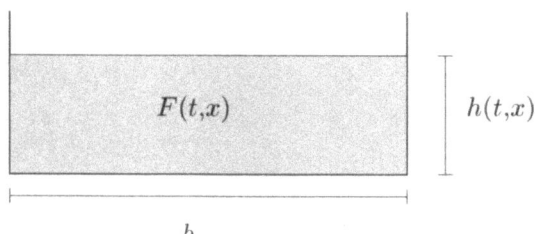

Figure 1.55 River geometry for the model problem

Moreover, we assume that there are no lateral in- and outflows and no friction. By choosing the length scale L, velocity scale $V = \sqrt{gL}$ and time scale $T = L/V$, we eventually get the system (in scaled variables)

$$\frac{\partial h}{\partial t} + \frac{\partial hv}{\partial x} = 0, \qquad \frac{\partial hv}{\partial t} + \frac{\partial}{\partial x}(hv^2 + h^2/2) = 0, \qquad (1.47)$$

or with $q = hv$, $\boldsymbol{u} = (h,q)$ and $\boldsymbol{G}(\boldsymbol{U}) = (q, q^2/h + h^2/2)$ and Riemann initial data

$$\frac{\partial U}{\partial t} + \frac{\partial}{\partial x} G(U) = 0, \qquad U(0,x) = \begin{cases} U_- & x < 0 \\ U_+ & x > 0 \end{cases} \qquad (1.48)$$

Typical Riemann initial conditions are shown in fig. 1.56.

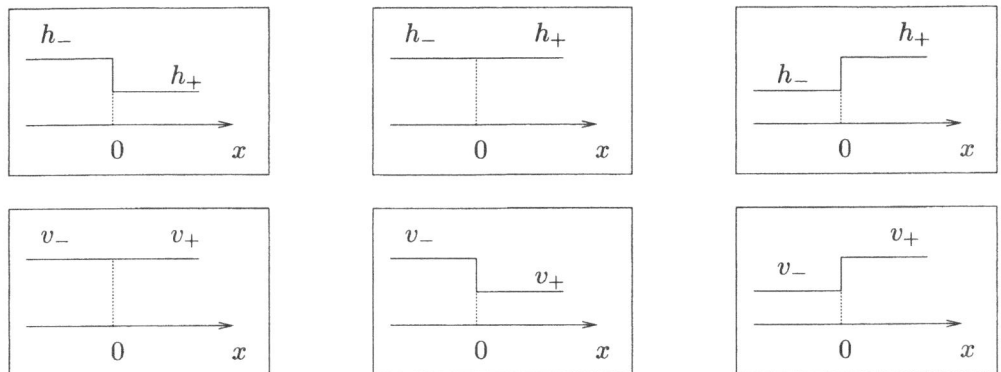

Figure 1.56 Possible Riemann initial values for water height and velocity

1.4.3   Rarefaction waves

We start with the construction of specific solutions of the system (1.47) called k-simple waves with $k = 1,2$. If $r_k(U)$ is an eigenvector of $\nabla G(U)$ with corresponding eigenvalue $\lambda_k(U)$, we proceed as follows:

1) we first solve $\gamma'(s) = r_k(\gamma(s))$ with some initial condition

2) we then solve $\frac{\partial w}{\partial t} + \lambda_k(\gamma(w))\frac{\partial w}{\partial x} = 0$ with initial conditions

3) finally, we set $U(t,x) = \gamma(w(t,x))$

The function U (the k-simple wave) is a solution of (1.47) because

$$\frac{\partial U}{\partial t} = \gamma'\frac{\partial w}{\partial t} = -r_k\lambda_k\frac{\partial w}{\partial x} == -\nabla G\, r_k\frac{\partial w}{\partial x} = -\nabla G\frac{\partial U}{\partial x} = -\frac{\partial}{\partial x}G(U).$$

In order to work out this program with rarefaction solutions w, we first determine the curves γ. Eigenvalues and eigenvectors of ∇G are given by

$$\lambda_{1/2} = v \mp \sqrt{h} = \frac{q}{h} \mp \sqrt{h}, \qquad r_{1/2} = (1,\lambda_{1/2})$$

and the curves in (h,q) space are (see fig. 1.57)

$$\gamma^{(1/2)}(h) = (h,\bar{\kappa}_{1/2}h \mp 2h^{3/2}), \qquad \bar{\kappa}_{1/2} = \bar{v} \pm 2\sqrt{h}$$

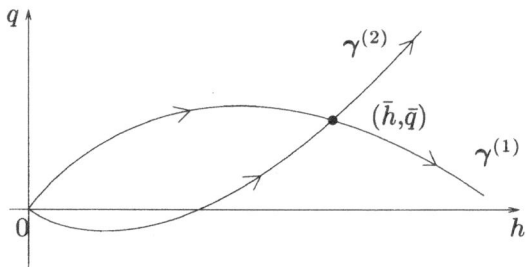

Figure 1.57 Integral curves of eigenvectors fields r_1 and r_2

In the second step, we solve the scalar equation for w which turns into a conservation law by introducing the flux function

$$g_k(w) = \int_0^w \lambda_k(\gamma^{(k)}(\sigma))\, d\sigma, \qquad w > 0.$$

Note that $\lambda_k(\gamma^{(k)}(\sigma)) = \frac{d}{d\sigma}\gamma_2^{(k)}(\sigma)$ and hence $g_k(w) = \gamma_2^{(k)}(w)$ so that g_1 is concave and g_2 is convex (the graphs of g_1, g_2 coincide with the curves in fig. 1.57). Consequently, the correct entropy solution of

$$\frac{\partial w}{\partial t} + \frac{\partial}{\partial x} g_k(w) = 0, \quad w(0,x) = \begin{cases} w_- & x < 0 \\ w_+ & x > 0 \end{cases}$$

is a rarefaction wave if $g_k'(w_+) > G_k'(w_-)$ which implies $w_- > w_+$ for $k = 1$ and $w_- < w_+$ in the case $k = 2$. In both cases, the property $g_k'' \neq 0$ together with $g_k' = \lambda_k(\gamma^{(k)})$ implies $r_k \cdot \nabla \lambda_k \neq 0$ (i.e. the pair (λ_k, r_k) is *genuinely nonlinear*). As specific example of a w-rarefaction wave, we consider $k = 1$, $\bar{h} = 1$, $\bar{v} = 0$, $w_- = 4$, and $w_+ = 1$. In this case, $\bar{\kappa}_1 = 2$ and $g_1(w) = 2w - 2w^{3/2}$ with derivative $g_1'(w) = 2 - 3\sqrt{w}$. Note that the rarefaction solution is based on

$$(g_1')^{-1}(\xi) = (2 - \xi)^2/9.$$

and has the complete form (see also fig. 1.58)

$$w(t,x) = \begin{cases} w_- = 4 & x/t < g_1'(w_-) = -4 \\ (2 - x/t)^2/9 & -4 < x/t < -1 \\ w_+ = 1 & x/t > g_1'(w_+) = -1 \end{cases}$$

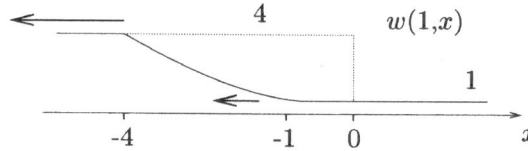

Figure 1.58 Solution of the Riemann problem for the auxiliary quantity w

Based on w, we now define the 1-rarefaction wave $U^0(x) = \gamma^{(1)}(w(0,x))$ (see fig. 1.59(a))

$$h(t,x) = w(t,x), \qquad v(t,x) = 2 - 2\sqrt{w(t,x)}$$

which solves the Riemann problem (1.48) with

$$U^0(x) = \gamma^{(1)}(w(0,x)) = \begin{cases} (4,-8) & x < 0 \\ (1,0) & x > 0 \end{cases}$$

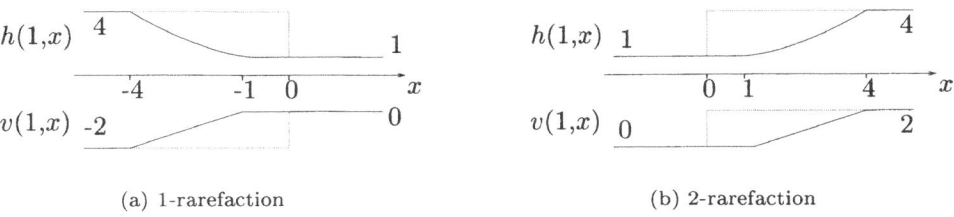

(a) 1-rarefaction   (b) 2-rarefaction

Figure 1.59 Rarefaction waves of the shallow water system

Similarly, a 2-rarefaction wave is obtained with $\bar{h} = 1$, $\bar{v} = 0$, $w_- = 1$, $w_+ = 4$, giving rise to $U_- = (1,0)$, and $U_+ = (4,8)$ (see fig. 1.59(b)).

1.4.4   Shock waves

Similar to the scalar case, one introduces weak solutions for systems of conservation laws: a function $U \in [\mathbb{L}_{loc}^\infty(\mathbb{R}^2)]^2$, $U(t,x) = 0$ for $t < 0$ is called *weak solution* of (1.48) if

$$\operatorname{div}_{\boldsymbol{y}}\mathcal{A}_i(U) = U_i^0\delta_{\Gamma_0}, \quad \text{in } \mathcal{D}'(\mathbb{R}^2), \quad i = 1,2$$

where $\mathcal{A}_i(U) = (U_i, G_i(U))$, $\boldsymbol{y} = (t,x)$ and $\Gamma_0 = \{0\} \times \mathbb{R}$.

Using the divergence relation (1.37), one shows that a piecewise classical solution is a weak solution if and only if $[\mathcal{A}_i(U)] \cdot \boldsymbol{n} = 0$ on the shock curve Γ_s. In terms of the slope $\dot{s}(t)$ of the shock curve, the condition reads

$$\dot{s}[U_i] = [G_i(U)], \qquad i = 1,2$$

and for the specific case of the shallow water system

$$\dot{s}[h] = [q], \qquad \dot{s}[q] = [q^2/h + h^2/2].$$

With the first condition one determines the shock speed $\dot{s} = [q]/[h]$ which then yields with the second relation

$$[q]^2 = [h][q^2/h + h^2/2]. \tag{1.49}$$

This relation implies that only selected pairs of states (q,h) and $(\bar{q},\bar{h})$ can be connected by a shock. Inserting $[q] = q - \bar{q}$, $[h] = h - \bar{h}$ and $[q^2/h + h^2/2] = q^2/h + h^2/2 - \bar{q}^2/\bar{h} + \bar{h}^2/2$, we can transform (1.49) into the relation

$$q = \bar{v}h \mp (h - \bar{h})\sqrt{\left(\frac{1}{h} + \frac{1}{\bar{h}}\right)/2} \tag{1.50}$$

Hence, the states (h,q) which can be connected with $(\bar{q},\bar{h})$ by a shock are located on two curves $q = q^{(1)}(h)$ and $q = q^{(2)}(h)$ depending on the sign $-$ or $+$ in (1.50) respectively (see fig. 1.60. The corresponding shock speeds are

$$\sigma_{1/2}(\boldsymbol{U},\bar{\boldsymbol{U}}) = \bar{v} \mp \sqrt{\frac{h}{2\bar{h}}(h + \bar{h})}.$$

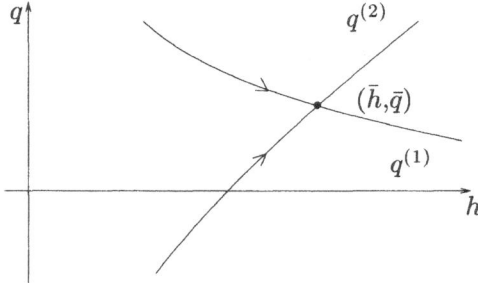

Figure 1.60 States which can be connected with $(\bar{q},\bar{h})$ by a weak solution with shock

In order to check which combination of states not only leads to a weak solution but also satisfies an entropy condition, we set $\boldsymbol{U}_- = (\bar{h},\bar{q})$ and consider successively $\boldsymbol{U}_+$ on the branches formed by the curves $q^{(1)}(h)$ and $q^{(2)}(h)$ in fig. 1.60. First, we consider the case in fig. 1.61(a). A simple calculation shows that on this branch

$$\lambda_1^- = \lambda_1(\boldsymbol{U}_-) < \sigma_1(\boldsymbol{U}_+,\boldsymbol{U}_-) < \lambda_1(\boldsymbol{U}_+) = \lambda_1^+$$

which leads to a configuration of characteristic ground curves as shown in fig. 1.61(b).

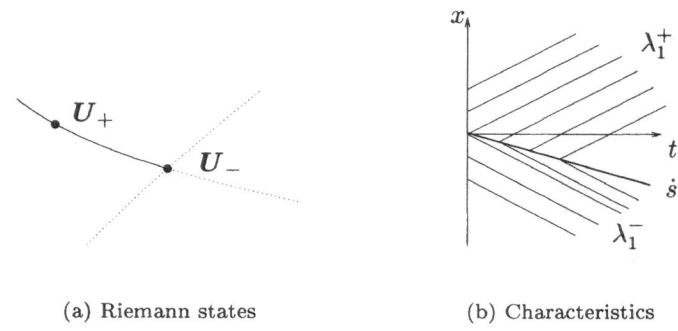

(a) Riemann states   (b) Characteristics

Figure 1.61 Situation in which shock solution is not entropy consistent

Obviously, the characteristics do not run into the shock and we discard this branch as being entropy violating. On the branch in fig. 1.62(a), on the other hand, we find

$$\lambda_1^+ < \sigma_1(U_+,U_-) < \min\{\lambda_1^-,\lambda_2^+\}$$

so that λ_1-characteristic ground curves run into the so called 1-shock (see fig. 1.62(b)).

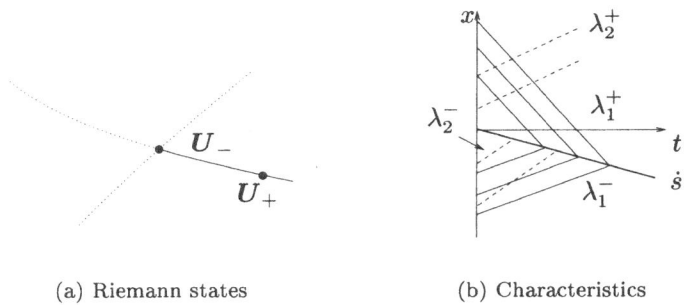

(a) Riemann states   (b) Characteristics

Figure 1.62 Situation of an entropy consistent 1-shock

Continuing with the branches of the curve $q^{(2)}$, we find that in the case of fig. 1.63(a)

$$\lambda_2^- < \sigma_2(U_+,U_-) < \lambda_2^+$$

leading to jumps which are not entropy consistent. Finally, a consideration for state U_+ on the branch in fig. 1.64(a) shows that

$$\max\{\lambda_1^-,\lambda_2^+\} < \sigma_2(U_+,U_-) < \lambda_2^-$$

so that λ_2-characteristics run into the entropy consistent 2-shock (see fig. 1.64(b)).

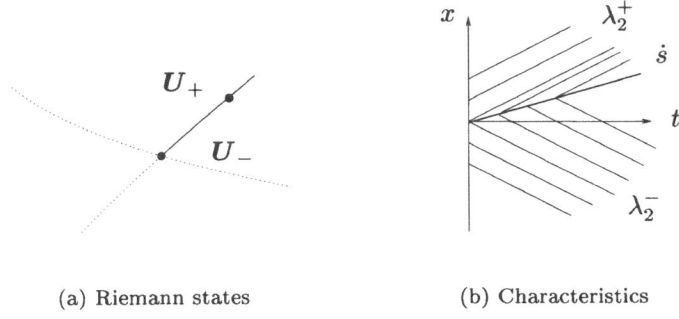

(a) Riemann states   (b) Characteristics

Figure 1.63 Situation in which shock solution is not entropy consistent

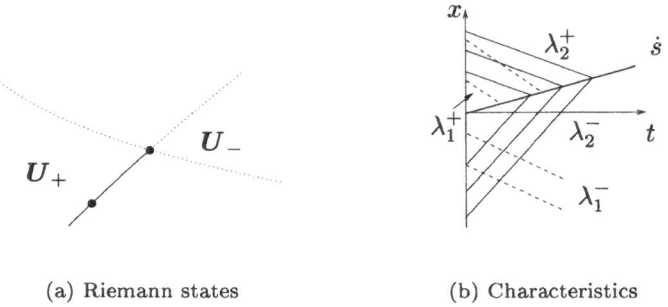

(a) Riemann states   (b) Characteristics

Figure 1.64 Situation of an entropy consistent 2-shock

Examples for 1- and 2-shocks are shown in fig. 1.65(a) and fig. 1.65(b) for $U_- = (1,0)$ and $U_+ = (2, -\sqrt{3/4})$ respectively $U_+ = (1/2, -\sqrt{3/8})$.

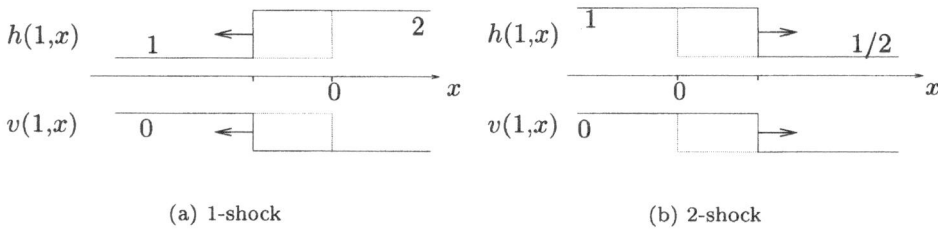

(a) 1-shock   (b) 2-shock

Figure 1.65 Shock solutions for the shallow water system

Summarizing our considerations we conclude that the states U_+ which can be connected with the state U_- by a shock wave are located on the lower branches of the $q^{(1)}$ and the $q^{(2)}$ curves which we denote by $S_1(U_-)$ and $S_2(U_-)$ respectively (see fig. 1.66).

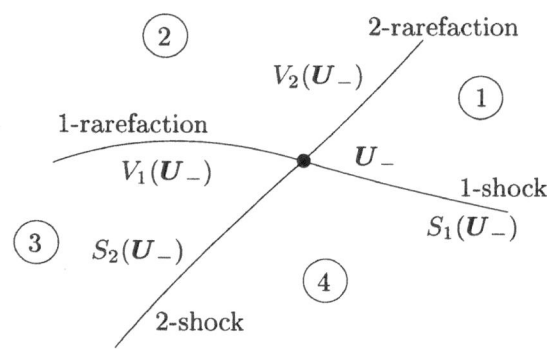

Figure 1.66 In regions 1 – 4 the solution of the Riemann problem is a combination of shocks and rarefaction waves

A connection between U_+ and U_- with an entropy consistent rarefaction wave is possible if U_+ is located on the upper branches of $\gamma^{(1)}$ or $\gamma^{(2)}$ (see fig. 1.57) which we denote $R_1(U_-)$ and $R_2(U_-)$. If U_+ is located on none of the branches S_1, S_2, R_1, and R_2, a connection with a single shock or rarefaction is not possible. For example, if U_+ is located in region 1 (fig. 1.66), there exists a unique $U_* \in S_1(U_-)$ such that $U_+ \in R_2(U_*)$ (see fig. 1.67).

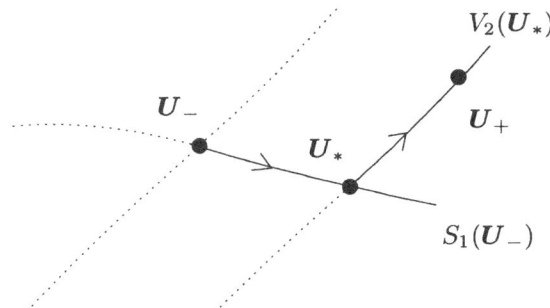

Figure 1.67 Connection of states U_- and U_+ via the intermediate state U_*

The corresponding (t,x) diagram and the solution at time $t = 1$ which is a combination of a 1-shock and a 2-rarefaction are shown in fig. 1.68.

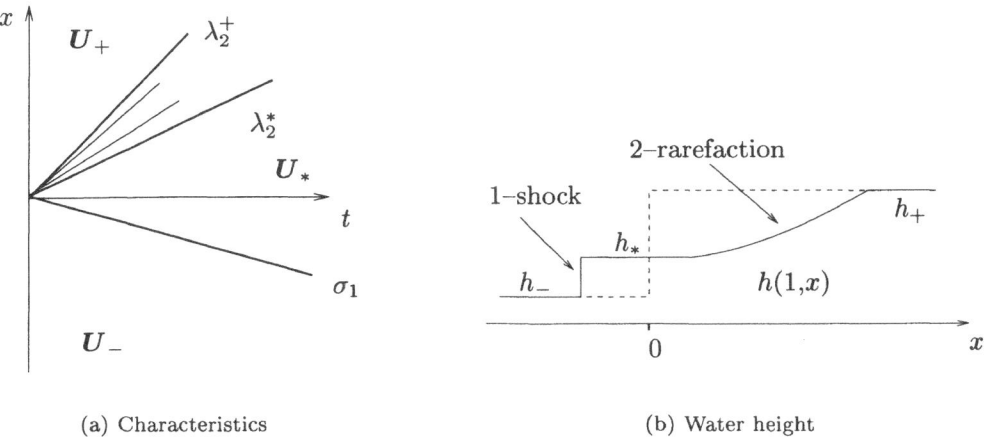

(a) Characteristics   (b) Water height

Figure 1.68 Combination of a 1-shock and a 2-rarefaction wave

For U_+ in region 2, two rarefaction waves are connected (fig. 1.69), region 3 corresponds to a 1-rarefaction and a 2-shock (fig. 1.70(a)). Finally, for U_+ in region 4, the solution is a combination of two shocks according to fig. 1.70(b).

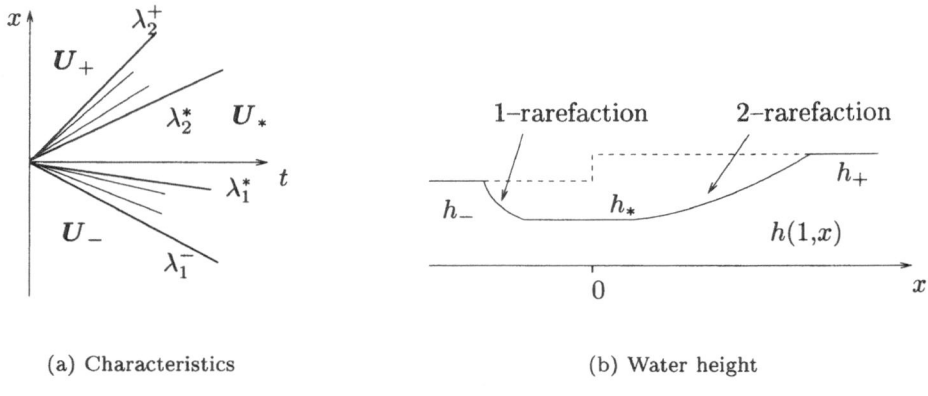

(a) Characteristics

(b) Water height

Figure 1.69 Combination of a 1-rarefaction and a 2-rarefaction wave

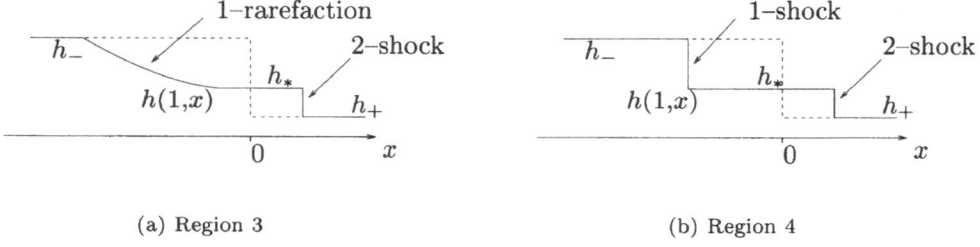

(a) Region 3

(b) Region 4

Figure 1.70 Solutions corresponding to U_+ in region 3 and 4 of fig. 1.66

For further details on Riemann problems for non-linear systems of conservation laws in one space dimension we refer, for example, to [7, 10, 6].

Bibliography

[1] C Cercignani. *The Boltzmann Equation and its Applications*. Springer, 1988.

[2] Alexandre J. Chorin and Jerrold E. Marsden. *A Mathematical Introduction to Fluid Mechanics*. Springer, 3rd edition, 1993.

[3] Lawrence C. Evans. *Partial Differential Equations*, volume 9 of *Graduate Studies in Mathematics*. American Mathematical Society, 1998.

[4] Avner Friedman. *Mathematics in Industrial Problems*, volume 16 of *The IMA Volumes in Mathematics and its Applications*. Springer, 1988.

[5] Edwige Godlewski and Pierre-Arnaud Raviart. *Hyperbolic systems of conservation laws*. Mathematiques & Applications. Ellipses, Paris, 1991.

[6] Edwige Godlewski and Pierre-Arnaud Raviart. *Numerical approximation of hyperbolic systems of conservation laws*, volume 118 of *Applied Mathematical Sciences*. Springer, New York, 1996.

[7] J. Kevorkian. *Partial differential equations. Analytical solution techniques*. Mathematics Series. Wadsworth & Brooks/Cole, Pacific Grove, Ca, 1990.

[8] J. Necas, J. Malek, M. Rokyta, and M. Ruzicka. *Weak and measure-valued solutions to evolutionary PDEs.*, volume 13 of *Applied Mathematics and Mathematical Computation*. Chapman & Hall, London, 1996.

[9] Peter Rentrop and Gerd Steinebach. Model and numerical techniques for the alarm system of river rhine. *Surv. Math. Ind.*, 6(4):245–265, 1997.

[10] Joel Smoller. *Shock waves and reaction-diffusion equations*, volume 258 of *Grundlehren der Mathematischen Wissenschaften*. Springer, New York, 2nd edition, 1994.

[11] Matthew Witten. *Hyperbolic partial differential equations : populations, reactors, tides and waves, theory and applications*. Pergamon Press, Oxford, 1983.

2   Central Schemes and Systems of Balance Laws

Introduction

The development of shock capturing schemes for the numerical approximation of the solution of conservation laws has been a very active field of research in the last two decades. There are several motivations for this effort. First, the challenge is a mathematical one. Solutions of conservation laws may develop jump discontinuities in finite time. To understand how to obtain numerical approximations that converge to the (discontinuous) solution has been a non trivial task. The mathematical theory of quasilinear hyperbolic systems of conservation laws has been used as a guideline in the development of modern shock capturing schemes. The concept of entropy condition and total variation diminishing are common to mathematical analysts and numerical analysts who deal with the problem. Another motivation for the development of shock capturing schemes is provided by the large number of applications. Complex flows in gas dynamics require the use of efficient and accurate schemes, which are able to deal with complex geometries. Unstructured mesh or adaptive mesh refinement become necessary so solve realistic problems. The schemes that are used for practical problems are usually different from the schemes for which theoretical results can be proven. For example, very little is known about the convergence property of high order schemes.

As a research field, I find it is one of the most fascinating in numerical analysis. In fact, it requires the knowledge of many areas of basic numerical analysis, from approximation theory to numerical methods for ordinary differential equations (many sophisticated concepts developed in the context of numerical methods for ODE's are now being imported and used for the development of numerical methods for PDE's), it is based on the mathematical theory of weak solutions for conservation laws, and it is strongly motivated by the several applications in many physical systems, from gas dynamics, MHD, reacting flows, semiconductor modeling, and so on.

A good introductory book which deals with wave propagation and shock capturing schemes (mainly upwind, in one space dimension) is the book by LeVeque [37]. A mathematically oriented reference book on the numerical solution of conservation laws is the book by Godlewski and Raviart [19]. Several schemes, mainly based on Riemann solver, with a lot of numerical examples are considered in the book by Toro [66]. A good review of modern numerical techniques for the treatment of hyperbolic systems of conservation laws is given in the lecture notes of a CIME course held in 1998 [14].

Most schemes for the numerical solution of conservation laws are based on the Godunov scheme, and on the numerical solution of the Riemann problem [34]. For numerical purpose it is often more convenient to resort to approximate solvers, such as the one proposed by Roe for gas dynamics [54].

Although such solution or its numerical approximation is known in many cases of physical relevance, there are other cases, such as gas in Extended Thermodynamics, or hydrodynamical models of semiconductors, in which the eigenvalues and eigenvectors of the matrix of the system are not known analytically, and for which the solution to the Riemann problem or its numerical approximation is hard to compute. In such cases it is desirable to use schemes that do not require the knowledge of the solution of the Riemann problem.

We call such schemes "Central schemes". The prototype central scheme is Lax-Friedrichs scheme. It is well known that Lax-Friedrichs scheme is more dissipative than first order upwind scheme, however it is simpler to use, since it does not require the knowledge of the sign of the flux derivative or the eigenvector decomposition of the system matrix (see for example [37] for a comparison between Lax-Friedrichs and upwind schemes).

Second order central schemes have been introduced in [50] and [57]. After that, central schemes have developed in several directions. We mention here the improvement of second order central scheme and the development of semidiscrete central scheme in one space dimension [32], the development of high order central schemes in one dimension [45, 11, 12, 38, 31], central schemes in several space dimensions on rectangular grids [6, 25, 28, 39, 40, 41, 43, 56, 57], and on unstructured grids [7, 8, 20], the development of central schemes for hyperbolic systems with source term [10, 46, 51], and the application of central schemes to geometrical optics [16], Hamilton-Jacobi equation [33, 44], incompressible flows [30, 42, 29], hydrodynamical models of semiconductors [1, 4, 5, 55, 67]. The above list is far from being complete. It is meant to give an idea of the recent spreading of central schemes in the numerical community.

The plan of the chapter is the following. In the next section we give a brief review on hyperbolic systems and their numerical approximation. Then we describe the popular Godunov method, based on the solution of the Riemann problem. Finally we present the second order Nessyahu-Tadmor scheme, with a few numerical examples. Section 2.2 is devoted to the development of high order central schemes in one space dimension. They are based on accurate non-oscillatory reconstruction of a function from cell-averages (Weighted Essentially Non-Oscillatory reconstruction), and on accurate evaluation of the integral of the flux, through the use of Runge-Kutta methods with Natural Continuous Extension. Section 2.3 is devoted to second and high order central schemes in two space dimensions. Some application to gas dynamics is presented. In the last section of the chapter we describe the treatment of systems with source terms, with particular emphasis on stiff sources. Finally, we describe possible developments of central schemes for systems with source.

2.1   Second order central schemes

2.1.1   Hyperbolic systems

Let us consider a system of equations of the form

$$\frac{\partial u}{\partial t} + \frac{\partial f(u)}{\partial x} = 0 \qquad (2.1)$$

where $u(x,t) \in \mathbb{R}^d$ is the unknown vector field, and $f : \mathbb{R}^d \to \mathbb{R}^d$ is assumed to be a smooth function. The system is hyperbolic in the sense that for any $u \in \mathbb{R}^d$, the Jacobian matrix $A = \nabla_u f(u)$ has real eigenvalues and its eigenvectors span $\mathbb{R}^d$.

Such system is linear if the Jacobian matrix does not depend on u, otherwise it is called quasilinear.

Linear hyperbolic systems are much easier to study. For these systems, the initial value problem is well posed, and the solution maintains the regularity of the initial data for any time.

Such systems can be diagonalized, and therefore they can be reduced to d linear scalar equations.

The situation is much different with quasilinear systems. For them the initial value problem is well posed locally in time. In general, the solution loses the regularity of the initial data after finite time. Even in the case of the single scalar equation, i.e. $d = 1$, the strong solution ceases to exist, and it is necessary to consider weak solutions. These have the general appearance of piecewise smooth functions, which contain jump discontinuities [35].

If we denote by x_Σ the position of the discontinuity Σ, and by V_Σ its velocity, then the jump conditions across Σ read

$$-V_\Sigma \llbracket u \rrbracket + \llbracket f \rrbracket = 0 \tag{2.2}$$

where for any quantity $h(x)$, $\llbracket h \rrbracket = h(x_\Sigma^+) - h(x_\Sigma^-)$ denotes the jump across the discontinuity interface Σ.

As an example, Figure (2.1) shows the solution of Burgers equation, for which $d = 1$ and $f = u^2/2$, with the initial condition

$$u(x,0) = 1 + \frac{1}{2}\sin(\pi x) \tag{2.3}$$

$x \in [-1,1]$, and periodic boundary conditions. A discontinuity forms at time $t = 2/\pi \approx 0.6366$. The figure shows the initial condition, and the solution at times $t = 0.5$ and $t = 0.8$. In the latter case, the parametric solution constructed by the characteristics is multi-valued. Single valued solution is restored by fitting a shock discontinuity at a position that maintains conservation [69].

Piecewise smooth solutions that satisfy the jump conditions are not unique. Entropy condition is used to guarantee uniqueness of the solution, at least in the scalar case.

Entropy condition states that for any convex function $\eta(u)$ there exists an entropy flux $\psi(u)$ such that the pair $[\eta,\psi]$ satisfies

$$\frac{\partial \eta}{\partial t} + \frac{\partial \psi(u)}{\partial x} \leq 0 \tag{2.4}$$

for any weak solution of the equation, and the equal sign holds for smooth solutions.

In the scalar case the entropy condition ensures that the weak solution is the unique viscosity solution, i.e. it is obtained as the limit

$$\lim_{\epsilon \to 0} u^\epsilon(x,t),$$

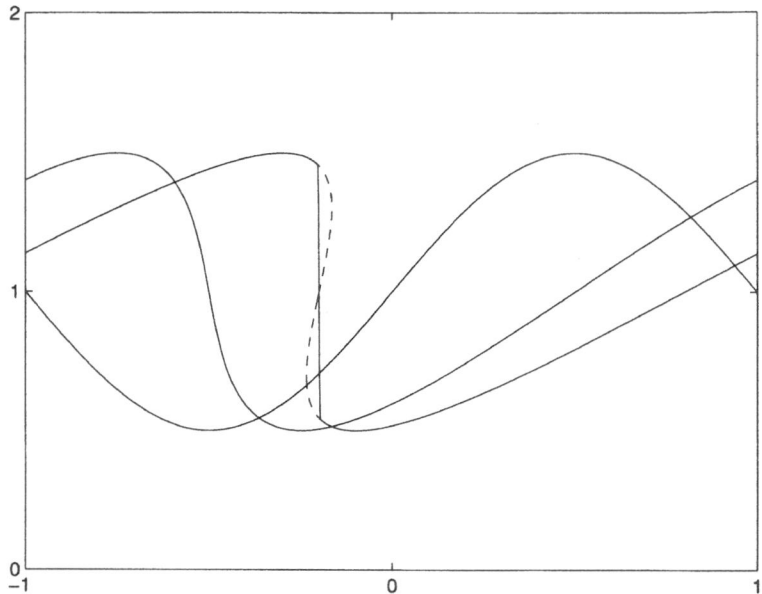

Figure 2.1 Burgers equation at different times ($t = 0,0.5,0.8$)

where u^ϵ satisfies the equation

$$\frac{\partial u^\epsilon}{\partial t} + \frac{\partial f(u^\epsilon)}{\partial x} = \epsilon \frac{\partial^2 u^\epsilon}{\partial x^2}.$$

For the relation between entropy condition and viscosity solutions in the case of systems see, for example, [35].

The mathematical theory of hyperbolic systems of conservation laws is used as a guideline for the construction of schemes for the numerical approximation of conservation laws. Consider, for example, the conservation property. Integrating Eq.(2.1) over an interval $[a,b]$ one has

$$\frac{d}{dt} \int_a^b u(x,t)\,dx = f(u(a,t)) - f(u(b,t))$$

if $u(a,t) = u(b,t)$ (for example if the boundary conditions are periodic) then the quantity $\int_a^b u(x,t)$ is conserved in time. Such conservation property is directly related to the jump condition (2.2).

It is important that a similar conservation property is maintained at a discrete level. In this way the schemes will provide the correct propagation speed for the discontinuities.

2.1.2   Conservative schemes

Most commonly used schemes for the numerical approximation of conservation laws are finite element, finite difference, and finite volume schemes. We shall mainly consider here finite volume schemes.

These are obtained by discretizing space into cells $I_j = [x_{j-1/2}, x_{j+1/2}]$, and time in discrete levels t_n. For simplicity we assume that the cells are all of the same size $\Delta x = h$, so that the center of cell j is $x_j = x_0 + jh$. This assumption is not necessary for finite volume schemes.

Integrating the conservation law on a cell in space-time $I_j \times [t_n, t_{n+1}]$ (see Figure (2.2)) one has

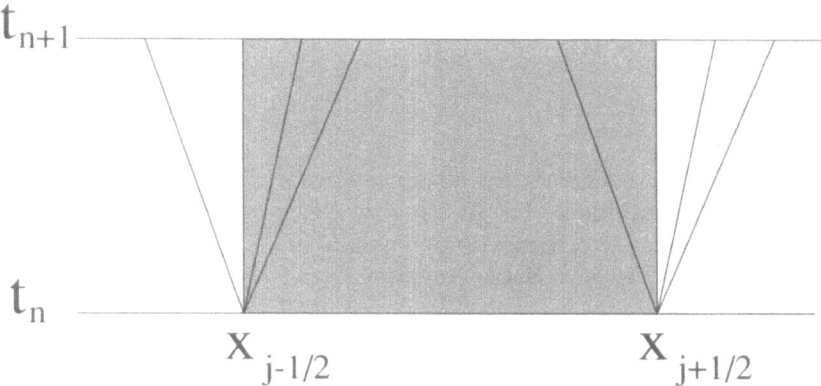

Figure 2.2 Integration over a cell and Godunov methods

$$\int_{x_{j-1/2}}^{x_{j+1/2}} u(x, t_{n+1}) = \int_{x_{j-1/2}}^{x_{j+1/2}} u(x, t_n) - \int_{t_n}^{t_{n+1}} (f(u(x_{j+1/2}, t)) - f(u(x_{j-1/2}, t))) \, dt.$$

$$(2.5)$$

This (exact) relation suggests the use of numerical scheme of the form

$$\bar{u}_j^{n+1} = \bar{u}_j^n - \frac{\Delta t}{h}(F_{j+1/2} - F_{j-1/2}),$$
$$(2.6)$$

where $\bar{u}_j^n$ denotes an approximation of the cell average of the solution on cell j at time t_n, and $F_{j+1/2}$, which approximates the time average of the flux on the boundary of the cell, is the so called *numerical flux*, and depends on the cell average of the cells surrounding point $x_{j+1/2}$. In the simplest case it is

$$F_{j+1/2} = F(\bar{u}_j, \bar{u}_{j+1})$$

with $F(u, u) = f(u)$ for consistency. Such schemes, called conservative schemes, satisfy a conservation property at a discrete level. This is essential in providing the correct propagation speed for discontinuities, which depend, through Eq. (2.2), uniquely on the conservation properties of the system.

Furthermore, Lax-Wendroff theorem ensures that if $u(x,t)$ is the limit of a sequence of discrete solutions $\bar{u}_j^n$ of a consistent conservative scheme, obtained as the discretization parameter h vanishes, then $u(x,t)$ is a weak solution of the original equation. Lax-Wendroff theorem assumes that the sequence of numerical solutions converges strongly to a function $u(x,t)$. Convergence of numerical schemes is studied through the TVD (Total Variation Diminishing) property. A discrete entropy condition is used to guarantee that the numerical solution converges to the unique admissible solution of Eq. (2.1). For a discussion on these issues see the book by LeVeque [37] of Godlewski and Raviart [19].

2.1.3   Godunov scheme

The prototype of upwind schemes for conservation laws is the Godunov scheme. It is based on two fundamental ideas. The first is that the solution is reconstructed from cell averages at each time step as a piecewise polynomial in x. In its basic form, the solution is reconstructed as a piecewise constant function

$$u(x,t_n) \approx R(x; \bar{u}^n) = \sum_j \bar{u}_j^n \chi_j(x) \tag{2.7}$$

where $\chi_j(x)$ is the indicator function of interval $I_j = [x_{j-1/2}, x_{j+1/2}]$. The second is that for a piecewise constant function, the solution of the system, for short time, can be computed as the solution of a sequence of Riemann problems.

A Riemann problem is a Cauchy problem for a system of conservation laws, where the initial condition is given by two constant vectors separated by an interface

$$u(x,0) = \begin{cases} u_- & x < 0 \\ u_+ & x > 0 \end{cases} \tag{2.8}$$

For the scalar equation the solution to the Riemann problem is known analytically. For a system of conservation laws, the solution consists of a similarity solution that depends on the variable x/t. In several cases of interests, such as gas dynamics with polytropic gas, the solution to the Riemann problem is known analytically [35].

Sometimes, for efficiency reason, it is more convenient to use approximate solutions to the Riemann problem [54].

Once the solution to the Riemann problem is known, it can be used for the construction of the Godunov scheme.

Let us denote by $u^*(u_-, u_+)$ the solution of the Riemann problem at $x = 0$. Then the *exact* solution of Eq.(2.1) can be computed from Eq.(2.5):

$$\bar{u}_j^{n+1} = \bar{u}_j^n - \frac{\Delta t}{\Delta x}(f(u^*(\bar{u}_{j-1}, \bar{u}_j)) - f(u^*(\bar{u}_j, \bar{u}_{j+1}))) \tag{2.9}$$

This relation is exact if the Riemann fan does not reach the boundary of the cell (see Figure (2.2)), i.e. if the following CFL condition is satisfied

$$\Delta t < \frac{\Delta x}{\rho(A)} \tag{2.10}$$

where $\rho(A) = \max_{1 \le i \le d} |\lambda_i(A)|$ is the spectral radius of the Jacobian matrix $A = \nabla_u f$, λ_i denoting the i-th eigenvalue.

Once the cell averages are computed at the new time t_{n+1}, then the solution at this time is again approximated by a piecewise constant solution of the form (2.7).

Godunov scheme is first order accurate, it is Total Variation Diminishing, and it satisfies a discrete entropy inequality. When applied to a linear system, Godunov method is equivalent to first order upwind scheme (see [37]).

Higher order version of Godunov scheme can be constructed. They are based on high order non oscillatory reconstruction and on the solution to the generalized Riemann problem.

High order non-oscillatory reconstruction is a crucial step, and we shall discuss it in detail when we deal with high order central schemes.

For the moment, we assume we are able to compute such reconstruction, of the form

$$u(x,t_n) \approx R(x; \bar{u}^n) = \sum_j R_j(x)\chi_j(x) \tag{2.11}$$

Then high order Godunov-type schemes are obtained by solving the system

$$\frac{d\bar{u}_j}{dt} = -\frac{f(u^*(u^-_{j+1/2}, u^+_{j+1/2})) - f(u^*(u^-_{j-1/2}, u^+_{j-1/2}))}{h} \tag{2.12}$$

where $u^-_{j+1/2} = R_j(x_{j+1/2})$, $u^+_{j+1/2} = R_{j+1}(x_{j+1/2})$. Because the values $u^-_{j+1/2}$ and $u^+_{j+1/2}$ depend on the reconstruction, which depends on the cell averages, it turns out that system (2.13) is a system of ordinary differential equations for the evolution of cell averages.

Such system may be solved by a suitable scheme for ODE's, for example a Runge-Kutta method, which maintains the accuracy of the spatial discretization (see [58] and references therein).

These methods are based on the (exact or approximate) solution to the Riemann problem. Such solution is not always available or inexpensive. As an alternative to these high order extension of the Godunov method, simpler schemes can be constructed, which make use of less expensive numerical flux function.

The general structure of such schemes is given by

$$\frac{d\bar{u}_j}{dt} = -\frac{F(u^-_{j+1/2}, u^+_{j+1/2}) - F(u^-_{j-1/2}, u^+_{j-1/2})}{h} \tag{2.13}$$

where $u^-_{j+1/2}$ and $u^+_{j+1/2}$ are defined as above, and the the numerical flux function $F(u^-, u^+)$ defines the scheme. The simplest choice of the numerical flux function is the so called Local Lax-Friedrichs flux:

$$F(u^-, u^+) = \frac{1}{2}(f(u^-) + f(u^+) - \alpha(u^+ - u^-)) \tag{2.14}$$

where $\alpha = \max(\rho(A(u^-)),\rho(A(u^+)))$, and $\rho(A)$ denotes the spectral radius of matrix A. The advantage of the local Lax-Friedrichs flux is that it does not require the knowledge of the solution to the Riemann problem, nor the exact knowledge of the eigenvalues and eigenvectors of the Jacobian matrix. Only an estimate of the largest eigenvalue is needed.

The disadvantage of this flux with respect to the Riemann solver is that it introduces a larger numerical dissipation.

Other flux functions are available. Common requirements that they have to satisfy are the following: they have to be

i) locally Lipschitz continuous in both argument

ii) nondecreasing in the first argument and non-increasing in the second argument (symbolically $F(\uparrow,\downarrow)$).

iii) consistent with the flux function, i.e. $F(u,u) = f(u)$.

Popular flux functions (for scalar equation), besides the local Lax Friedrichs described above, are the Godunov flux

$$F(a,b) = \begin{cases} \min_{a \le u \le b} f(u) & \text{if } a \le b \\ \max_{b \le u \le a} f(u) & \text{if } a > b \end{cases}$$

and the Engquist-Osher flux [15]

$$F(a,b) = \int_0^a \max(f'(u),0)\,du + \int_0^b \min(f'(u),0)\,du + f(0)$$

Godunov flux is the least dissipative, and the Lax-Friedrichs the most of the three.

Notice, however, that the numerical dissipation is proportional to the jump $u^+ - u^-$, which is extremely small for high order schemes and smooth solution. For a scheme of order p, in fact, it is $u^+ - u^- = O(h^p)$. Therefore the numerical dissipation becomes large only near discontinuities, i.e. where it is most needed. As a result, the difference between exact Riemann flux and local Lax-Friedrichs flux is very pronounced for low order schemes, but it is not so dramatic for very high order schemes [58].

We shall come back to discuss about high order schemes later.

Now let us consider again a piecewise polynomial reconstruction and let us integrate the equation (2.1) on a staggered grid, as shown in Figure (2.3).

Integrating on the staggered grid one obtains

$$\int_{x_j}^{x_{j+1}} u(x,t_{n+1}) = \int_{x_j}^{x_{j+1}} u(x,t_n) - \int_{t_n}^{t_{n+1}} (f(u(x_{j+1},t)) - f(u(x_j,t)))\,dt \qquad (2.15)$$

Once again, this formula is *exact*. In order to convert it into a numerical scheme one has to approximate the staggered cell average at time t_n, and the time integral of the flux on the border of the cells.

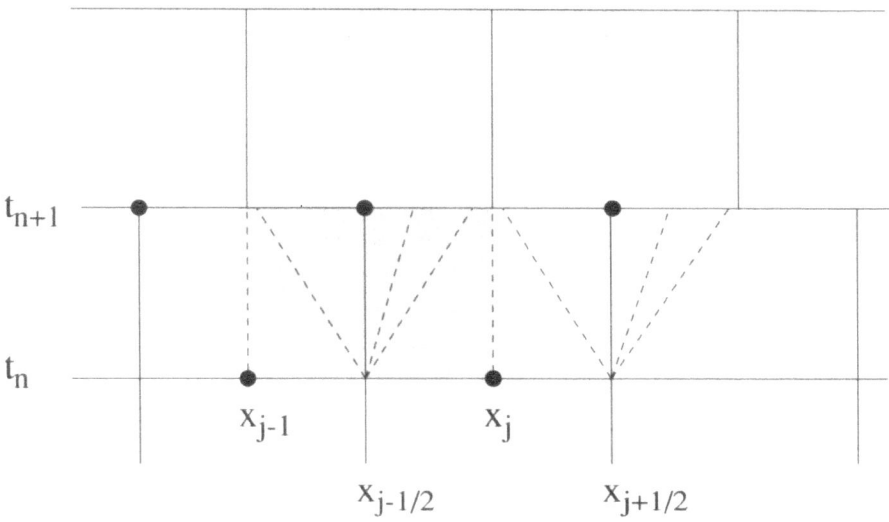

Figure 2.3 Integration on a staggered grid

Let us assume that the function $u(x,t_n)$ is reconstructed form cell averages as a piecewise polynomial function. Then the function is smooth at the center of the cell and its discontinuities are at the edge of the cell. If we integrate the equation on a staggered cell, then there will be a fan of characteristics propagating from the center of the staggered cell, while the function on the edge (dashed vertical lines in the figure) will remain smooth, provided the characteristic fan does not intersects the edge of the cell, i.e. provided a suitable CLF condition of the form

$$\Delta t < \frac{\Delta x}{2\rho(A)} \tag{2.16}$$

is satisfied.

The simplest central scheme is obtained by piecewise constant reconstruction of the function, and by using a first order quadrature rule in the evaluation of the integrals. The resulting scheme is

$$\bar{u}_{j+1/2}^n = \frac{1}{2}(\bar{u}_j^n + \bar{u}_{j+1}^n) - \lambda(f(\bar{u}_{j+1}^n) - f(\bar{u}_j^n)) \tag{2.17}$$

where $\lambda = \Delta t/\Delta x$ denotes the mesh ratio. Such scheme is just Lax-Friedrichs scheme on a staggered grid. The advantage of the Lax-Friedrichs scheme over upwind scheme is that it is much simpler to apply, since no knowledge is required of the characteristic structure of the system. However, the disadvantage is that it is much more dissipative than the upwind scheme.

As an example, let us consider the numerical solution of the Riemann problem for the simple linear equation

$$u_t + u_x = 0, \quad u(x,0) = \begin{cases} 1 & x < 0 \\ 0 & x > 0 \end{cases} \tag{2.18}$$

The result for upwind and Lax Friedrichs schemes are shown in Figure (2.4), obtained with 100 grid points on the interval $[-0.2,1.8]$, $t_{\max} = 1.4$ and $\Delta t/\Delta x = 0.8$ for the upwind scheme and $\Delta t/\Delta x = 0.4$ for the staggered Lax-Friedrichs. Both schemes are exact for the linear equation when the maximum mesh ratio (1 for upwind and 1/2 for LxF) is used. Notice that the first order upwind scheme performs slightly better than the staggered Lax-Friedrichs, in spite of the fact that it uses half time steps (due do the more severe stability condition of the latter). A comparison of the two schemes in terms of their modified equation can be seen, for example, in [37].

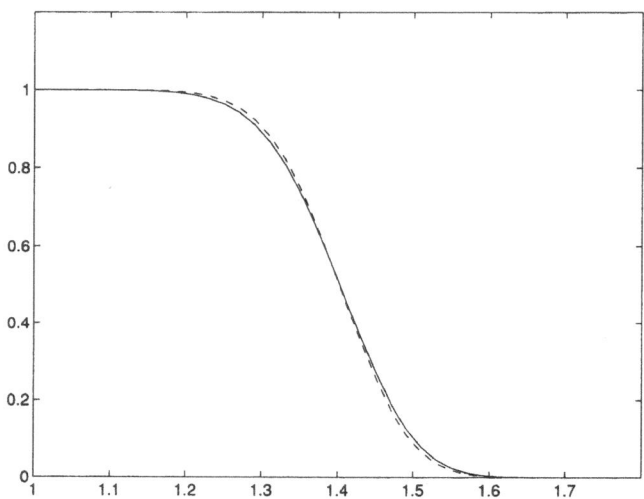

Figure 2.4 Comparison between staggered Lax-Friedrichs (continuous line) and first order upwind (dashed line). $\Delta t/\Delta x$ is 0.8 for upwind scheme and 0.4 for Lax-Friedrichs scheme

2.1.4   The Nessyahu-Tadmor scheme

A second order scheme is obtained by using a piecewise linear approximation for the reconstruction of the function, and a second order quadrature rule (for example the midpoint rule) for the computation of the time integral of the flux on the edges of the cell.

Such scheme has been proposed by Nessyahu and Tadmor [50], and independently by Sanders and Weiser [57].

The staggered cell average is given by

$$\begin{aligned}
\bar{u}^n_{j+1/2} &= \frac{1}{h} \int_{x_j}^{x_{j+1}} R(x; \bar{u}^n) \, dx = \frac{1}{h} \left(\int_{x_j}^{x_{j+1/2}} L_j(x) \, dx + \int_{x_{j+1/2}}^{x_{j+1}} L_{j+1}(x) \, dx \right) \\
&= \frac{1}{2}(\bar{u}^n_j + \bar{u}^n_{j+1}) - \frac{1}{8}(u'_{j+1} - u'_j),
\end{aligned}$$

where $L_j(x) = \bar{u}_j^n + u_j'(x - x_j)/h$ is the linear reconstruction in cell I_j. Here u_j'/h denotes a first order approximation of the derivative of the function in the cell. The value of the field u at the node of the midpoint rule, $u_j^{n+1/2}$, can be computed by first order Taylor expansion, which is equivalent to forward Euler scheme,

$$u_j^{n+1/2} = \bar{u}_j^n - (\lambda/2)f_j'.$$

Here f_j'/h denotes a first order approximation of the space derivative of the flux.

In order to prevent spurious oscillations in the numerical solution, it is essential that these derivatives are computed by using a suitable *slope limiter*. Several choices are possible for the slope limiter. The simplest one is the MinMod limiter. It is defined according to

$$MM(a,b) = \begin{cases} \min(a,b) & \text{if } a < 0 \text{ AND } b < 0 \\ \max(a,b) & \text{if } a > 0 \text{ AND } b > 0 \\ 0 & \text{if } ab < 0 \end{cases}.$$

Such simple limiter, however, degrades the accuracy of the scheme near extrema. A better limiter is the so called UNO (Uniform Non Oscillatory) limiter, proposed by Harten et al. [23].

Such limiter can be written as

$$u_j' = MM(d_{j-1/2} + \frac{1}{2}MM(D_{j-1},D_j), d_{j+1/2} - \frac{1}{2}MM(D_j,D_{j+1})), \qquad (2.19)$$

where

$$d_{j+1/2} = u_{j+1} - u_j, \quad D_j = u_{j+1} - 2u_j + u_{j-1}.$$

Other limiters are possible (see [50]).

The quantity f_j' can be computed either by applying a slope limiter to $f(\bar{u}_j^n)$ or by using the relation

$$f_j' = A(\bar{u}_j^n)u_j'.$$

After one time step, one finds the solution $\{\bar{u}_{j+1/2}^{n+1}\}$ on the staggered cells. Then one repeats a similar step, and after the second step, the solution at time t_{n+2}, $\{\bar{u}_j^{n+1}\}$, is determined on the original grid.

Theoretical properties of the scheme, such as TVD property and the so called "cell entropy inequality" are discusses in [50].

As an example we show here some numerical tests performed by the NT scheme. Figure (2.5) represents the computation on the linear equation $u_t + u_x = 0$, performed by using the two limiters described above. The initial condition is $u(x,0) = 1 + 0.5\sin(2\pi x)$, and the boundary conditions are periodic in $[0,1]$. The solution is computed after 5 periods. The clipping effect due to the MinMod limiter is evident, while the NT scheme with UNO limiter maintains good accuracy after a long time.

The computation shown in Figure (2.1) was performed by NT scheme with 1000 grid points.

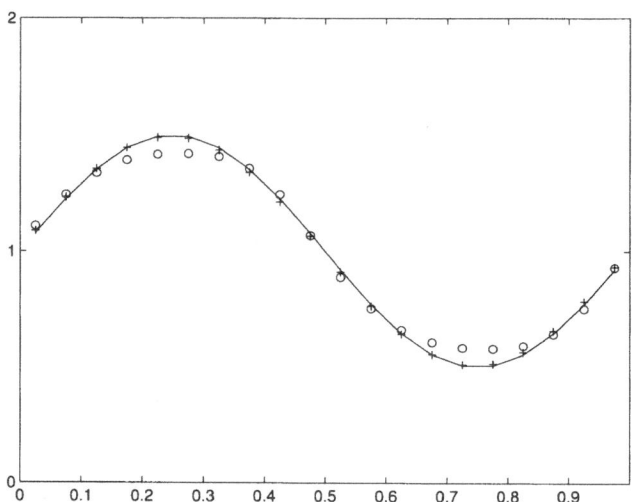

Figure 2.5 NT scheme applied to the linear equation. Mesh ratio $\Delta t/\Delta x = 0.45$. $t_{\max} = 5$.
$N = 20$. MM limiter (circles) and UNO limiter (+). The line represents the exact solution.

More interesting test concerns the solution of the Euler equation of gas dynamics.

Such equations are of the form (2.1), with

$$u = \begin{pmatrix} \rho \\ \rho v \\ E \end{pmatrix}, \quad f = \begin{pmatrix} \rho v \\ \rho v^2 + p \\ v(E + p) \end{pmatrix}, \tag{2.20}$$

where ρ, v, p, and E denote, respectively, the density, velocity, pressure and energy density per unit volume of the gas. For a polytropic gas, energy and pressure are related by

$$p = (\gamma - 1)(E - \frac{1}{2}\rho v^2),$$

where $\gamma = c_p/c_v$ is the polytropic constant, c_p and c_v denote the specific heat at constant pressure and volume, respectively. For a monoatomic gas it is $\gamma = 5/3$, and for a biatomic gas, like air, it is $\gamma = 7/5$. Typical test problems for gas dynamics are the Sod and Lax tests. These are Riemann problems defined by the initial condition

$$\begin{cases} (\rho_l,(\rho v)_l,E_l) = (1,0,2.5), & x < 0.5, \\ (\rho_r,(\rho v)_r,E_r) = (0.125,0,0.25), & x > 0.5. \end{cases} \quad \text{Sod problem} \tag{2.21}$$

$$\begin{cases} (\rho_l,(\rho v)_l,E_l) = (0.445,0.311,8.928), & x < 0.5, \\ (\rho_r,(\rho v)_r,E_r) = (0.5,0,1.4275), & x > 0.5. \end{cases} \quad \text{Lax problem} \tag{2.22}$$

The polytropic constant is $\gamma = 1.4$ and the final time is $t_{\max} = 0.16$.

The result of NT scheme applied to the Sod problem is shown in Figure (2.6).

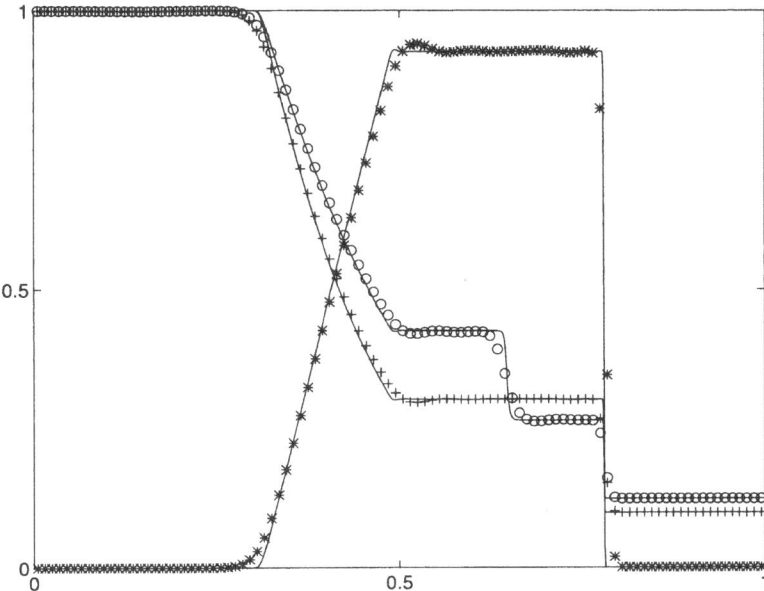

Figure 2.6 NT scheme applied to the Sod problem of gas dynamics. $N = 100$. $t_{\max} = 0.16$. UNO limiter. Courant number 0.475. Density (o), velocity (*), and pressure (+). The line is the reference solution computed by 1000 grid points.

2.2 High order central schemes

In most applications, second order schemes have been successfully used. There is, however, a great interest in constructing and using high order (or it would be more appropriate to say high resolution) schemes.

It may seem strange to look for high order schemes for the numerical solution of quasilinear hyperbolic systems, since the solution itself is not regular, and in general presents jump discontinuities. However, we shall show that there are several reasons for using high order schemes.

Although the mathematical theory of quasilinear hyperbolic systems is developed in the space of bounded variation functions, in most practical cases the solutions are more regular than that. In fact in many practical cases they are piecewise smooth functions, very well represented by piecewise polynomials. This is the reason why numerically computed convergence rate are better than the sharpest theoretical estimate: the convergence rate is computed on piecewise smooth solutions, while the theoretical estimates are computed on BV functions. Pointwise error estimates for the convergence of viscosity solution to the scalar conservation law are considered in [63, 64].

High order schemes give good accuracy of the solution in the smooth regions. They also provide sharp resolution of the discontinuities. These two benefits justify the use of high resolution schemes.

A warning, however, is necessary. Most schemes for the treatment of discontinuous solutions are non linear, and with variable order. They adjust themselves, so that near nonlinear discontinuities the right amount of numerical dissipation is introduced in order to prevent the formation and amplification of spurious oscillations.

The loss of accuracy near the discontinuities may affect the accuracy even in the smooth regions, since small perturbations formed near the shock may propagate in the inner flow. Such effect is usually small, and it is presently subject of investigation [17].

There is a more subtle reason for desiring high resolution central schemes: economy. We shall see that it is possible to design high resolution central schemes that are very easy to use, and that can be easily adapted to a large variety of systems, without any knowledge of the solution of the Riemann problem. These are the Central WENO schemes, that we describe in the next sections, and their semidiscrete analogue, i.e. the WENO schemes with local Lax-Friedrichs flux function. Although such schemes are more dissipative than state-of-the-art Riemann-based flux functions, the difference in the resolution is sensible at the level of low (first and second) order schemes, while the it is almost negligible for high resolution schemes (except for special cases, such as, for example, contact discontinuities). For this reason, it is sometimes more convenient to use a third order central scheme than a second order upwind scheme.

In the construction of high order central schemes, let us start from relation (2.15). We assume that the solution at time t_n is reconstructed as a piecewise polynomial function as in (2.11). Then, by approximating the integral of the flux by a quadrature formula, one obtains the general scheme

$$\bar{u}_{j+1/2}^{n+1} = \frac{1}{\Delta x} \int_{x_j}^{x_{j+1}} R(x; \bar{u}^n)\, dx \tag{2.23}$$

$$+ \lambda \sum_{i=0}^{s} \gamma_l [f(\hat{u}(x_j, t_n + \beta_l \Delta t)) - f(\hat{u}(x_{j+1}, t_n + \beta_l \Delta t))] \tag{2.24}$$

The parameters γ_l and β_l are the weights and the nodes of the particular quadrature formula, and $\hat{u}$ are the intermediate values of the field at the nodes, predicted by Runge-Kutta method (see below). For a fourth-order method one can use, for example, Simpson's rule.

The staggered cell-averages at time t_n, $\bar{u}_{j+1/2}^n$, are given by

$$\bar{u}_{j+1/2}^n = \frac{1}{h} \int_{x_j}^{x_{j+1}} R(x; \bar{u}^n) dx = \frac{1}{h} \left[\int_{x_j}^{x_{j+1/2}} R_j(x) dx + \int_{x_{j+1/2}}^{x_{j+1}} R_{j+1}(x) dx \right]. \tag{2.25}$$

2.2.1   Time evolution: Runge-Kutta methods with Natural Continuous Extension

Consider the Cauchy problem

$$\begin{cases} u' = F(t, u(t)) \\ u(t_0) = u_0. \end{cases}$$

We adapt to our context the notation used in [70] according to which the solution at the $(n + 1)$-th time step obtained with a ν stage Runge-Kutta scheme can be written as

$$u^{n+1} = u^n + \Delta t \sum_{i=1}^{\nu} b_i K^{(i)}, \qquad (2.26)$$

where the $K^{(i)}$'s are the Runge-Kutta fluxes

$$K^{(i)} = F\left(t^n + \Delta t c_i, u^n + \Delta t \sum_{j=1}^{i-1} a_{ij} K^{(j)}\right),$$

and the c_i are given by $c_i = \sum_j a_{ij}$. The method is completely determined by the vector b and the matrix a, which is lower triangular with zero diagonal for explicit schemes.

In our case we are solving a sequence of Cauchy problems in order to obtain the predicted intermediate values required for the quadrature of the fluxes. At the j-th grid point we have

$$\begin{cases} u_j'(\tau) &= F(\tau, u_j(\tau)) = -f_x\left(u(x_j, t^n + \tau)\right) \\ u_j(\tau = 0) &= R(x_j; \bar{u}^n). \end{cases} \qquad (2.27)$$

Thus the computation of the i-th Runge-Kutta flux $K^{(i)}$ requires the evaluation of the x-derivative of f at the intermediate time $t = t^n + c_i \Delta t$. The predicted point-values of u_j at time $t = t^n + c_i \Delta t$ are used to compute the point-values, $f(u_j)$. These predicted values of f are then used for reconstructing an interpolant from which the point-values of the derivative $(f_x)_j$ are computed. To maintain high accuracy and control over oscillations in the evaluation of f_x required in (2.27), the reconstruction of the interpolant of both f and u is essentially the same and is described in §2.2.2.

Following [11], we use here *Natural Continuous Extensions* of Runge-Kutta schemes (NCE) which provide a uniform accuracy of the solution in the time interval $[t^n, t^{n+1}]$ (consult [70]). Each ν-stage Runge-Kutta method of order p has a NCE u of degree $\bar{d} \leq p$ in the sense that there exist ν polynomials $b_i(\theta)$, $i = 1, \cdots, \nu$ of degree at most $\bar{d}$, such that

1. $u(t^n + \theta \Delta t) := u^n + \Delta t \sum_{i=1}^{\nu} b_i(\theta) K^{(i)} \qquad 0 \leq \theta \leq 1$;

2. $u(t^n) = u^n \qquad$ and $\qquad u(t^n + \Delta t) = u^{n+1}$;

3. $\max_{t^n \leq t \leq t^n + \Delta t} \left| w^{(l)}(t) - u^{(l)}(t) \right| = O((\Delta t)^{\bar{d}+1-l}), \qquad 0 \leq l \leq \bar{d}$,

where u^n is the numerical solution computed with the RK scheme at time level t^n, and $w(t)$ is the exact solution of the equation with $w(t^n) = u^n$. Note that the polynomials $b_i(\theta)$ depend only on the Runge-Kutta method chosen and not on the particular ODE being solved. In the next sections we shall show how to use Runge-Kutta methods with NCE in order to guarantee high order accuracy in time.

We end this section listing the NCE's which are of interest for our schemes.

1) **RK1**   (for *second-order* scheme)
   $\bar{d} = \nu = p = 1$
   $b_1(\theta) = \theta;$

2) **RK2**   (for *third-order* scheme)
   $\bar{d} = \nu = p = 2$
   $b_1(\theta) = (b_1 - 1)\theta^2 + \theta$
   $b_2(\theta) = b_2\theta^2;$

3) **RK3**   (Not used since no NCE of degree $\bar{d} = 3$ exists in this case)   (2.28)
   $\bar{d} = 2, \nu = p = 3$
   $b_i(\theta) = 3(2c_i - 1)b_i\theta^2 + 2(2 - 3c_i)b_i\theta$ $i = 1,2,3$

4) **RK4**   (for *fourth-order* scheme)
   $\bar{d} = 3, \nu = p = 4$
   $b_1(\theta) = 2(1 - 4b_1)\theta^3 + 3(3b_1 - 1)\theta^2 + \theta$
   $b_i(\theta) = 4(3c_i - 2)b_i\theta^3 + 3(3 - 4c_i)b_i\theta^2$ $i = 2,3,4$

No NCE of degree $\bar{d} = 4$ exists in this case.

In our case, we used the following set of coefficients. For our third order method:

$$b = \begin{pmatrix} 1/2 \\ 1/2 \end{pmatrix} \qquad\qquad a = \begin{pmatrix} 0 & 0 \\ 1 & 0 \end{pmatrix} \qquad\qquad (2.29)$$

For our fourth order method:

$$b = \begin{pmatrix} 1/6 \\ 1/3 \\ 1/3 \\ 1/6 \end{pmatrix} \qquad\qquad a = \begin{pmatrix} 0 & 0 & 0 & 0 \\ 1/2 & 0 & 0 & 0 \\ 0 & 1/2 & 0 & 0 \\ 0 & 0 & 1 & 0 \end{pmatrix} \qquad\qquad (2.30)$$

We proceed in the next section by presenting the Central-WENO (CWENO) reconstruction which supplies the required elements, R_j, for the overall piecewise-polynomial interpolant (2.11).

2.2.2   The Central-CWENO (CWENO) Reconstruction

In this section, we present our new Central-WENO (CWENO) piecewise-parabolic reconstruction which will be then utilized to construct a fourth-order method.

Let u be the exact solution at time t^n, and $\bar{u}_j^n$ the numerical approximation of its cell average on the cell I_j. Starting from the data $\{\bar{u}_j^n\}$ we apply the reconstruction scheme, obtaining the function $R(x; \bar{u}^n)$. By construction, $R(x; \bar{u}^n) = \sum_j R_j(x)\chi_j(x)$, where in this case $R_j(x)$ is a polynomial of degree 2. We require that our reconstruction satisfies the following properties:

1. **Accuracy**

   i) Cell averages We shall require that Eq. (2.25) provides a high order accurate approximation of the staggered cell averages at time t^n, i.e.

   $$\bar{u}_{j+1/2}^n = \frac{1}{h}\int_{x_j}^{x_{j+1}} u(x,t^n)\,dx + O(h^s),$$

   where s denotes the spatial order of the method. This requirement is satisfied by imposing that the polynomial reconstruction from cell-averages must satisfy:

   $$\frac{1}{2h}\int_{x_j}^{x_{j+1/2}} R(x;\bar{u}^n)dx = \frac{1}{2h}\int_{x_j}^{x_{j+1/2}} u(x,t^n)dx + O(h^s).$$

   $$\frac{1}{2h}\int_{x_j}^{x_{j+1}} R(x;\bar{u}^n)dx = \frac{1}{2h}\int_{x_j}^{x_{j+1}} u(x,t^n)dx + O(h^s), \qquad (2.31)$$

   ii) Pointwise values We look for a reconstruction, $\tilde{R}(x;\bar{u}^n)$, that satisfies the following requirements

   $$\tilde{R}(x_j;\bar{u}^n) = u(x_j,t^n) + O(h^s). \qquad (2.32)$$

   iii) Space derivatives of the flux A separate reconstruction $\hat{R}(x;f^n)$ of f will be used for the derivatives of the fluxes, $\partial_x f(u(x,t))$, for which we require

   $$\hat{R}'(x_j;f^n) = \partial_x f(u(x_j,t^n)) + O(h^{s-1}). \qquad (2.33)$$

2. **Conservation**
   For the function reconstruction R we require

   $$\frac{1}{h}\int_{I_j} R_j(x)dx = \bar{u}_j. \qquad (2.34)$$

3. **Non-oscillatory Reconstruction**
   Avoid spurious oscillations in the sense of ENO/WENO reconstruction ([23], [47]).

As we shall see, in general different reconstructions are obtained for R and $\tilde{R}$, since the accuracy requirements are different.

Here we shall describe in detail fourth order CWENO.

In each cell, I_j, we reconstruct three polynomials of degree 2, $P_{j-1}(x),P_j(x),P_{j+1}(x)$. Each of these polynomials is constructed by posing the following interpolation requirements:

$$\begin{cases} \frac{1}{h} \int_{I_{k-1}} P_k(x)dx = \bar{u}_{k-1}, \\ \\ \frac{1}{h} \int_{I_k} P_k(x)dx = \bar{u}_k, \qquad\qquad\qquad k = j-1, j, j+1 \qquad (2.35) \\ \\ \frac{1}{h} \int_{I_{k+1}} P_k(x)dx = \bar{u}_{k+1}. \end{cases}$$

The reconstruction is created by considering a convex combination of the above polynomials, $P_k(x)$,

$$R_j(x) = w_j^{-1} P_{j-1}(x) + w_j^0 P_j(x) + w_j^1 P_{j+1}(x), \qquad (2.36)$$

where the weights $w_j^k, k = -1, 0, 1$, satisfy $w_j^k \geq 0$, and $\sum_{k=-1}^1 w_j^k = 1$.

A basic fact of interpolation theory states that each polynomial P_k, $k = j-1, j, j+1$, is uniformly third order accurate in the cell j [58],

$$P_k(x) - u(x, t_n) = O(h^3), \quad k = j-1, j, j+1, \quad x \in I_j.$$

If we consider a convex combination of these three polynomials, we have two more degrees of freedom, which may be used to impose more order conditions. For example we can impose that the pointwise value of the polynomial is fifth order accurate at a given point, or we can impose that the half-cell average of it is fifth order accurate. Classical WENO schemes are based on imposing fifth order accuracy at the edge of the cell [26].

The stencil used in the reconstruction of the second degree polynomial $R_j(x)$ contains five points. Note that this convex combination retains the interpolation requirement (2.34) for $R_j(x)$ in I_j, but otherwise does not fulfill any other interpolation requirements in the neighboring cells.

Since $deg(P_k(x)) = 2$, $k = j-1, j, j+1$, one can rewrite

$$P_k(x) = \tilde{u}_k + \tilde{u}'_k(x - x_k) + \frac{1}{2}\tilde{u}''_k(x - x_k)^2, \quad k = j-1, j, j+1. \qquad (2.37)$$

The reconstructed point-values, $\tilde{u}_k$, and the reconstructed discrete first and second derivatives, $\tilde{u}'_k, \tilde{u}''_k$, are uniquely determined by the interpolation requirements (2.35), as

$$\begin{aligned} \tilde{u}_k &= \bar{u}_k - \frac{\bar{u}_{k-1} - 2\bar{u}_k + \bar{u}_{k+1}}{24}, \\ \tilde{u}'_k &= \frac{\bar{u}_{k+1} - \bar{u}_{k-1}}{2h}, \qquad\qquad\qquad\qquad\qquad (2.38) \\ \tilde{u}''_k &= \frac{\bar{u}_{k+1} - 2\bar{u}_k + \bar{u}_{k-1}}{h^2}, \qquad k = j-1, j, j+1. \end{aligned}$$

All that is left in order to end the reconstruction is to determine the weights, $w_j^k, k = -1, 0, 1$. Two ingredients are taken into account in the construction: accuracy and non-oscillatory requirements.

The two requirements will be satisfied by the procedure described below. Following the notations of [26], in order to guarantee convexity, $\sum_{k=-1}^1 w_j^k = 1$, the weights, w_j^k, are written as

$$w_j^k = \frac{\alpha_j^k}{\alpha_j^{-1} + \alpha_j^0 + \alpha_j^1}, \qquad k = -1, 0, 1, \tag{2.39}$$

where

$$\alpha_j^k = \frac{C_k}{(\epsilon + IS_j^k)^p}, \qquad k = -1, 0, 1. \tag{2.40}$$

The constants, C_k, ϵ, p, and the smoothness indicators, IS_j^k, will be determined below.

Since we can not satisfy all the accuracy requirements simultaneously, we split the computation into two parts and by that we are led to two different sets of constants C_k. The first set corresponds to the accuracy requirement in the reconstruction of the cell-averages (2.31), the second set of constants corresponds to the accuracy requirements in the reconstruction of the pointwise values, and the third set corresponds to the accuracy requirements in the computation of the flux derivatives (2.33). Due to cancellation, any symmetric choice of coefficients will result in a fourth-order approximation of the point-values in the center of the cells ($s = 4$ in (2.32)). In order to satisfy (2.32) for $s = 5$ one has to use non-positive constants, namely a *non-convex* combination of the stencils. The treatment of negative weights has been recently investigated by Shu [59]. We shall describe later the use of this technique.

Clearly, the use of different sets of constants imposes absolutely no problems on the implementation of the algorithm.

A straightforward computation results with the desired constants which are displayed in Table 2.1.

	C_{-1}	C_0	C_1	accuracy
cell-averages	3/16	5/8	3/16	h^5
derivatives	1/6	2/3	1/6	h^4
point-values	$-9/80$	49/40	$-9/80$	h^6
	any symmetric combination			h^4

Table 2.1   The constants of the Central-WENO reconstruction

Note that fifth order requirement for the pointwise reconstruction will actually produce sixth order approximation, but using negative weights. By setting

$$\alpha_j^k = C_k, \qquad k = -1, 0, 1 \tag{2.41}$$

instead of (2.40), one obtains a linear high order scheme, which would give good accuracy in the smooth regions, and would produce oscillations near discontinuities. The nonlinear weights are used to prevent oscillations. The basic idea is that one should weight the stencils so that only stencils that do not contain the discontinuity will be used in practice in the reconstruction. This is obtained by formula (2.40) and by suitable smoothness indicators, which are a quantitative measure of the local roughness of the solution.

Several different ways to determine the smoothness indicators were suggested in the literature (see, e.g., [47] and [26]). Here we use the one taken from [26], which amounts to a measure on the L^2-norms of the derivatives:

$$\text{IS}_j^k = \sum_{l=1}^{2} \int_{x_{j-1/2}}^{x_{j+1/2}} h^{2l-1} (P_k^{(l)})^2 dx, \qquad k = -1, 0, 1, \tag{2.42}$$

where $P_k^{(l)}$ denotes the l-th derivative of $P_k(x)$. An explicit integration of (2.42) yields

$$
\begin{aligned}
\text{IS}_j^{-1} &= \frac{13}{12}(\bar{u}_{j-2} - 2\bar{u}_{j-1} + \bar{u}_j)^2 + \frac{1}{4}(\bar{u}_{j-2} - 4\bar{u}_{j-1} + 3\bar{u}_j)^2, \\
\text{IS}_j^0 &= \frac{13}{12}(\bar{u}_{j-1} - 2\bar{u}_j + \bar{u}_{j+1})^2 + \frac{1}{4}(\bar{u}_{j-1} - \bar{u}_{j+1})^2, \\
\text{IS}_j^1 &= \frac{13}{12}(\bar{u}_j - 2\bar{u}_{j+1} + \bar{u}_{j+2})^2 + \frac{1}{4}(3\bar{u}_j - 4\bar{u}_{j+1} + \bar{u}_{j+2})^2.
\end{aligned}
\tag{2.43}
$$

In smooth regions, a Taylor expansion of (2.43) gives

$$\text{IS}_j^k = (\tilde{u}'h)^2 + \frac{13}{12}(\tilde{u}''h^2)^2 + O(h^6), \qquad k = -1, 0, 1. \tag{2.44}$$

Hence, $\text{IS}_j^k = O(h^2)$, and in critical points it is $O(h^4)$. In non-smooth regions, $\text{IS}_j^k = O(1)$, and by that the normalized weight of the corresponding stencil will be negligible. Therefore, our reconstruction follows the WENO methodology by automatically avoiding the information coming from non-smooth regions which are the cause for spurious oscillations.

The remaining parameters to be determined in (2.40) are ϵ and p. The constant ϵ was inserted in the denominator in order to prevent it from vanishing. In [26] an $\epsilon = 10^{-6}$ and $p = 2$ were selected. We use the same values.

In short, our reconstruction from cell average routine accepts in input the values $\{\bar{u}_j\}$ of cell averages at time t^n, and produces in output the point values u_j, u'_j, u''_j which completely determine the reconstruction polynomial $R_j(x) = u_j + u'_j(x - x_j)/h + (1/2)u''_j(x - x_j)/h^2$.

A few modifications are needed to compute the reconstruction from point values for the flux $f_j = f(u_j)$ which is needed in the Runge-Kutta step.

Here the candidate polynomials P_k satisfy the interpolation requirements (compare with (2.35)):

$$
\begin{cases}
P_k(x_{k-1}) = f_{k-1}, \\
P_k(x_k) = f_k, \\
P_k(x_{k+1}) = f_{k+1}.
\end{cases}
\qquad k = j-1, j, j+1 \tag{2.45}
$$

Thus the reconstructed point values $\tilde{f}_k$ and the reconstructed first and second derivatives $\tilde{f}'_k$ and $\tilde{f}''_k$ are given by:

$$\tilde{f}_k = f_k, \quad \tilde{f}'_k = \frac{f_{k+1} - f_{k-1}}{2h}, \quad \tilde{f}''_k = \frac{f_{k+1} - 2f_k + f_{k-1}}{h^2}, \quad k = j-1, j, j+1.$$

For the evaluation of the intermediate values, all that will be needed is a pointwise reconstruction of the space derivative f'_j which is given by:

$$f'_j = w_j^{-1}(\tilde{f}'_{j-1} + h\tilde{f}''_{j-1}) + w_j^0 \tilde{f}'_j + w_j^1(\tilde{f}'_{j+1} - h\tilde{f}''_{j+1}). \tag{2.46}$$

The computation of the weights in (2.46) is the same as above (2.39)–(2.44). Here, we use the second set of constants C_k's of Table 2.1, and the computation of the smoothness indicators IS_k^j in (2.43) involves the point values f_j, instead of the cell averages.

Three different families of smoothness indicators have to be computed: one for the cell average, one for the pointwise values and one for the flux derivative. The latter, in principle, has to be computed at each stage of the Runge-Kutta method. Numerical experiment show that it is enough to compute it only once per time step, at the beginning of each step.

We shall only construct a fourth order scheme, therefore we shall use only positive weights. For the pointwise reconstruction we use the same set of constants used for the calculation of the cell average. In this way only two families of smoothness indicators have to be computed at each time step.

Fifth order accuracy space could be obtained by using the technique described at the end of the section.

2.2.3 Numerical results

In this section we test the fourth-order scheme. We start from a single scalar equation where we numerically compute the order of accuracy of our schemes. We also demonstrate on a model problem how the smoothness indicators trigger the selection of the correct stencil when discontinuities are present.

The construction of the smoothness indicators is more delicate for systems of equations than in the scalar case. A naive component by component extension of the scalar scheme does not yield the best results. Better results are obtained by using a global smoothness indicator which is the same for all components of vector u.

We then apply our scheme to some classical test problems of gas dynamics. Our results show that the use of the same smoothness indicator for all the components produces better results and is computationally cheaper compared with the componentwise indicator.

Scalar equation

As an application of the central WENO schemes we study the performance of our schemes by applying them to the Burgers equation.

$$u_t + (\tfrac{1}{2}u^2)_x = 0,$$
$$u(x,t = 0) = 1 + \tfrac{1}{2}\sin(\pi x),$$
$$\text{periodic boundary conditions on [-1,1],}$$
$$\text{integration times: } T = 0.33 \text{ and } T = 1.5.$$

Here $T = 0.33$ is used for convergence tests, and $T = 1.5$ for the shock capturing test (the shock develops at $T_s = 2/\pi$).

Burgers equation, $u_0 = 1 + 1/2\sin(\pi x)$

N	L^1 error	L^1 order	L^∞ error	L^∞ order
20	0.2926E-02		0.9462E-02	
40	0.2459E-03	3.5728	0.1139E-02	3.0549
80	0.1419E-04	4.1150	0.8631E-04	3.7216
160	0.6821E-06	4.3787	0.4461E-05	4.2742
320	0.3227E-07	4.4017	0.2296E-06	4.2800
640	0.1766E-08	4.1916	0.1269E-07	4.1779

Table 2.2 Convergence test for the fourth order CWENO scheme. $T = 0.33$, $\lambda = 0.66 * 2$, $\epsilon = 10^{-6}$, $p = 2$.

Convergence tests are useful when testing a nonlinear scheme. First they are used to check the correctness of the code. Second, because the schemes are nonlinear, one should check that the nonlinear weights do not destroy the accuracy.

If the expected accuracy is not reached, then it is useful to perform the check on a smooth solution with a linear scheme, where the weights equal the constant C_k. In this way one verifies whether the lack of accuracy is due to the nonlinear weights or to a programming error.

The computational parameters used in the following tests are $\epsilon = 10^{-6}$, $p = 2$. The mesh ratio was chosen as $\lambda = 0.66\lambda_{max}$ for the nonlinear (Burgers) equation. The parameter λ_{max} was computed in order to satisfy stability conditions and it depends on the scheme. We used $\lambda_{max} = 2/7$.

First we perform an accuracy test, computing the order of accuracy at $T = 0.33$, well before the shock formation time $T = 2/\pi$. The results appear in Table 2.2. We observe the correct order of accuracy.

The shock-capturing properties of the CWENO scheme are illustrated in Figure 2.7 for the Burgers equation. The pictures on the left refer to the solution before shock formation ($T = 0.5$) and the pictures on the right refer to the solution after shock formation ($T = 1.5$). The bottom part of the picture shows the weights w_j^k computed in the reconstruction from cell averages. In particular, the central weight corresponds to w_j^0, while the left weight corresponds to w_j^{-1}. Before the shock formation, the weights remain close to their equilibrium values (given in Table 1). An abrupt change can be seen after the shock formation. Here the stencils that would yield oscillations are assigned almost a zero weight. Thus the solution is oscillation-free even after the shock forms. The shock transition occurs within two cells.

Systems of equations

We apply our schemes to the system of Euler equations of gas dynamics for a polytropic gas with constant $\gamma = 1.4$. The variables ρ, v, E, and p, denote the density, velocity, total energy per unit volume and the pressure, respectively. We consider the test problems (2.21) and (2.22).

Burgers equation. Solution and weights

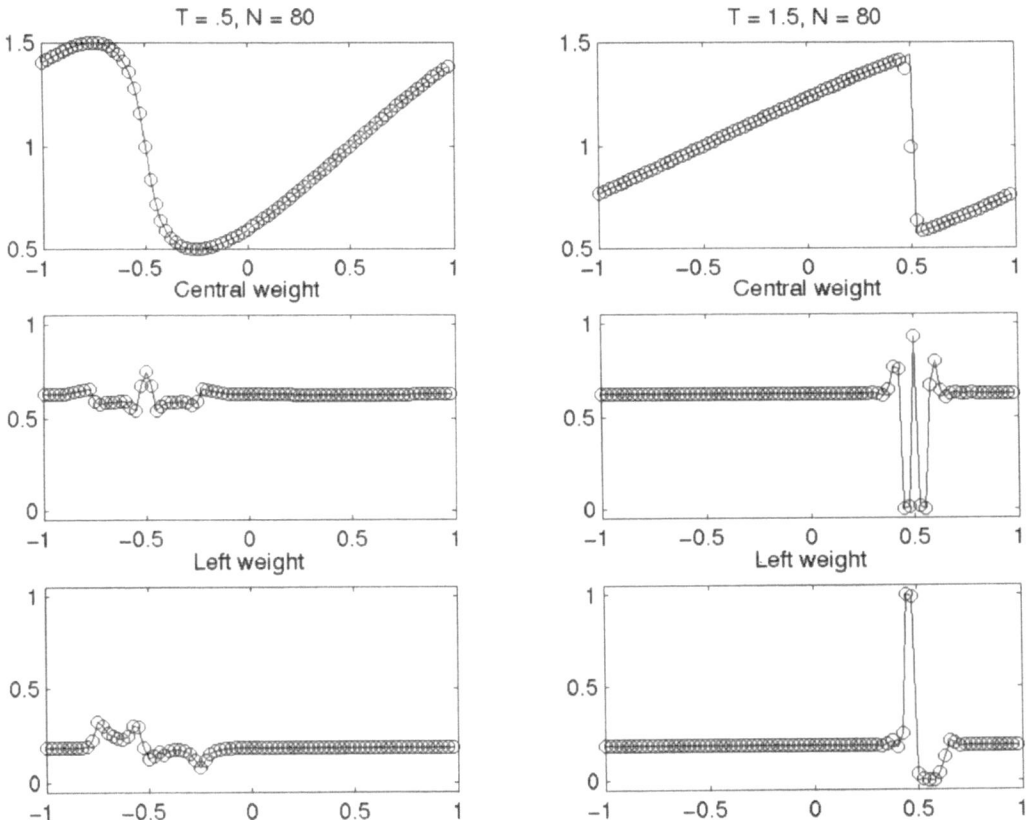

Figure 2.7 Weights computed from cell averages, $m = 4$

In both cases the computational domain is $[0,1]$; we integrate the equations up to $T = 0.16$, i.e. before the perturbations reach the boundary of the computational region. The mesh ratio here is $\lambda = 0.1$, which is the optimal one for Lax initial data.

The number of cells, N, was taken as $N = 100$ in order to compare with the upwind literature, and also as $N = 400$ to show the behavior of the weights: when the solution is well resolved, the weights coincide almost everywhere with the linear weights.

We apply one smoothness indicator for all the components:

$$\mathrm{IS}_j^k = \frac{1}{d} \sum_{r=1}^{d} \frac{1}{||\bar{u}_r||_2} \left(\sum_{l=1}^{2} \int_{x_{j-1/2}}^{x_{j+1/2}} h^{2l-1} \left(P_{j+k,r}^{(l)} \right)^2 \, dx \right), \quad k = -1, 0, 1. \tag{2.47}$$

Here d is the number of equations, and $P_{k,r}$ denotes the k-th polynomial for the r-th component. The quantity $||\bar{u}_r||_2$ is a scaling factor, and it is defined as the L^2 norm of the cell averages of the r-th component of u, namely:

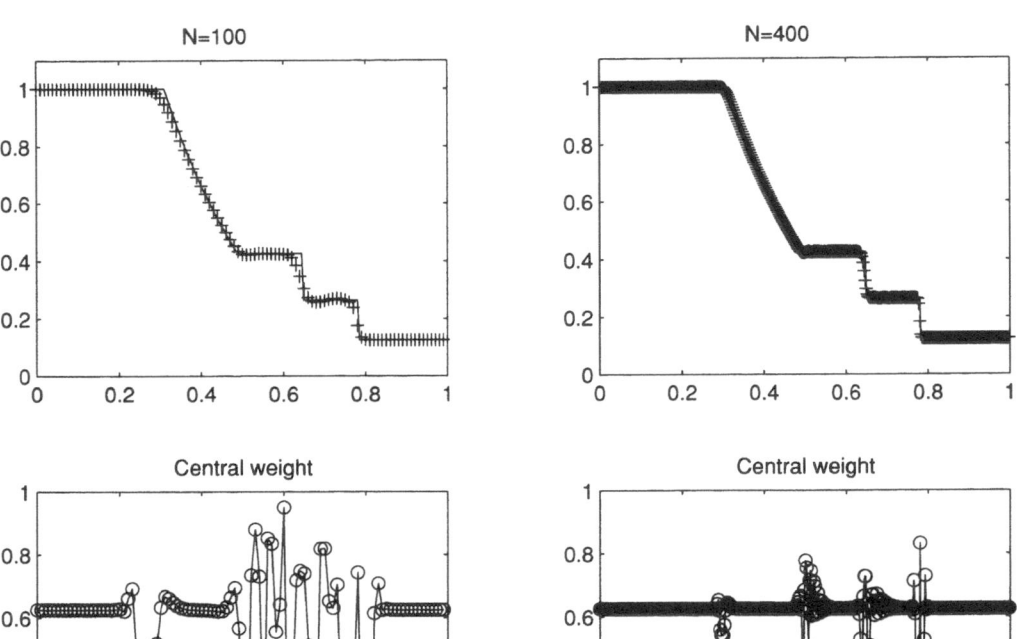

Figure 2.8 Fourth order CWENO. Density (top) and Central weights (bottom) for Sod's problem: $\lambda = 0.1$, $T = 0.16$. The weight shown is computed once per time step during the reconstruction from cell-averages

$$\|\bar{u}_r\|_2 = \left(\sum_{\text{all } j} |\bar{u}_{j,r}|^2 h \right)^{1/2} .$$

This choice is not crucial if all components are of the same order. However it would be more appropriate to use $\|\bar{u}_r\|_2^2$ as a scaling factor, so that the smoothness indicators will be scale-invariant with respect to a change in the physical units of the component of the field vector u.

The integral in (2.47) can be exactly integrated (see (2.43)).

This strategy will be called "Global smoothness indicators": its results on Sod's problem are shown in Figure 2.8. A comparison between componentwise and global smoothness indicators is shown in the case of the Lax test (see Figure 2.9).

To summarize, the "Global smoothness indicators" strategy is much less expensive and gives better results than the "Componentwise smoothness indicators" strategy.

Componentwise and Global Smoothness Indicators

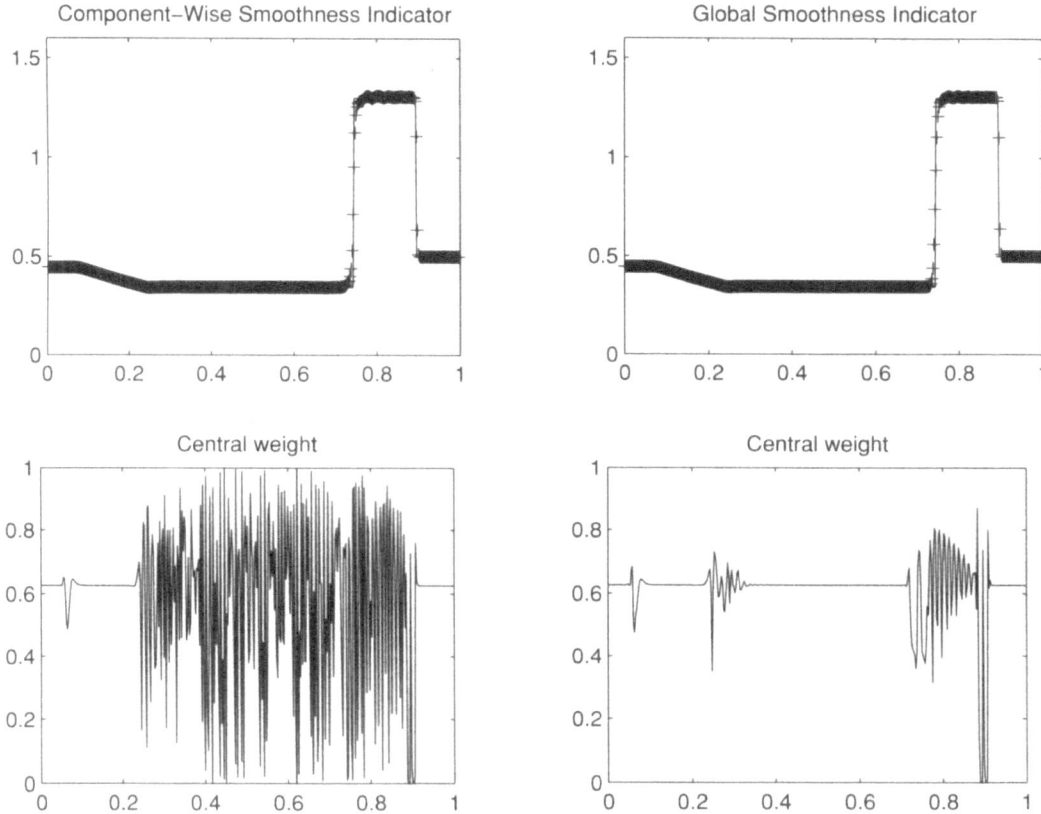

Figure 2.9 Fourth order CWENO. Density (top) and Central weights (bottom) for Lax' problem: $\lambda = 0.1$, $T = 0.16$, $N = 400$. The weight shown is computed once per time step during the reconstruction from cell-averages for the density

2.2.4   Improvements

The Centered WENO schemes described before are not optimal for two reasons.

First, because of the restriction on the uniform accuracy of NCE of Runge Kutta methods, a fourth order RK scheme has been used to compute the predictor values for the fourth order CWENO scheme. This requires the computation of the stage values, and then the evaluation of the flux at the nodes of the quadrature formula.

Second, only fourth order CWENO is obtained with three parabola, because when looking for a fifth order scheme, negative weights appear. Here we mention how it is possible to optimize CWENO schemes, by using fourth order RK scheme for constructing a fourth order CWENO, without the additional evaluation of the flux at the nodes of the quadrature formula, and how to use a fifth order Runge-Kutta method for a fifth order scheme [52].

Let us write Eq.(2.1) in the form

$$\frac{\partial u}{\partial t} = F(u), \tag{2.48}$$

where $F = -\partial f/\partial x$. By applying a ν-stage Runge-Kutta to Eq. (2.48), one obtains an expression for the approximation of the pointwise value of $u(x,t+\Delta t)$

$$u(x,t+\Delta t) = u(x,t) + \sum_{i=1}^{\nu} b_i K^{(i)}(x), \tag{2.49}$$

$$K^{(i)}(x) = F\left(u(x,t) + \Delta t \sum_{l=1}^{i-1} a_{il} K^{(l)}(x)\right), \quad i = 1, \dots, \nu \tag{2.50}$$

By integrating Eqs. (2.49) and (2.50) between x_j and x_{j+1} one has

$$\bar{u}_{j+1/2}^{n+1} = \bar{u}_{j+1/2}^{n} + \Delta t \sum_{i=1}^{\nu} b_i \bar{K}_{j+1/2}^{(i)} \tag{2.51}$$

$$\bar{K}_{j+1/2}^{(i)} = -\frac{1}{h}[f(u_{j+1}(t+c_i\Delta t)) - f(u_j(t+c_i\Delta t))], \quad i = 1, \dots, \nu \tag{2.52}$$

where $u_j(t+c_i\Delta t)$ denotes the stage value of the field u at grid point x_j. The scheme can be therefore summarized as follows

- Compute the stage values (predictor)

$$u_j^{(i)} = u_j^n + \Delta t \sum_{l=1}^{i-1} a_{il} F(u_j^{(l)}) \tag{2.53}$$

- compute the new cell average (corrector)

$$\bar{u}_{j+1/2}^{n+1} = \bar{u}_{j+1/2}^{n} - \lambda \sum_{l=1}^{\nu} b_l [f(u_{j+1}^{(l)}) - f(u_j^{(l)})] \tag{2.54}$$

CWENO reconstruction will be used to compute

- cell average, $h\bar{u}_{j+1/2}^{n} = \int_{x_j}^{x_{j+1/2}} R_j(x)\,dx + \int_{x_{j+1/2}}^{x_{j+1}} R_{j+1}(x)\,dx$

- pointwise values of the field u_j^n at point x_j, $u_j^n = \tilde{R}_j(x_j)$

- flux derivative, $F(u_j^{(l)}) \approx -\hat{R}'(x_j)$

Notice that, as usual, three different CWENO reconstructions will be used, according to the objective.

By using this technique, fourth order RK predictor can be used to obtain a fourth order CWENO scheme. No additional flux evaluation is needed, since the fluxes used in the corrector are computed at the stage values, and therefore $\nu - 1$ of them have already been computed during the predictor stage.

In order to obtain a fifth order scheme, then the constants $\mathbb{C} = (-9/80, 49/40, -9/80)$ have to be used for the reconstruction of the pointwise values (see Table (2.1)).

Note that negative weights appear in the reconstruction of pointwise values from cell averages. Straightforward use of WENO reconstruction with negative weights would produce spurious oscillations (and possibly blow-up of the numerical solution). This problem can be overcome using a technique recently developed by Shu [59]. First, use two sets of positive constants, C_i^+, C_i^-, such that $C_i = C_i^+ - C_i^-$, $i = -1, 0, 1$. This is done in such a way that the constants are well bounded away from zero. Shu proposes

$$C_i^+ = \frac{1}{2}(\theta|C_i| + C_i), \quad C_i^- = \frac{1}{2}(\theta|C_i| - C_i),$$

and uses $\theta = 3$. Then compute

$$\alpha_\pm^i = \frac{C_i^\pm}{(IS^i + \epsilon)^2}, \quad i = -1, 0, 1$$

Finally, compute the two sets of weights

$$w_+^i = \frac{\alpha_+^i}{\sum_{l=-1}^1 \alpha_+^i} \sum_{l=-1}^1 C_i^+, \quad w_-^i = \frac{\alpha_-^i}{\sum_{l=-1}^1 \alpha_-^i} \sum_{l=-1}^1 C_i^-,$$

The final reconstruction is obtained by

$$R_j(x) = \sum_{i=-1}^1 (w_{j+}^i - w_{j-}^i)P_j^i.$$

2.2.5 Semidiscrete central schemes

One of the drawbacks of NT and many other central schemes is staggering. There is a non-staggered version of the scheme, which is developed in [24].

The central schemes we have considered so far are based on a simultaneous discretization in space and time. There is a procedure to derive semidiscrete central schemes, which are discretized only in space. The discretization in time can then be performed by a suitable ODE solver (method of lines). A second order semidiscrete central scheme has been derived by Kurganov and Tadmor [32]. The derivation is based on a different treatment of the the region which lies in the characteristic fan from the region of the mesh which lies in the smooth region. The authors derive a fully discrete second order scheme, which is less dissipative that Nessyahu-Tadmor scheme, and then, by letting the time step vanish, they deduce a semidiscrete scheme. When applied to conservation laws, the scheme is basically equivalent to a second order monotone finite volume scheme with local Lax-Friedrichs flux function. The authors apply the scheme to several problems containing a source term, such as convection diffusion equations. However, they are not interested in the treatment of very stiff sources. A third order extension of the scheme is carried out in [31], and applied to several balance laws. As we shall remark later, semidiscrete central schemes can be used as building blocks for the construction of high order schemes for systems with stiff source.

2.3 Multidimensional central schemes

In this section we describe how to derive second and higher order central schemes, and we show some applications to gas dynamics.

Central schemes can be extended to problems in several dimensions. Second order central schemes on rectangular grids have been considered by Jiang and Tadmor [25], Sanders and Weiser [57], and by Arminjon et al. [6]. They have been extended to unstructured grids by Arminjon et al. [7]. High order central schemes in two dimensions have been considered in the papers [39], [40], [41]

Consider the two-dimensional system of conservation laws

$$u_t + f(u)_x + g(u)_y = 0, \tag{2.55}$$

subject to the initial values

$$u(x,y,t=0) = u_0(x,y),$$

and to boundary conditions, which we do not specify at this point. The flux functions f and g are smooth vector valued functions, $f,g : \mathbb{R}^d \to \mathbb{R}^d$. The system (2.55) is assumed to be hyperbolic in the sense that for any unit vector $(n_x,n_y) \in \mathbb{R}^2$, the matrix $n_x \nabla_u f + n_y \nabla_u g$ has real eigenvalues and its eigenvectors form a basis of $\mathbb{R}^d$. In order to integrate numerically (2.55), we introduce a rectangular grid which for simplicity will be assumed to be uniform with mesh sizes $h = \Delta x = \Delta y$ in both directions. We will denote by $I_{i,j}$ the cell centered around the grid point $(x_i,y_j) = (i\Delta x, j\Delta y)$, i.e., $I_{i,j} = [x_i - h/2, x_i + h/2] \times [y_j - h/2, y_j + h/2]$. Let Δt be the time step and denote by $u_{i,j}^n$ the approximated point-value of the solution at the (i,j)-th grid point at time $t^n = n\Delta t$. Finally, let $\bar{u}_{i,j}^n$ denote the cell average of a function u evaluated at the point (x_i,y_j),

$$\bar{u}_{i,j}^n = \frac{1}{h^2} \int_{I_{i,j}} u(x,y,t^n) \, dx \, dy.$$

Given the cell-averages $\{\bar{u}_{i,j}^n\}$ at time t^n, Godunov-type methods provide the cell-averages at the next time-step, t^{n+1}, in the following way: first, a piecewise-polynomial reconstruction is computed from the data $\{\bar{u}_{i,j}^n\}$ resulting with

$$u^n(x,y) = \sum_{i,j} R_{i,j}(x,y)\chi_{i,j}(x,y). \tag{2.56}$$

Here, $R_{i,j}(x,y)$ is a suitable polynomial (which has to satisfy conservation, accuracy and non-oscillatory requirements), while $\chi_{i,j}(x,y)$ is the characteristic function of the cell $I_{i,j}$. Thus, in general, the function $u^n(x,y)$ will be discontinuous along the boundaries of each cell $I_{i,j}$.

In order to proceed, the reconstruction, $u^n(x,y)$, is evolved according to some approximation of (2.55) for a time step Δt. We will use the fact that the solution remains smooth at the vertical edges of the staggered control volume, $I_{i+1/2,j+1/2} \times [t^n, t^{n+1}]$, provided that the time-step Δt satisfies the CFL condition

$$\Delta t < \frac{h}{2} \frac{1}{\max(|\sigma_x|, |\sigma_y|)}.$$

Here, $I_{i+1/2,j+1/2} = [x_i, x_{i+1}] \times [y_j, y_{j+1}]$ (see Figure 2.10; the edges at which the solution remains smooth are denoted by dotted vertical lines), and σ_x and σ_y are the largest (in modulus) eigenvalues of the Jacobian of f and g, respectively.

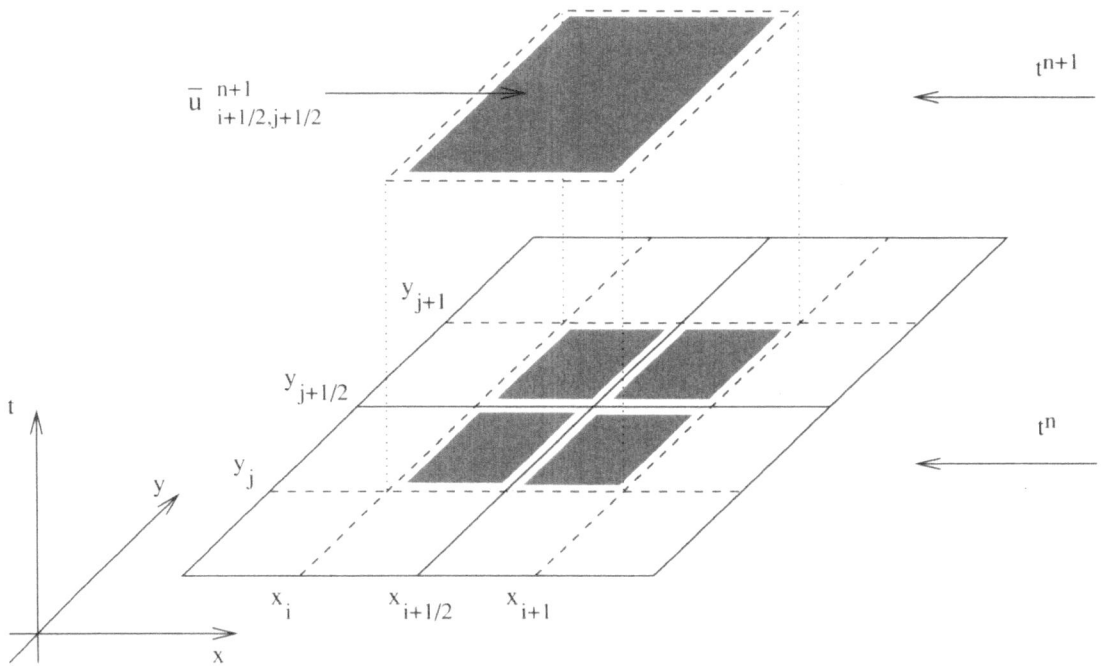

Figure 2.10 The Two-Dimensional Stencil

An exact integration of the system (2.55) with data $u^n(x,y)$ over the control volume $I_{i+1/2,j+1/2} \times [t^n, t^{n+1}]$ results with

$$\bar{u}^{n+1}_{i+\frac{1}{2},j+\frac{1}{2}} = \frac{1}{h^2} \int\int_{I_{i+\frac{1}{2},j+\frac{1}{2}}} u^n(x,y)\, dx\, dy \qquad (2.57)$$

$$- \frac{1}{h^2} \int_{\tau=t^n}^{t^{n+1}} \left\{ \int_{y=y_j}^{y_{j+1}} [f(u(x_{i+1},y,\tau)) - f(u(x_i,y,\tau))]\, dy \right\} d\tau$$

$$- \frac{1}{h^2} \int_{\tau=t^n}^{t^{n+1}} \left\{ \int_{x=x_i}^{x_{i+1}} [g(u(x,y_{j+1},\tau)) - g(u(x,y_j,\tau))]\, dx \right\} d\tau.$$

The first integral on the RHS of (2.57) is the cell-average of the function $u^n(x,y)$ on the staggered cell $I_{i+1/2,j+1/2}$. Given the reconstructed function $u^n(x,y)$, (2.56), this term can be computed exactly: it will consist of a contribution of four terms, resulting from averaging $R_{i+1,j+1}(x,y)$, $R_{i,j+1}(x,y)$, $R_{i+1,j}(x,y)$, and $R_{i,j}(x,y)$, on the corresponding quarter-cells.

The advantage of the central framework appears in the evaluation of the time integrals appearing in (2.57). Since the solution remains smooth on the segments $(x_i, y_j) \times [t^n, t^{n+1}]$, we can evaluate the time integrals with a quadrature rule using only nodes lying in these segments.

A second order scheme is obtained by approximating the integral of the flux f as

$$\int_{\tau=t^n}^{t^{n+1}} \int_{y=y_j}^{y_{j+1}} f(x_i, y, \tau) \, dy \, d\tau \approx \frac{h \Delta t}{2} \left(f(u_{i,j}^{n+1/2}) + f(u_{i,j+1}^{n+1/2}) \right) \tag{2.58}$$

and likewise for the integral of the flux g. By applying the same discretization used for the Nessyahu-Tadmor scheme, one obtains the two dimensional counterpart, which has been introduced in [6] and, independently, in [25]. First, in each cell (i,j) the field u is reconstructed by a piecewise linear approximation,

$$L_{i,j}(x,y) = \bar{u}_{i,j}^n + u_{i,j}' \frac{x - x_i}{h} + u_{i,j}^{\backslash} \frac{y - y_j}{h}$$

where u'/h and $u^{\backslash}/h$ denote, respectively, first order approximations of x- and y-partial derivatives, and can be computed by using a suitable slope limiter, as in the one dimensional case. The resulting scheme has a compact form similar to the one dimensional one, and can be written as

$$
\begin{aligned}
\bar{u}_{i+1/2,j+1/2}^{n+1} = \ & \bar{u}_{i+1/2,j+1/2}^n \\
& - \frac{\lambda}{2}(f(u_{i+1,j}^{n+1/2}) + f(u_{i+1,j+1}^{n+1/2}) - f(u_{i,j}^{n+1/2}) - f(u_{i,j+1}^{n+1/2})) \\
& - \frac{\lambda}{2}(g(u_{i,j+1}^{n+1/2}) + g(u_{i+1,j+1}^{n+1/2}) - g(u_{i,j}^{n+1/2}) - g(u_{i+1,j}^{n+1/2}))
\end{aligned}
\tag{2.59}
$$

where $\lambda = \Delta t / h$, and

$$
\begin{aligned}
\bar{u}_{i+1/2,j+1/2}^n = \ & \frac{1}{4}(\bar{u}_{i,j}^n + \bar{u}_{i+1,j}^n + \bar{u}_{i+1,j+1}^n + \bar{u}_{i,j+1}^n) + \frac{1}{16}(u_{i,j}' - u_{i+1,j}' \\
& + u_{i,j+1}' - u_{i+1,j+1}' + u_{i,j}^{\backslash} - u_{i,j+1}^{\backslash} + u_{i+1,j}^{\backslash} - u_{i+1,j+1}^{\backslash})
\end{aligned}
\tag{2.60}
$$

and the predictor values are evaluated as

$$u_{i,j}^{n+1/2} = \bar{u}_{i,j}^n - \frac{\lambda}{2} f_{i,j}' - \frac{\lambda}{2} f_{i,j}^{\backslash} \tag{2.61}$$

Once again, the first order approximation $f_{i,j}'$ and $g_{i,j}^{\backslash}$ can be computed either by a slope limiter acting on $f(\bar{u}_{i,j})$ and $g(\bar{u}_{i,j})$ or by

$$f' = A(\bar{u}_{i,j}) u_{i,j}', \quad g^{\backslash} = B(\bar{u}_{i,j}) u_{i,j}^{\backslash},$$

where A and B are the Jacobian matrices $A = \nabla_u f$, $B = \nabla_u g$. The above second order central scheme is very simple to use, and it has been successfully applied to a large variety of problems, such as double Mach reflection [25] and MHD equations [65]. Modifications and improvements of the above second order central scheme have been considered by several authors (see [28, 56, 43]).

Two dimensional central schemes have been extended to unstructured grids [7]. Two dimensional central schemes on unstructured grids are very flexible, and they provide good resolution even for complex flows. Theoretical convergence results for such schemes are available [8, 20].

Here we consider the development of high order central schemes in two space dimensions. For a fourth-order method one can use Simpson's rule for the time integrals

$$\int_{t^n}^{t^{n+1}} f(u(x_i,y_j,\tau))d\tau = \tag{2.62}$$

$$= \frac{\Delta t}{6} \left[f\left(u_{i,j}^n\right) + 4f\left(u_{i,j}^{n+1/2}\right) + f\left(u_{i,j}^{n+1}\right) \right] + O\left((\Delta t)^5\right),$$

and the following centered quadrature rule for the integrals in space,

$$\int_{x_i}^{x_{i+1}} f(x)dx = \frac{h}{24}[-f(x_{i+2}) + 13f(x_{i+1}) + 13f(x_i) - f(x_{i-1})] + O(h^5). \tag{2.63}$$

In this way, the quadrature rule for approximating the integrals of the fluxes involves only nodes on the segments $(x_i,y_j) \times [t^n,t^{n+1}]$.

The quadrature in time, (2.62), requires the prediction of the values of the solution at later times. In the case of Simpson's rule, one has to generate the values of $u_{i,j}$ at times $t^{n+1/2}$, t^{n+1}. (The point-value $u_{i,j}^n$ can be obtained directly from the reconstruction $u_{i,j}(t^n) = u^n(x_i,y_j)$). Once again we use the smoothness of the numerical solution along the segments $(x_i,y_j) \times [t^n,t^{n+1}]$ to consider the sequence of Cauchy problems

$$\begin{cases} v_{i,j}'(\tau) = F(\tau,v_{i,j}(\tau)) := -f_x(v(x_i,y_j,t^n + \tau)) - g_y(v(x_i,y_j,t^n + \tau)), \\[2mm] v_{i,j}(\tau = 0) = u^n(x_i,y_j). \end{cases} \tag{2.64}$$

In order to obtain the intermediate values at $t^{n+1/2}$ and t^{n+1}, all that is required is to solve (2.64) up to these times using a Runge-Kutta scheme. When more than one intermediate value is required (as in the case of Simpson's rule), it is possible to solve (2.64) once with the largest time required, and then reconstruct the other values with the required accuracy using the Natural Continuous Extension (NCE), [70]. More details will follow below.

2.3.1 Multidimensional CWENO reconstruction

In this section we will describe in detail our new reconstruction step. We start with the reconstruction from cell-averages, (2.56), which is needed at the beginning of each time-step. We then proceed with the reconstruction from point-values which is used for evaluating the fluxes in the RK step (2.64). This section ends with a discussion of the modifications to the algorithm which are required for solving systems of equations.

2.3.2 The Reconstruction from Cell Averages

In every cell $I_{i,j}$ we reconstruct a bi-quadratic polynomial, $R_{i,j}(x,y)$, which is written as a convex combination of nine bi-quadratic polynomials, $P_{i,j}(x,y)$, centered in the cells around $I_{i,j}$,

$$R_{i,j}(x,y) = \sum_{l,k=-1}^{1} w_{i,j}^{l,k} P_{i+l,j+k}(x,y). \tag{2.65}$$

The bi-quadratic polynomials $P_{i,j}(x,y)$, which serve as the building blocks for the reconstruction (2.65), interpolate the data $\{\bar{u}^n\}$ in the sense of cell averages (see below). They approximate the function $u(x,y)$ whose cell averages are $\{\bar{u}^n\}$ with third order accuracy. The combination (2.65) is designed to increase accuracy *and* to avoid spurious oscillations.

The weights $w_{i,j}^{l,k}$ in (2.65) are computed using a nonlinear algorithm which satisfy a positivity requirement, $w_{i,j}^{l,k} \geq 0$, and a conservation requirement, $\sum_{l,k=-1}^{1} w_{i,j}^{l,k} = 1$.

For simplicity of notation, let us introduce the 3×3 matrices:

$$(\Omega_{i,j})_{l,k} = w_{i,j}^{l,k}, \qquad l,k = -1,0,1.$$

Thus each matrix $\Omega_{i,j}$ contains the nine non-constant weights needed to compute the reconstruction on the cell $I_{i,j}$. Note that the first index, l, is associated with the x-variable, while the index k is associated with the y-variable.

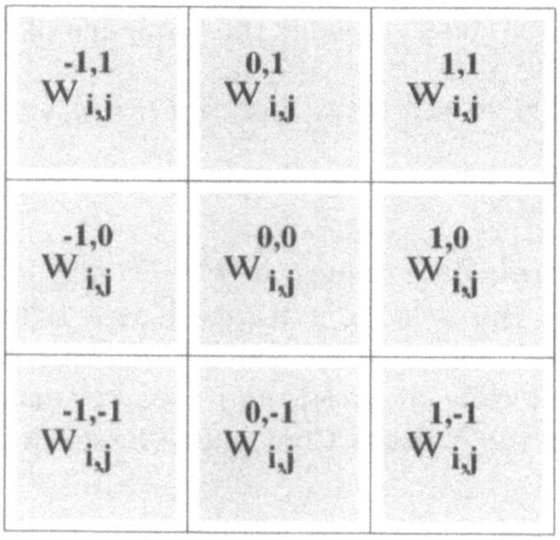

Figure 2.11 The Weight Matrix $\Omega_{i,j}$

Let $I_{i,j}^m$, $m = 1,\dots,4$, denote the four quarters of the cell $I_{i,j}$, with $I_{i,j}^1$ being the upper-right quarter while the other three quarters are numbered clock-wise (see Figure 2.12). In order to obtain a fourth-order computation of the first term on the RHS of (2.57), the reconstructed polynomial, $R_{i,j}(x,y)$, must recover the averages over the four quarter cells with fourth-order accuracy,

$$\bar{R}_{i,j}^{(m)} := \frac{1}{4h^2} \int_{I_{i,j}^m} R_{i,j}(x,y,t^n) dx dy = \frac{1}{4h^2} \int_{I_{i,j}^m} u(x,y,t^n) + O(h^4), \quad m = 1,\dots,4, \tag{2.66}$$

where $u(x,y,t^n)$ denotes the exact solution of the equation at time t^n. On the other hand, the derivatives of the fluxes should be recovered with third-order accuracy. We therefore need to accurately evaluate the intermediate values, $u(x,y,t^n + \beta_i \Delta t)$, with $\beta_1 = 1/2$ and $\beta_2 = 1$, (see (2.26), (2.28) and (2.62)), and in particular, we need an

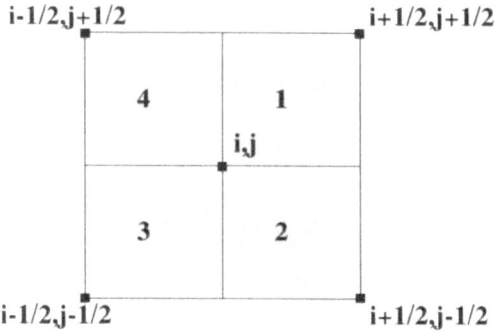

Figure 2.12 The Quarter Cells

accurate reconstruction of the point-values of the solution at the integer grid points (i,j) at time t^n.

The output of the reconstruction routine from cell averages at the beginning of the time step must therefore provide a fourth-order approximation of

(a) the four quarter-cell averages

$$\bar{R}_{i,j}^{(m)} = \frac{1}{4h^2} \sum_{l,k=-1}^{1} w_{i,j}^{l,k} \int_{I_{i,j}^m} P_{i+l,j+k}(x,y)\, dx dy, \qquad m = 1,\dots,4. \qquad (2.67)$$

(b) the point-values at the integer grid-points

$$R_{i,j}(x_i,y_j) = \sum_{l,k=-1}^{1} w_{i,j}^{l,k} P_{i+l,j+k}(x_i,y_j). \qquad (2.68)$$

The reconstruction routine from point values called at each evaluation of the Runge-Kutta fluxes must provide a third order approximation of the derivatives of the flux at the integer grid points

$$R_{i,j}^x(x_i,y_j) = \sum_{l,k=-1}^{1} w_{i,j}^{l,k} \partial_x P_{i+l,j+k}(x_i,y_j), \qquad (2.69)$$

$$R_{i,j}^y(x_i,y_j) = \sum_{l,k=-1}^{1} w_{i,j}^{l,k} \partial_y P_{i+l,j+k}(x_i,y_j), \qquad (2.70)$$

where the polynomials $P_{i+l,j+k}$ interpolate the data $f(u(\cdot,\cdot))$ in (2.69), while in (2.70) the polynomials $P_{i+l,j+k}$ interpolate the data $g(u(\cdot,\cdot))$. Generally, the weights $w_{i,j}^{l,k}$ in (2.67) and in (2.68) will be different from the weights in (2.69) and (2.70) due to the different accuracy requirements.

We would like to stress that there is no need to explicitly compute all the coefficients of the polynomial $R_{i,j}(x,y)$. All that is needed are the point-values and the quarter-cell averages of these polynomials, or their derivatives at the grid points.

We are now ready to present the construction of the fundamental bi-quadratic polynomials, $P_{i,j}(x,y)$.

2.3.3   The Bi-Quadratic Polynomials

In this section we explicitly give the coefficients of the interpolating polynomials $P_{i,j}(x,y)$, which serve as the building block for the reconstruction of $R_{i,j}(x,y)$ in (2.65). In each cell, $I_{i,j}$, we write the polynomial $P_{i,j}(x,y)$ as

$$\begin{aligned}
P_{i,j}(x,y) \;=\; & b_0 + b_1(x - x_i) + b_2(y - y_j) + b_3(x - x_i)(y - y_j) + \qquad (2.71) \\
& + b_4(x - x_i)^2 + b_5(y - y_j)^2 + b_6(x - x_i)^2(y - y_j) + \\
& + b_7(x - x_i)(y - y_j)^2 + b_8(x - x_i)^2(y - y_j)^2,
\end{aligned}$$

where for simplicity we have omitted the indices (i,j) from the coefficients $\{b_m\}$. The nine coefficients b_m are uniquely determined by the interpolation conditions

$$\frac{1}{h^2} \int_{x_i - \frac{h}{2} + lh}^{x_i + \frac{h}{2} + lh} \int_{y_j - \frac{h}{2} + kh}^{y_j + \frac{h}{2} + kh} P_{i,j}(x,y)\, dy\, dx = \bar{u}_{i+l,j+k}, \qquad l,k = -1,0,1,$$

i.e., the polynomials $P_{i,j}(x,y)$ interpolate the data $\{\bar{u}_{i,j}\}$ in the sense of cell-averages. The resulting expressions of the coefficients are

$$b_0 = \bar{u} - \frac{h^2}{24}(\hat{u}_{xx} + \hat{u}_{yy}) + \frac{h^4}{24^2}\hat{u}_{xxyy}, \quad b_1 = \hat{u}_x - \frac{h^2}{24}\hat{u}_{xyy},$$

$$b_2 = \hat{u}_y - \frac{h^2}{24}\hat{u}_{xxy}, \qquad\qquad\qquad b_3 = \hat{u}_{xy},$$

$$b_4 = \tfrac{1}{2}\hat{u}_{xx} - \frac{h^2}{48}\hat{u}_{xxyy}, \qquad\qquad b_5 = \tfrac{1}{2}\hat{u}_{yy} - \frac{h^2}{48}\hat{u}_{xxyy},$$

$$b_6 = \tfrac{1}{2}\hat{u}_{xxy}, \qquad\qquad\qquad\qquad b_7 = \tfrac{1}{2}\hat{u}_{xyy},$$

$$b_8 = \tfrac{1}{4}\hat{u}_{xxyy},$$

where the following notation for divided differences was used,

$$\hat{u}_{x_{i,j}} = \frac{\bar{u}_{i+1,j} - \bar{u}_{i-1,j}}{2h}, \qquad \hat{u}_{y_{i,j}} = \frac{\bar{u}_{i,j+1} - \bar{u}_{i,j-1}}{2h},$$

$$\hat{u}_{xx_{i,j}} = \frac{\bar{u}_{i+1,j} - 2\bar{u}_{i,j} + \bar{u}_{i-1,j}}{h^2}, \qquad \hat{u}_{yy_{i,j}} = \frac{\bar{u}_{i,j+1} - 2\bar{u}_{i,j} + \bar{u}_{i,j-1}}{h^2},$$

$$\hat{u}_{xy_{i,j}} = \frac{\bar{u}_{i+1,j+1} - \bar{u}_{i+1,j-1} - \bar{u}_{i-1,j+1} + \bar{u}_{i-1,j-1}}{4h^2},$$

$$\hat{u}_{xyy_{i,j}} = \frac{(\bar{u}_{i+1,j+1} - 2\bar{u}_{i+1,j} + \bar{u}_{i+1,j-1}) - (\bar{u}_{i-1,j+1} - 2\bar{u}_{i-1,j} + \bar{u}_{i-1,j-1})}{2h^3},$$

$$\hat{u}_{xxy_{i,j}} = \frac{(\bar{u}_{i+1,j+1} - 2\bar{u}_{i,j+1} + \bar{u}_{i-1,j+1}) - (\bar{u}_{i+1,j-1} - 2\bar{u}_{i,j-1} + \bar{u}_{i-1,j-1})}{2h^3},$$

$$\begin{aligned}
\hat{u}_{xxyy_{i,j}} = \frac{1}{h^4}\Big[& (\bar{u}_{i+1,j+1} - 2\bar{u}_{i+1,j} + \bar{u}_{i+1,j-1}) - 2(\bar{u}_{i,j+1} - 2\bar{u}_{i,j} + \bar{u}_{i,j-1}) + \\
& + (\bar{u}_{i-1,j+1} - 2\bar{u}_{i-1,j} + \bar{u}_{i-1,j-1})\Big].
\end{aligned}$$

2.3.4   The Weights

The weights $w_{i,j}^{l,k}$ in the reconstruction (2.65) are computed following the WENO/CWENO ideas presented in [47], [26] and [38]. The goal is to choose weights such that

(a) In smooth regions maximum accuracy is obtained.

(b) In non-smooth regions, information coming from non-smooth stencils is switched off in order to prevent the onset of spurious oscillations.

In order to achieve these goals, the weights $w_{i,j}^{l,k}$ are written as

$$w_{i,j}^{l,k} = \frac{\alpha_{i,j}^{l,k}}{\sum_{l,k=-1}^{1} \alpha_{i,j}^{l,k}}, \tag{2.72}$$

where

$$\alpha_{i,j}^{l,k} = \frac{C^{l,k}}{(\epsilon + \mathrm{IS}_{i,j}^{l,k})^p}. \tag{2.73}$$

Here, $C^{l,k}$ are the constants which are chosen in order to maximize accuracy in smooth regions, $\mathrm{IS}_{i,j}^{l,k}$ are the "smoothness indicators" (see below), p is a constant and ϵ is introduced in order to prevent division by zero. Following previous works (e.g. [38]), in all our numerical experiments we use $p = 2$ and $\epsilon = 10^{-6}$.

The "smoothness indicators", $\mathrm{IS}_{i,j}^{l,k}$, are designed to measure the smoothness (or, more precisely, the roughness) of the polynomials $P_{i+l,j+k}$ in the cell $I_{i,j}$. This is done by evaluating a suitable function of the norms of the derivatives of the polynomial on the cell $I_{i,j}$, namely

$$\mathrm{IS}_{i,j}^{l,k} = \int_{I_{i,j}} \left(|\partial_x P_{i+l,j+k}|^2 + |\partial_y P_{i+l,j+k}|^2 + h^2 |\partial_{xx}^2 P_{i+l,j+k}|^2 + h^2 |\partial_{yy}^2 P_{i+l,j+k}|^2 \right) \, dx \, dy. \tag{2.74}$$

The integrals in (2.74) can be computed exactly, but they involve a large number of function evaluations. We compute the integrals with a Gaussian quadrature with four nodes on the rectangle $I_{i,j}$.

All is left to compute are the constants $C^{l,k}$ in (2.73).

We seek the values of a set of constants, $C^{l,k}$, such that the integral of the reconstruction on each quarter-cell is fourth-order accurate.

We start with the upper-right quarter cell and use symmetry considerations to label $C^{l,k}$ as $q_1, \ldots q_6$, such that, $C^{1,1} = q_1$, $C^{1,0} = C^{0,1} = q_2$, $C^{-1,1} = C^{1,-1} = q_3$, $C^{-1,0} = C^{0,-1} = q_4$, $C^{-1,-1} = q_5$ and $C^{0,0} = q_6$, (see Figure 2.13). Since $C^{l,k} \geq 0$ and $\sum_{l,k=-1}^{1} C^{l,k} = 1$, we have

$$q_6 = 1 - q_1 - 2q_2 - 2q_3 - 2q_4 - q_5, \qquad q_m \geq 0, \quad m = 1, \cdots, 6.$$

Imposing the accuracy requirements (2.66) for the upper-right quarter cell, results with the following system

$$\begin{cases} q_2 = -q_1 + q_4 + q_5, \\ q_3 = \frac{3}{16} - q_4 - q_5, \end{cases}$$

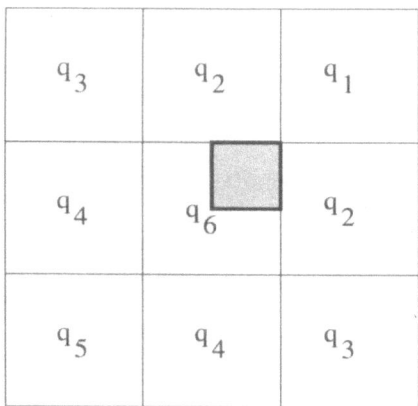

Figure 2.13 The Nine Weights

while q_1 , q_4 and q_5 remain arbitrary.

One possible solution is to choose $q_1 = q_4 = q_5 = \frac{1}{16}$, from which it follows that $q_2 = q_3 = \frac{1}{16}$ and $q_6 = \frac{1}{2}$. This gives a symmetric combination which can be therefore used for all four quarter-cell averages (and not only for the upper-right quarter cell),

$$\mathbb{C} = \begin{pmatrix} 1/16 & 1/16 & 1/16 \\ 1/16 & 1/2 & 1/16 \\ 1/16 & 1/16 & 1/16 \end{pmatrix}. \tag{2.75}$$

By symmetry, this specific choice of $\mathbb{C}$ also gives fourth-order accuracy for the computation of point-values at the center of the cell.

As in the one dimensional case, the smoothness indicators are computed only once, at the beginning of each time step.

2.3.5   The Reconstruction of Flux Derivatives

In order to compute each RK flux in (2.64), it is necessary to evaluate the function

$$F(u)_{i,j} := -f_x(u) - g_y(u)\Big|_{i,j}, \tag{2.76}$$

where u is evaluated at each intermediate time $t_i = t^n + \Delta t\, c_i$ of the RK scheme, (2.26). It is therefore necessary to compute the intermediate values of u :

$$u_{i,j}^{(l)} = u_{i,j}^n + \Delta t \sum_{k=1}^{l-1} a_{l,k} K^{(k)}, \quad l = 1, \dots, \nu. \tag{2.77}$$

Given the intermediate values in (2.77) we can evaluate $f(u_{i,j}^{(l)})$ and $g(u_{i,j}^{(l)})$, which can be then used to compute the discrete derivatives of f and g required in (2.76). These derivatives can be calculated using a procedure which is equivalent to the reconstruction procedure that was used earlier. This time, however, we require that the point-values of the derivative of the reconstruction will be third-order accurate. For simplicity, assume that we start with the function $u_{i,j}$. As before, we write the final reconstruction as a convex combination of interpolating polynomials (compare with (2.65)),

$$R_{i,j}(x,y) = \sum_{l,k=-1}^{1} \hat{w}_{i,j}^{l,k} \hat{P}_{i+l,j+k}(x,y). \tag{2.78}$$

This time, the polynomials interpolate the data in the sense of point-values

$$\hat{P}_{i,j}(x_{i+l}, y_{j+k}) = \hat{u}_{i+l,j+k} \quad l,k = -1,0,1.$$

where $\hat{u}_{i+l,j+k}$ denotes either $f(u_{i+l,j+k})$ or $g(u_{i+l,j+k})$.

The non-constant coefficients in (2.78) are (compare with (2.72)),

$$\hat{w}_{i,j}^{l,k} = \frac{\hat{\alpha}_{i,j}^{l,k}}{\sum_{l,k=-1}^{1} \hat{\alpha}_{i,j}^{l,k}},$$

where for the derivative in the x-direction, one has

$$\hat{\alpha}_{i,j}^{l,k} = \frac{\hat{C}_x^{l,k}}{(\epsilon + \mathrm{IS}_{i,j}^{l,k})^p},$$

and a similar expression holds for the derivative in the y-direction. The smoothness indicators are the same as those computed at the beginning of the time-step. Since we are interested in an accurate reconstruction of the derivatives in (2.76), the constants $\hat{C}_x^{l,k}$ must be chosen in order to satisfy

$$|\partial_x R_{i,j} - u_x(x_i, y_j)| = O(h^3).$$

A straightforward computation results with the possible choice of $\hat{C}_x^{l,k}$ as

$$\hat{C}_x^{l,k} = \begin{pmatrix} 0 & 0 & 0 \\ 1/6 & 2/3 & 1/6 \\ 0 & 0 & 0 \end{pmatrix}. \tag{2.79}$$

For the y-derivative one can choose the transpose of (2.79), $\hat{C}_y^{l,k} = (\hat{C}_x^{l,k})^t$. With this choice, the mixed terms of the bi-quadratic polynomials do not play any role and the differentiation formulas become very simple:

$$\begin{aligned} \left.\frac{\partial R_{i,j}}{\partial x}\right|_{(x_i,y_j)} &= \sum_{l=-1}^{1} \hat{w}_{i,j}^{l,0} \left.\frac{\partial \hat{P}_{i+l,j}}{\partial x}\right|_{(x_i,y_j)} \\ \left.\frac{\partial R_{i,j}}{\partial y}\right|_{(x_i,y_j)} &= \sum_{k=-1}^{1} \hat{w}_{i,j}^{0,k} \left.\frac{\partial \hat{P}_{i,j+k}}{\partial x}\right|_{(x_i,y_j)} \end{aligned} \tag{2.80}$$

2.3.6 The Algorithm

We would like to summarize the different stages of the algorithm obtained in the previous sections. Given $\bar{u}_{i,j}^n$, compute $\bar{u}_{i+1/2,j+1/2}^{n+1}$ according to (2.57), i.e.,

$$\bar{u}_{i+1/2,j+1/2}^{n+1} = \mathcal{I}_1 + \mathcal{I}_2,$$

where

$$\mathcal{I}_1 = \frac{1}{h^2} \int \int_{I_{i+\frac{1}{2},j+\frac{1}{2}}} u^n(x,y)\, dx\, dy,$$

and

$$\mathcal{I}_2 = \quad - \frac{1}{h^2} \int_{\tau=t^n}^{t^{n+1}} \left\{ \int_{y=y_j}^{y_{j+1}} [f\left(u(x_{i+1},y,\tau)\right) - f\left(u(x_i,y,\tau)\right)]\, dy \right\}\, d\tau$$

$$- \frac{1}{h^2} \int_{\tau=t^n}^{t^{n+1}} \left\{ \int_{x=x_i}^{x_{i+1}} [g\left(u(x,y_{j+1},\tau)\right) - g\left(u(x,y_j,\tau)\right)]\, dx \right\}\, d\tau.$$

$\mathcal{I}_1$ is the sum of the four quarter-cell averages defined in (2.67),

$$\mathcal{I}_1 = \bar{R}_{i,j}^{(1)} + \bar{R}_{i,j+1}^{(2)} + \bar{R}_{i+1,j+1}^{(3)} + \bar{R}_{i+1,j}^{(4)}$$

where the polynomials $P_{i+l,j+k}(x,y)$ appearing in (2.67) are given by (2.71) and the weights $w_{i,j}^{l,k}$ are given by (2.72).

The integrals in $\mathcal{I}_2$ are replaced by the quadrature (2.62) and (2.63).

$$\int_{t^n}^{t^{n+1}} f(u(x_i,y_j,z))dz =$$

$$= \frac{\Delta t}{6} \left[f\left(u_{i,j}^n\right) + 4f\left(u_{i,j}^{n+1/2}\right) + f\left(u_{i,j}^{n+1}\right) \right] + O\left((\Delta t)^5\right),$$

and

$$\int_{x_i}^{x_{i+1}} f(x)dx = \frac{h}{24}[-f(x_{i+2}) + 13f(x_{i+1}) + 13f(x_i) - f(x_{i-1})] + O(h^5).$$

The time quadrature required the prediction of the intermediate values, which can be obtained with the RK scheme, (2.26), and their NCE (2.28). This ODE solver requires on the RHS the values of the derivatives of the fluxes given by (2.80), which are evaluated at the integer grid-points (and therefore utilizes the point-values recovered by (2.68)).

2.3.7 Systems of Equations

There are not that many modifications required in order to solve systems of equations instead of solving scalar equations. Basically, one has to extend the algorithm to systems using a straightforward component-wise approach.

The only delicate point is the computation of the smoothness indicators. A component-wise evolution of the smoothness indicators where each component may rely on a different stencil has some disadvantages, as already pointed out in the one dimensional case. The simplest and most robust way to compute the smoothness indicators is to apply *global smoothness indicators*: all components have the same indicator, which is computed as an average of the smoothness indicators of each component,

$$S_{i,j}^{l,k} = \frac{1}{d} \sum_{m=1}^{d} \left\{ \int_{I_{i,j}} \left(|\partial_x P_{i+l,j+k}^m|^2 + |\partial_y P_{i+l,j+k}^m|^2 + \right. \right. \tag{2.81}$$

$$\left. \left. h^2 |\partial_{xx}^2 P_{i+l,j+k}^m|^2 + + h^2 |\partial_{yy}^2 P_{i+l,j+k}^m|^2 \right) dx dy \right\} \left(\|\bar{u}^{(m)}\|_2 + \epsilon \right)^{-1}.$$

Here $P_{i,j}^m$ denotes the m-th component of the vector valued interpolation polynomial, centered on the cell $I_{i,j}$, and:

$$\|\bar{u}^{(m)}\|_2^2 = \sum_{i,j} |\bar{u}_{i,j}^{(m)}|^2 h^2,$$

where (m) denotes the m-th component of the vector $\bar{u}_{i,j}$. Therefore, the Global Smoothness Indicator is an average of all componentwise Smoothness Indicators, each of which is normalized with respect to the norm of the corresponding field.

2.3.8 Numerical Examples

In order to show the non-oscillatory and high resolution properties of the fourth order CWENO scheme, we consider three test problems. The first is a convergence test performed on the linear equation

$$u_t + u_x + u_y = 0$$

with initial condition $u(x,y,0) = \sin^2(2\pi x)\sin^2(2\pi y)$, with periodic boundary conditions. The convergence results of this test are reported in Table 2.3.

Linear advection, Nonlinear weights

N	L^1 error	L^1 order	L^∞ error	L^∞ order
10	8.763E-03	-	2.464E-02	-
20	5.092E-04	4.10	1.632E-03	3.92
40	3.001E-05	4.08	8.747E-05	4.22
80	1.828E-06	4.04	4.836E-06	4.18
160	1.135E-07	4.01	2.802E-07	4.11

Table 2.3 $T = 1$, $\lambda = 0.45$, $u_0(x) = \sin^2(\pi x)\sin^2(\pi y)$

It is evident that the scheme shows an accuracy of fourth order, as expected.

The second test consists in the solution of the 2D Burgers equation, with flux $f(u) = g(u) = -u^2/2$. The initial data, taken from [25], consists in a two dimensional Riemann problem, and is given by

0.8	0.5
-1	-0.2

The configuration is centered at $(1/2,1/2)$, and the computational region is $[0,1] \times [0,1]$. The boundary conditions are $\partial u/\partial n = 0$. Such conditions are perfectly justified until the signal reaches the boundary. The flux is Burgers', $f(u) = g(u) = -u^2/2$. The number of grid points in each direction is $N = 80$, and $\lambda = 0.25$.

The solution at $T = 0.5$ is shown in Figure 2.14. The figure shows the absence of spurious oscillations in a problem involving shock interaction. All discontinuities are very sharp.

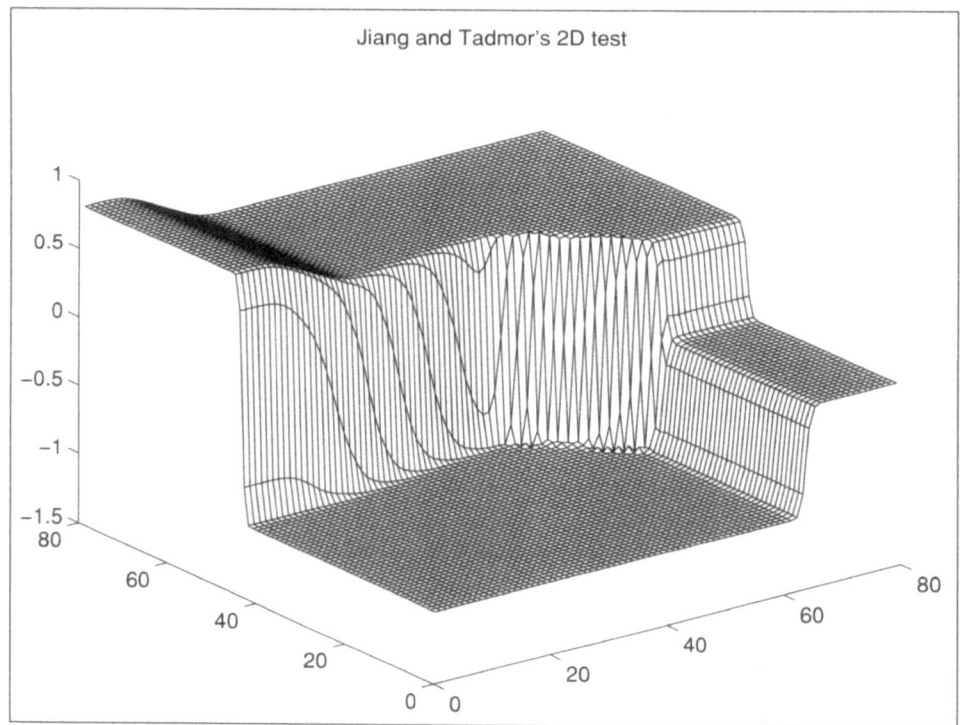

Figure 2.14 Burgers equation. Solution at $T = 0.5$. $N = 80$, $\lambda = 0.25$

The third test is described below.

Two-Dimensional Gas Dynamics Equations

We consider the system of equations for gas dynamics in 2D

$$U_t + F(U)_x + G(U)_y = 0,$$

where

$$U = \begin{pmatrix} \rho \\ \rho u \\ \rho v \\ E \end{pmatrix}, \qquad F = \begin{pmatrix} \rho u \\ \rho u^2 + p \\ \rho u v \\ u(E + p) \end{pmatrix}, \qquad G = \begin{pmatrix} \rho v \\ \rho u v \\ \rho v^2 + p \\ v(E + p) \end{pmatrix}.$$

Here, ρ is the density, u and v are the two components of the velocity, $E = \rho e + \frac{1}{2}\rho(u^2 + v^2)$ is the total energy per unit volume, and e is the internal energy of the gas. The system is closed by defining the pressure p through the equation of state. For a polytropic gas $p = \rho e(\gamma - 1)$, where the constant γ is the ratio of specific heats. In all tests considered, $\gamma = 1.4$.

$$\begin{pmatrix} \rho \\ u \\ v \\ p \end{pmatrix} = \begin{pmatrix} 2 \\ -0.75 \\ 0.5 \\ 1 \end{pmatrix} \quad \bigg| \quad \begin{pmatrix} \rho \\ u \\ v \\ p \end{pmatrix} = \begin{pmatrix} 1 \\ -0.75 \\ -0.5 \\ 1 \end{pmatrix}$$

$$\begin{pmatrix} \rho \\ u \\ v \\ p \end{pmatrix} = \begin{pmatrix} 1 \\ 0.75 \\ 0.5 \\ 1 \end{pmatrix} \quad \bigg| \quad \begin{pmatrix} \rho \\ u \\ v \\ p \end{pmatrix} = \begin{pmatrix} 3 \\ 0.75 \\ -0.5 \\ 1 \end{pmatrix}$$

Table 2.4 Initial condition for Configuration 5 (Figure 2.3.8

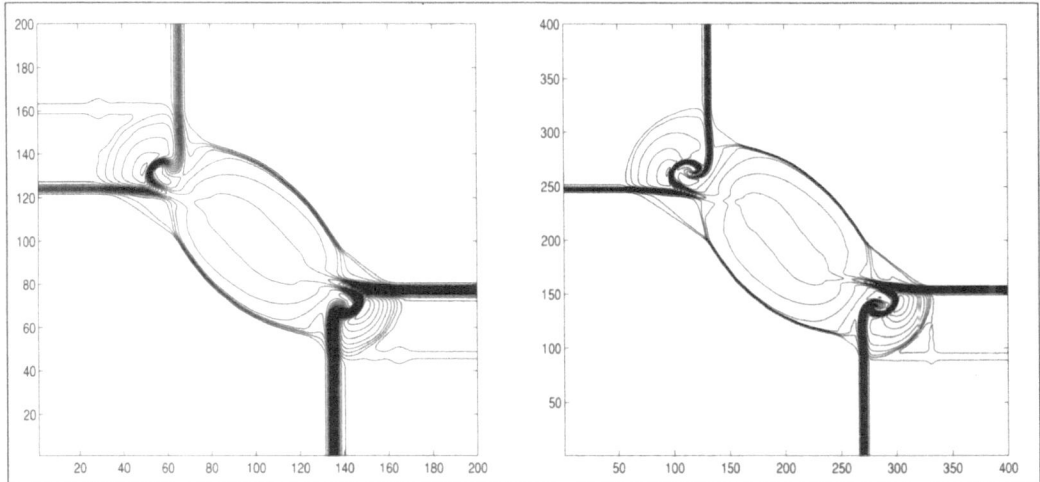

Figure 2.15 Two dimensional Riemann Problem. Solution at time $T = 0.23$ for the initial data reported in Table 2.4. Comparison with two different grids. On the left, the grid is 200×200, on the right, the grid is 400×400

The test problems shown here are two dimensional Riemann problems, taken from [36].

The initial state is given by a piecewise constant data in the four quadrants of $\mathbb{R}^2$. The initial condition for the first one is given in Table 2.4 (denoted as Configuration 5 in [36]). These initial data result in four interacting contact discontinuities.

The results of the computation are shown in Figure 2.3.8, at time T=0.23. On the left, we show the density obtained with a 200×200 grid, while on the right we show the density on a 400×400 grid. The mesh ration one half the one chosen in [36], due to our more restrictive CFL, namely $\lambda = 0.5 * 0.2494$.

We first note that there is a very strong increase in resolution as the cell dimensions are halved due to the high-order accuracy of the scheme. When we compare the results obtained on the fine grid with the corresponding ones in Figure 5 of [36], we find that the two pictures are almost identical. Although the positive schemes used by Lax and Liu in [36] are only second order accurate, we believe that our results are quite striking. In fact, while the positive scheme makes use of the Jacobian and the matrix of eigenvectors of the system of gas dynamics, our scheme requires only the definition of the fluxes. Still, the physics of the problem, apparently, is perfectly caught.

$$
\begin{pmatrix} \rho \\ u \\ v \\ p \end{pmatrix} = \begin{pmatrix} 1.0222 \\ -0.6179 \\ 0.1 \\ 1 \end{pmatrix} \quad \begin{pmatrix} \rho \\ u \\ v \\ p \end{pmatrix} = \begin{pmatrix} 0.5313 \\ 0.1 \\ 0.1 \\ 0.4 \end{pmatrix}
$$

$$
\begin{pmatrix} \rho \\ u \\ v \\ p \end{pmatrix} = \begin{pmatrix} 0.8 \\ 0.1 \\ 0.1 \\ 1 \end{pmatrix} \quad \begin{pmatrix} \rho \\ u \\ v \\ p \end{pmatrix} = \begin{pmatrix} 1 \\ 0.1 \\ 0.8276 \\ 1 \end{pmatrix}
$$

Table 2.5 Initial condition for Configuration 16 (Figure 2.16)

We end our discussion showing the results obtained for Configuration 16 of [36]. The initial condition can be found in Table. 2.3.8. The resulting solution is composed of two contact discontinuities, a rarefaction and a shock wave. We show the results for the density at $T = 0.2$ on a 400×400 grid in Figure 2.16. The CFL number is $\lambda = 0.5 * 0.2494$. These results should be compared with the corresponding ones in [36], Figure 16. We note that the shock is sharp, and the resolution of the two contact discontinuities is also good. Moreover, there are no spurious oscillations even though the wave pattern is complex.

2.3.9 Improvements

The same improvements considered in one dimension can be extended in the two dimensional case. We discuss here only the better use of Runge-Kutta schemes, and we do not consider the problem of the negative weights.

System (2.55) can be written in the form (2.48)

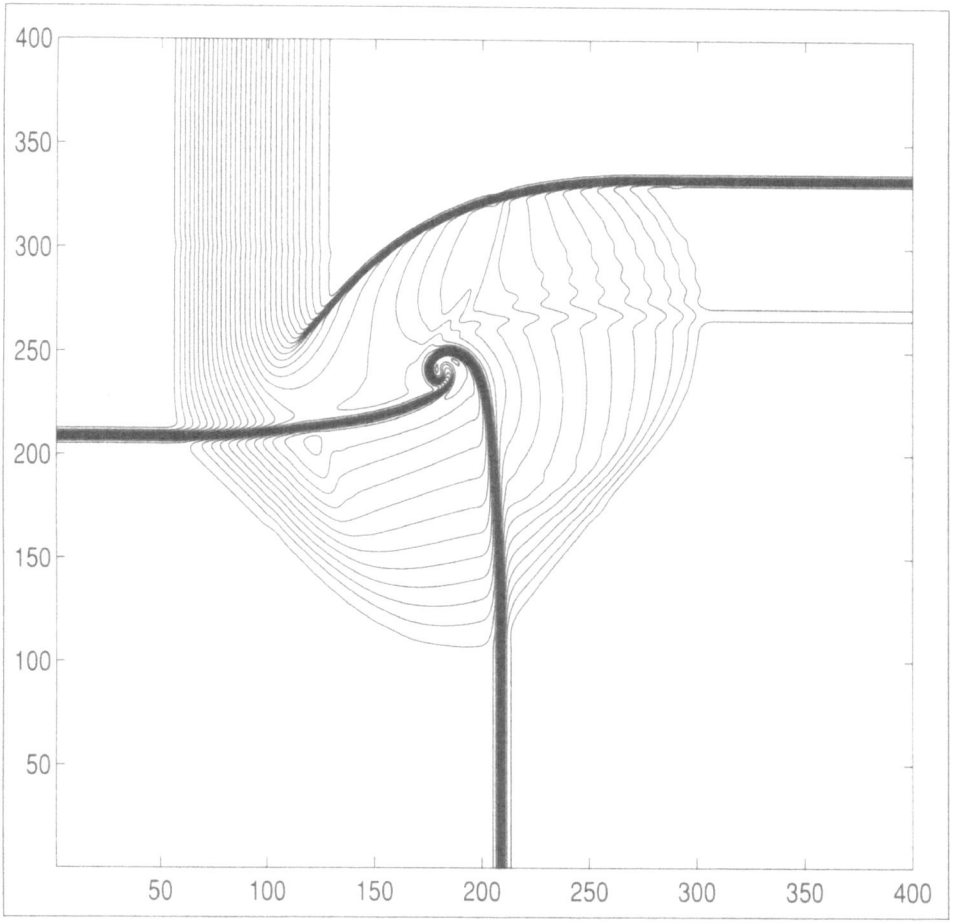

Figure 2.16 Two dimensional Riemann problem. 400×400 grid. Solution at time $T = 0.2$ for the initial data reported in Figure 2.3.8

$$\frac{\partial u}{\partial t} = F(u)$$

where $F(u) = -\partial f / \partial x - \partial g / \partial y$. By applying a ν-stage explicit Runge-Kutta scheme pointwise at the different stages, and taking the average of the solution at time t_{n+1}, one obtains

$$u^{(k)}(x,y) = u^n(x,y) + \Delta t \sum_{l=1}^{k-1} a_{kl} F(u^{(l)}(x,y))$$

$$
\begin{aligned}
\bar{u}^{n+1}(x + h/2, y + h/2) = {} & \bar{u}^n(x + h/2, y + h/2) \\
& - \lambda \sum_{k=1}^{\nu} b_k \Big[\int_y^{y+h} (f(u^{(k)}(x+h,\tilde{y})) - f(u^{(k)}(x,\tilde{y}))) \, d\tilde{y} \\
& \quad - \int_x^{x+h} (g(u^{(k)}(\tilde{x},y+h)) - g(u^{(k)}(\tilde{x},y))) \, d\tilde{x} \Big]
\end{aligned}
\tag{2.82}
$$

The above relation can be discretized on a grid, by performing a WENO reconstruction of the function. Using the WENO reconstruction, the staggered cell average can be computed. About the computation of the integral of the flux appearing in (2.82), one can use either the quadrature formula (2.63) or a Gaussian quadrature formula with nodes inside the cell. The latter approach appears more robust, but is more expensive, since it requires more function evaluations.

2.4   Treatment of the source

Several models in mathematical physics are described by quasilinear hyperbolic systems with source term. The source may be of relaxation type, or it may represent a diffusion term. In this case it is more appropriate to talk about balance laws, rather than conservation laws.

Hyperbolic systems with relaxation appear in discrete velocity models in kinetic theory [18], gas with vibrational degrees of freedom [68], gas in Extended Thermodynamics [49], hydrodynamical models for semiconductors [2, 3], shallow water equations [62], and in several other physical models.

In this section we consider systems of balance laws of the form

$$\frac{\partial u}{\partial t} + \frac{\partial f(u)}{\partial x} = -\frac{1}{\epsilon}g(u) \tag{2.83}$$

The parameter ϵ represents the relaxation time. If it is very small than we say the the relaxation term is *stiff*.

Several schemes have been proposed for the treatment of systems with source term. The most common technique is based on fractional step methods.

In the simplest version, one step for (2.83) is obtained by solving, for a time step Δt, the sequence of the relaxation and convection equations. Given the numerical approximation of the solution, u^n, the new approximation u^{n+1} is obtained as

$$\frac{\partial \tilde{u}}{\partial t} = -\frac{1}{\epsilon}g(\tilde{u}), \quad \tilde{u}(x,0) = u^n(x) \tag{2.84}$$

$$\frac{\partial \hat{u}}{\partial t} + \frac{\partial f(\hat{u})}{\partial x} = 0, \quad \hat{u}(x,0) = \tilde{u}(x,\Delta t) \tag{2.85}$$

At the end of the two steps, one sets $u^{n+1}(x) = \hat{u}(x,\Delta t)$. Such splitting provides a first order (in time) approximation of the solution. A second order scheme can be obtained by using the so called *Strang splitting*: it is obtained by solving (2.84) for a time step $\Delta t/2$, then equation (2.85) for a time step Δt, and finally equation (2.84) for a time step $\Delta t/2$ again. Of course, each step (2.84) and (2.85) has to be solved at least with second order accuracy.

This approach can be generalized to obtain high order methods.

There are several cases, however, when the splitting approach is not very effective. When the relaxation term is stiff, for example, Strang splitting loses second order accuracy, and more sophisticated splitting techniques are necessary to obtain a second order scheme.

In [27], for example, a second order method for systems with stiff source is considered. In [13], a second order scheme for hyperbolic systems with relaxation is derived, which maintain second order accuracy both in the stiff and non stiff limit. These methods have been developed in the context of upwind schemes.

The same approach used in [13] can not be straightforwardly extended to central schemes.

A different approach has been proposed by Bereux and Sainsaulieu in [10], where a system with a possibly stiff relaxation term is considered. Their scheme maintains second order accuracy for stiff source.

A natural way to treat source term in central scheme is to include the source in the integration over the cell is space-time. The schemes we consider here are of this form.

Integrating Eq.(2.83) over the space-time (see Figure 2.3) one has (assume $\epsilon = 1$)

$$\int_{x_j}^{x_{j+1}} u(x,t_{n+1}) = \int_{x_j}^{x_{j+1}} u(x,t_n) - \int_{t_n}^{t_{n+1}} (f(u(x_{j+1},t)) - f(u(x_j,t)))\, dt \quad (2.86)$$

$$+ \int_{t_n}^{t_{n+1}} \int_{x_j}^{x_{j+1}} g(x,t)\, dx\, dt. \quad (2.87)$$

Numerical schemes are obtained by a suitable discretization of the integrals.

Here we only consider second order schemes. We shall discuss later how to derive higher order schemes.

For a second order scheme we use a piecewise linear reconstruction in each cell, as in the Nessyahu-Tadmor scheme, and midpoint rule for the computation of the flux integral.

Different schemes are obtained, according to the discretization of the integral of the source term. If the source is not stiff, then a fully explicit time discretization can be used, resulting in the following scheme

$$u_{j+1/2}^{n+1} = \bar{u}_{j+1/2}^n + \frac{\Delta t}{\Delta x}(f(u_j^{n+1/2}) - f(u_{j+1}^{n+1/2}))$$

$$+ \frac{\Delta t}{2}(g(u_j^{n+1/2}) + g(u_{j+1}^{n+1/2})) \quad (2.88)$$

where

$$\bar{u}_{j+1/2}^n = \frac{1}{2}(u_j^n + u_{j+1}^n) + \frac{1}{8}(u_j' - u_{j+1}') \quad (2.89)$$

and the predictor values $u_j^{n+1/2}$ are computed by

$$u_j^{n+1/2} = u_j^n - \frac{\lambda}{2}f_j' + \frac{\Delta t}{2}g(\bar{u}_j^n). \quad (2.90)$$

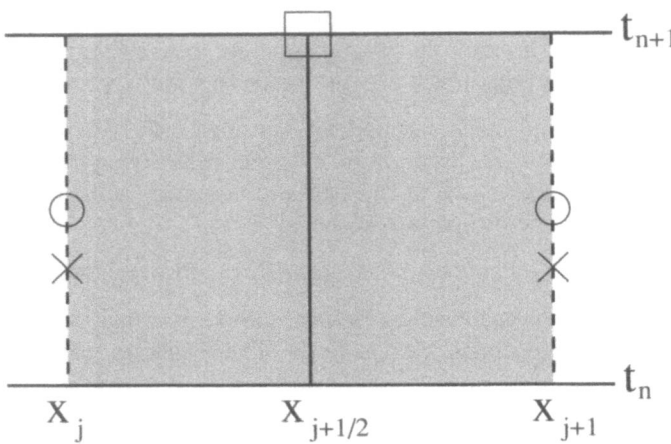

Figure 2.17 Nodes in space time for the second order Uniform Central Scheme

Note that this scheme is fully explicit, therefore it is subject to stability restriction due to both the flux and the source term. If the stability restriction of the source term is more severe than the one due to the flux, then not only efficiency, but also accuracy of the calculation will be affected. In this case, in fact, one has to use a Courant number much smaller than the one allowed by the CFL restriction. It is well known that in this case the numerical dissipation will be larger than necessary, and the accuracy of the scheme will be poor. This problem can be partially circumvented, for moderately stiff source, by the use of semidiscrete schemes. (For a discussion of this issue see, for example, [32]). When the stiffness increases it is better to treat the source implicitly. This can be done at the stage of the predictor, as

$$u_j^{n+1/2} = u_j^n - \frac{\lambda}{2} f_j' + \frac{\Delta t}{2} g(u_j^{n+1/2}) \qquad (2.91)$$

The time discretization used for the source is the midpoint implicit scheme, which is a Gauss-collocation scheme with one level. Such scheme is A-stable, but not L-stable, and therefore it is not suitable for very stiff source (see [21]). A numerical scheme which is stable and accurate even for very stiff source has been proposed in [46]. It is obtained by using two predictor stages, one for the flux, and one which is needed for the source. The scheme can be written in the form

$$
\begin{aligned}
u_j^{n+1/2} &= u_j^n - \frac{\lambda}{2} f_j' + \frac{\Delta t}{2} g(u_j^{n+1/2}) \\
u_j^{n+1/3} &= u_j^n - \frac{\lambda}{3} f_j' + \frac{\Delta t}{3} g(u_j^{n+1/3}) \\
u_{j+1/2}^{n+1} &= \bar{u}_{j+1/2}^n - \lambda (f(u_{j+1}^{n+1/2}) - f(u_j^{n+1/2})) \\
&\quad + \frac{\Delta t}{8} (3g(u_j^{n+1/3}) + 3g(u_{j+1}^{n+1/3}) + 2g(u_{j+1/2}^{n+1}))
\end{aligned}
\qquad (2.92)
$$

where $\bar{u}_{j+1/2}^n$ is computed by 2.89. We shall call the above scheme Uniformly accurate Central Scheme of order 2 (UCS2).

This scheme has the following properties. When applied to hyperbolic systems with relaxation, it is second order accurate in space and time both in the non stiff case (i.e. $\epsilon = 1$) and in the stiff limit (i.e. $\epsilon = 0$).

A small degradation of accuracy is observed for intermediate values of the relaxation parameter ϵ.

The time discretization used in the above scheme is a particular case of Runge-Kutta Implicit-Explicit (IMEX) scheme. Such schemes are particularly important when one has to solve systems that contain the sum of a non stiff (possibly expensive to compute) and a stiff term. These systems may be convection-diffusion equations, or hyperbolic systems with stiff relaxation. In these cases it is highly desirable to use a scheme which is explicit in the non stiff term, and implicit in the stiff term.

Runge-Kutta IMEX schemes have been studied in [9] and in [53].

Let us consider a generic system

$$y' = f(t,y) + \frac{1}{\epsilon}g(t,y) \qquad (2.93)$$

which may represent a system of ordinary differential equations or a discretization of a system of partial differential equations (method of lines).

An Implicit-Explicit Runge-Kutta scheme for system (2.93) is of the form

$$Y_i = y_0 + h\sum_{j=1}^{i-1} \tilde{a}_{ij} f(t_0 + \tilde{c}_j h, Y_j) + h\sum_{j=1}^{\nu} a_{ij} \frac{1}{\epsilon} g(t_0 + c_j h, Y_j), \qquad (2.94)$$

$$y_1 = y_0 + h\sum_{i=1}^{\nu} \tilde{w}_i f(t_0 + \tilde{c}_i h, Y_i) + h\sum_{i=1}^{\nu} w_i \frac{1}{\epsilon} g(t_0 + c_i h, Y_i). \qquad (2.95)$$

The matrices $\tilde{A} = (\tilde{a}_{ij})$, $\tilde{a}_{ij} = 0$ for $j \geq i$ and $A = (a_{ij})$ are $\nu \times \nu$ matrices such that the resulting scheme is explicit in f, and implicit in g. An IMEX Runge-Kutta scheme is characterized by these two matrices and the coefficient vectors $\tilde{c} = (\tilde{c}_1, \dots, \tilde{c}_\nu)^T$, $\tilde{w} = (\tilde{w}_1, \dots, \tilde{w}_\nu)^T$, $c = (c_1, \dots, c_\nu)^T$, $w = (w_1, \dots, w_\nu)^T$. They can be represented by a double *tableau* in the usual Butcher notation,

$$\begin{array}{c|c} \tilde{c} & \tilde{A} \\ \hline & \tilde{w}^T \end{array} , \qquad \begin{array}{c|c} c & A \\ \hline & w^T \end{array} .$$

A sufficient condition to guarantee that f is always evaluated explicitly is that the scheme for g is diagonally implicit i.e. $a_{ij} = 0$, for $j > i$, and that the first column of a is zero.

Stability analysis of the scheme is studied in [46], and the analysis of other IMEX RK schemes is performed in [53]. In particular, the concept of A-stable scheme for a single complex linear equation is extended to IMEX schemes. Here we report the result found in [46].

By applying the generic IMEX scheme to the linear equation (2.93) with $f(y) = \lambda_1 y$, $g(y) = \lambda_2 y$, and $y_0 = 1$, one obtains the function of absolute stability

$$y_1 = \mathcal{R}(z_1, z_2) \equiv 1 + (z_1 \tilde{w}^T + z_2 w^T)(I - z_1 \tilde{A} - z_2 A)^{-1} e, \qquad (2.96)$$

where $z_1 = \lambda_1 h$, $z_2 = \lambda_2 h$, $e = (1, \dots, 1)^T \in \mathbb{R}^\nu$. The function $\mathcal{R}(z_1, z_2)$ is the *function of absolute stability*.

The *region of absolute stability* S_A associated to scheme (2.94, 2.95) is defined as

$$S_A = \{(z_1, z_2) \in \mathbb{C}^2 : |\mathcal{R}(z_1, z_2)| \le 1\}.$$

It is evident that the region does not contain the set $\mathbb{C}^- \times \mathbb{C}^-$. Our goal is to show that there exist two regions of the complex plane, $S_1 \subset \mathbb{C}$, $S_2 \subset \mathbb{C}$, with the following properties:

$$S_A \supset S_1 \times S_2, \qquad (2.97)$$
$$S_2 \supset \mathbb{C}^- \equiv \{z \in \mathbb{C} : \mathrm{Re}(z) \le 0\}, \qquad (2.98)$$

and to compute them. In particular, one is interested in the largest set S_1 for which $S_2 \supset \mathbb{C}^-$. Such region is defined by

$$S_1 = \{z_1 \in \mathbb{C} : \sup_{z_2 \in \mathbb{C}^-} |\mathcal{R}(z_1, z_2)| \le 1\}.$$

The region S_1 for scheme UCS2 is reported in Figure (2.18).

We remark that this stability analysis is valid only for the scalar equation. The extension to systems of equations is presently under investigation.

To show the effectiveness of the scheme, we apply it to the monoatomic gas in Extended Thermodynamics.

Monoatomic gas in Extended Thermodynamics   As an application of the scheme UCS2, we consider the a Riemann problem for a gas in Extended Thermodynamics. In usual gas dynamics, a gas is described by the conservation laws for mass, momentum, end energy. Pressure, stress tensor, and heat flux are related to the conservative quantities by constitutive relations (i.e. the equation of state for the pressure and the Navier-Stokes-Fourier relations that express stress and heat flux in terms of the gradient of velocity and temperature. In Extended Thermodynamics stress tensor and heat flux are independent thermodynamical variables, for which additional balance laws can be written [49]. In one dimension the gas is described by a system of five balance equations, the usual three conservation equations for mass, momentum, and energy, and two additional equations for the balance of stress σ (just one component in one dimension) and heat flux q. The system is of the form

$$u_t + f(u)_x = -\frac{1}{\epsilon} g(u)$$

with

$$u = \begin{pmatrix} \rho \\ \rho v \\ \frac{1}{2}\rho v^2 + \frac{3}{2}p \\ \frac{2}{3}\rho v^2 + \sigma \\ \rho v^3 + 5vp + 2\sigma v + 2q \end{pmatrix}, \; g = \begin{pmatrix} 0 \\ 0 \\ 0 \\ \rho \sigma \\ \frac{2}{3}\rho(2q + 3v\sigma) \end{pmatrix}$$

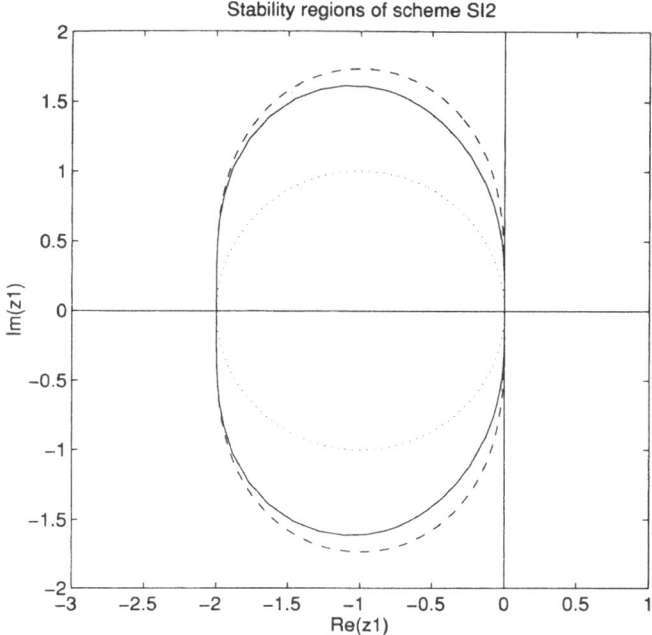

Figure 2.18 Stability region S_1 in the explicit parameter for scheme UCS2 (continuous line). The dotted and the dashed lines are the stability region of explicit Euler and explicit second-order two-level Runge-Kutta respectively.

$$f = \begin{pmatrix} \rho v \\ \rho v^2 + p + \sigma \\ \frac{1}{2}\rho v^3 + \frac{5}{2}vp + \sigma v + q \\ \frac{2}{3}\rho v^3 + \frac{4}{3}vp + \frac{7}{3}v\sigma + \frac{8}{15}q \\ \rho v^4 + 5\frac{p^2}{\rho} + 7\frac{\sigma p}{\rho} + \frac{32}{5}qv + v^2(8p + 5\sigma) \end{pmatrix}$$

As $\epsilon \to 0 \Rightarrow \sigma \to 0$, $q \to 0$ and the equations reduce to the Euler equations for monoatomic gas.

Such system is hyperbolic in a suitable region of the field vector u, which contains the equilibrium manifold $\sigma = 0$, $q = 0$.

No simple expression for the eigenvalues and eigenvector is known for such system, therefore it would be expensive to use characteristic based methods.

We consider the following Riemann problem.

$$\rho = 1, \quad u = 0, \quad p = \frac{5}{3}, \quad \sigma = 0, \quad q = 0 \quad \text{for } x < 0.5, \tag{2.99}$$

$$\rho = \frac{1}{8}, \quad u = 0, \quad p = \frac{1}{6}, \quad \sigma = 0, \quad q = 0 \quad \text{for } x > 0.5. \tag{2.100}$$

The numerical solution to the Riemann problem is shown in Figure (2.4). By comparison, the solution to the Euler equations corresponding to the same initial conditions is reported. The two solutions are very close for density and velocity, because the value of the relaxation parameter is very small.

Stress tensor and heat flux are mainly concentrated at the shock. Two schemes are compared: UCS2 and a scheme based on second order upwind + Strang splitting. It is evident that UCS2 gives a better resolution than the splitting scheme, because of the loss of accuracy of the latter for small values of the relaxation parameter.

The comparison is shown between the numerical solution of the Euler equations and of the gas in Extended Thermodynamics.

Further developments

The scheme UCS2 shown above is not optimal, since it makes uses of three implicit stages in order to obtain second order accuracy. More efficient time discretization can be obtained. A recently developed central scheme by Pareschi guarantees uniform accuracy in the stiff and non stiff limit with two explicit and two implicit stages [51].

There are several topics we did not have space to deal with in this chapter. We shall briefly mention them here.

We did not say anything about boundary conditions. In all calculations we used trivial boundary conditions, either periodic, or flat. Such boundary conditions can be easily treated by using enough ghost cells out of the computational domain. The cells will be then filled with a periodic replica of the computed solution (for periodic B.C.) or with constant values (for zero flux B.C.) The latter are correct until the signal reaches the boundary.

A case which is particularly important in the applications is given by reflecting boundary conditions for gas dynamics. This case is treated in the usual way, i.e. pressure and density are symmetric with respect to the boundary, and velocity is antisymmetric. Such boundary conditions are assigned both at even and odd time steps. At odd time step, the value of the field variables on the boundary cell can be computed if enough ghost cells are filled.

The treatment of more complex boundary conditions goes beyond the scope of the book.

About the derivation of schemes for hyperbolic systems with stiff source, we remark that semidiscrete central schemes can can be used as building blocks for second order IMEX time discretization of hyperbolic systems with source term. If one is interested in higher order schemes, however, this approach can not be straightforwardly used. Higher order semidiscrete central scheme based on WENO reconstruction are basically equivalent to WENO finite volume schemes developed and used by Shu [58]. However, the integration of a source term in space-time may pose some difficulties for stiff problems.

Higher order IMEX time discretization can be used by adopting *finite difference space discretization*, rather than *finite volume*. Finite difference schemes have the same structure of finite volume. They have a conservative form, and the basic unknown is the *pointwise value* rather that the cell average of the unknown field $u(x,t)$. They have the advantage of being more efficient that their finite volume counterpart, especially in multidimensional computation. However, they have the disadvantage of requiring a uniform (or smoothly varying) grid, and therefore they can not be extended to unstructured grids. An extensive treatment of high order finite difference schemes is given in [60, 26, 58].

The development of high order finite volume schemes for systems with stiff source is still an open problem.

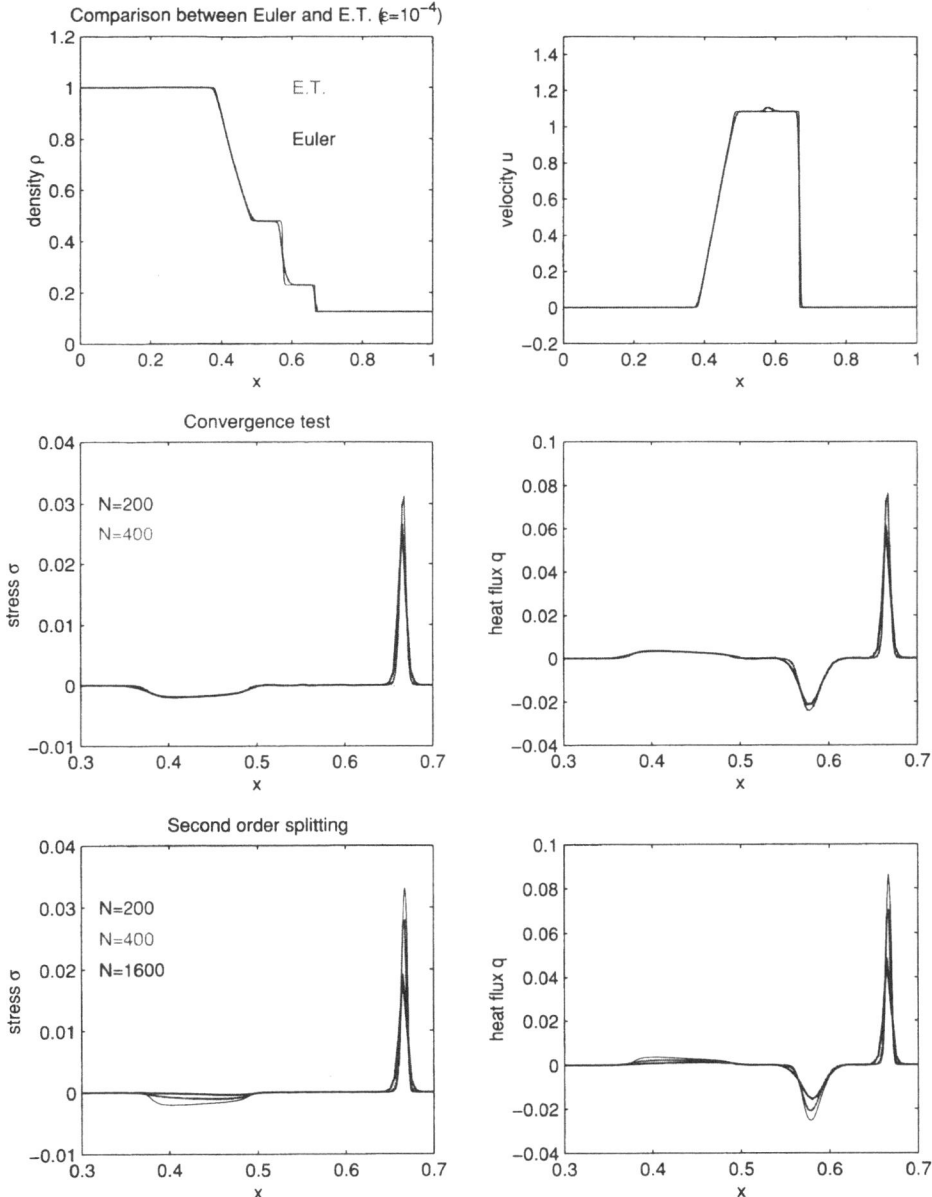

Figure 2.19 Solution of the Riemann problem in Extended Thermodynamics, and comparison with the solution to the Euler equations. Initial data 2.99, with $\varepsilon = 10^{-4}$. Mesh ratio $\Delta t = \Delta x/9$. Top pictures: density (left) and velocity (right) computed with 200 grid points with UCS2 for Extended Thermodynamics (gray line) and NT scheme for Euler equation (black line). Bottom and middle pictures: Shear stress (left) and heat flux (right) for Extended Thermodynamics model with $\varepsilon = 10^{-4}$. Note that the x-scale has been magnified. Middle pictures: UCS2 scheme with 200 (gray line) and 400 (black line). Bottom pictures: upwind+splitting scheme with 200 (gray line) and 400 (black line). The thin line in the bottom pictures is the reference solution obtained by UCS2 with 1600 grid points.

Bibliography

[1] A. M. ANILE, M. JUNK, V. ROMANO AND G. RUSSO, *Cross-validation of numerical schemes for extended hydrodynamical models of semiconductors*, Math. Models Methods Appl. Sci. **10**, 833–861 (2000).

[2] A. M. ANILE AND S. PENNISI, *Thermodynamic derivation of the hydrodynamical model for charge transport in semiconductors*, Phys. Rev B **46**, 13186–13193 (1992).

[3] A. M. ANILE AND O. MUSCATO, *Improved hydrodynamical model for carrier transport in semiconductors* Phys. Rev B **51**, 16728–16740 (1995).

[4] A. M. ANILE, N. NIKIFORAKIS AND R. M. PIDATELLA, *Assessment of a high resolution centered scheme for the solution of hydrodynamical semiconductor equations*, SIAM J. Sci. Comput. **22**, 1533–1548 (2000).

[5] A. M. ANILE, V. ROMANO AND G. RUSSO, *Extended hydrodynamical model of carrier transport in semiconductors*, SIAM J. Appl. Math. **61**, 74–101 (2000).

[6] P. ARMINJON AND M.-C. VIALLON, *Généralisation du schéma de Nessyahu-Tadmor pour une équation hyperbolique deux dimensions d'espace. (French) [Generalization of the Nessyahu-Tadmor scheme for hyperbolic equations in two space dimensions]* C. R. Acad. Sci. Paris Sér. I Math. **320**, 85–88 (1995).

[7] P. ARMINJON, M.-C. VIALLON AND A. MADRANE, *A finite volume extension of the Lax-Friedrichs and Nessyahu-Tadmor schemes for conservation laws on unstructured grids.* Int. J. Comput. Fluid Dyn. **9**, 1–22 (1997).

[8] P. ARMINJON AND M.-C. VIALLON, *Convergence of a finite volume extension of the Nessyahu-Tadmor scheme on unstructured grids for a two-dimensional linear hyperbolic equation*, SIAM J. Numer. Anal. **36**, 738–771 (1999).

[9] U. ASCHER, S. RUUTH, AND R. J. SPITERI, *Implicit-explicit Runge-Kutta methods for time dependent Partial Differential Equations*, Appl. Numer. Math. **25**, 151–167 (1997).

[10] F. BEREUX AND L. SAINSAULIEU, *A Roe-type Riemann solver for hyperbolic systems with relaxation based on time-dependent wave decomposition*, Numer. Math. **77**, 143–185 (1997).

[11] F. BIANCO, G. PUPPO AND G. RUSSO, *High Order Central Schemes for Hyperbolic Systems of Conservation Laws*, SIAM J. Sci. Comput. **21**, 294–322 (1999).

[12] F. BIANCO, G. PUPPO AND G. RUSSO, *High Order Central Schemes for Hyperbolic Systems of Conservation Laws*, HYP-98 Seventh International Conference on Hyperbolic Problems, Theory, Numerics, Applications, ETH Zuerich, Switzerland, February 9-13, 1998.

[13] R. E. CAFLISCH, S. JIN AND G. RUSSO, *Uniformly accurate schemes for hyperbolic systems with relaxation*, SIAM J. Numer. Anal. **34**, 246–281 (1997).

[14] B. COCKBURN, C. JOHNSON, SHU C.-W. AND E. TADMOR, *Advanced Numerical Approximation of Nonlinear Hyperbolic Equations*, Lecture Notes in Mathematics **1697** (editor: A. Quarteroni), Springer, Berlin, 1998.

[15] B. ENGQUIST AND S.OSHER, *One sided difference approximations for nonlinear conservation laws*, Math. Comp. **36**, 321–351 (1981).

[16] B. ENGQUIST AND O. RUNBORG, *Multiphase computations in geometrical optics*, J. Comput. Appl. Math. **74** 175-192 (1996).

[17] B. ENGQUIST AND B. SJÖGREEN, *The convergence rate of finite difference schemes in the presence of shocks.* SIAM J. Numer. Anal. **35**, 2464–2485 (1998).

[18] R. GATIGNOL, *Théorie cinétique des gaz à répartition discrète de vitesses*, Lectures Notes in Physics, **36**, Springer-Verlag, Berlin-New York, 1975.

[19] E. GODLEWSKI AND P.-A. RAVIART, *Numerical approximation of hyperbolic systems of conservation laws.* Applied Mathematical Sciences, **118**, Springer-Verlag, New York, 1996.

[20] B. HAASDONK, B. KRÖNER AND D. ROHDE, *Convergence of a staggered Lax-Friedrichs scheme for nonlinear conservation laws on unstructured two-dimensional grids*, Numer. Math. **88**, 459–484 (2001).

[21] E. HAIRER AND G. WANNER, *Solving ordinary differential equations, Vol.2 Stiff and differential-algebraic problems*, Springer-Verlag, New York, 1987.

[22] E. HARABETIAN AND R. PEGO, Nonconservative hybrid shock capturing schemes. J. Comput. Phys. **105**, 1–13 (1993).

[23] A. HARTEN, B. ENGQUIST, S. OSHER AND S. CHAKRAVARTHY, *Uniformly High Order Accurate Essentially Non-oscillatory Schemes III*, J. Comput. Phys. **71**, 231–303 (1987).

[24] G.-S. JIANG, D. LEVY, C.-T.LIN, S. OSHER AND E. TADMOR, *High-Resolution Non-Oscillatory Central Schemes with Non-Staggered Grids for Hyperbolic Conservation Laws*, SIAM J. Numer. Anal. **35**, 2147–2168 (1998).

[25] G.-S. JIANG AND E. TADMOR, *Nonoscillatory Central Schemes for Multidimensional Hyperbolic Conservation Laws*, SIAM J. Sci. Comput. **19**, 1892–1917 (1998).

[26] G.-S. JIANG AND C.-W. SHU, *Efficient Implementation of Weighted ENO Schemes*, J. Comput. Phys. **126**, 202–228 (1996).

[27] S. JIN, *Runge-Kutta methods for hyperbolic systems with stiff relaxation terms*, J. Comp. Phys. **122**, 51–67 (1995).

[28] T. KATSAOUNIS AND D. LEVY, *A modified structured central scheme for 2D hyperbolic conservation laws*, Appl. Math. Lett. **12**, 89–96 (1999).

[29] R. KUPFERMAN, *A numerical study of the axisymmetric Couette-Taylor problem using a fast high-resolution second-order central scheme*, SIAM J. on Sci. Comput. **20**, 858–877 (1998).

[30] R. KUPFERMAN AND E. TADMOR, *A fast high-resolution second-order central scheme for incompressible flows*, Proc. Nat. Acad. Sci. U.S.A. **94**, 4848–4852 (1997).

[31] A. A. KURGANOV AND D. LEVY, *A third-order semi-discrete central scheme for conservation laws and convection-diffusion equation*, SIAM J. Sci. Comput. **22**, 1461–1488 (2000).

[32] A. KURGANOV AND E. TADMOR, *New High-Resolution Central Schemes for Nonlinear Conservation Laws and Convection-Diffusion Equations*, J. Comput. Phys. **160**, 214–282 (2000).

[33] A. KURGANOV AND E. TADMOR, *New High-Resolution Semi-Discrete Central Schemes for Hamilton-Jacobi Equations*, J. Comput. Phy. **160**, 720-742 (2000).

[34] P.D.LAX, *Hyperbolic systems of conservation laws II*, Comm. Pure Appl. Math. **10**, 105–119 (1957).

[35] P.D.LAX, *Hyperbolic systems of conservation laws and the mathematical theory of shock waves*, SIAM, Philadelphia, 1973.

[36] P. D. LAX AND X. D. LIU, *Solutions of Two-Dimensional Riemann Problems of Gas Dynamics by Positive Schemes*, SIAM J. Sci. Comp. **19**, 319–340 (1998).

[37] R. J. LEVEQUE, *Numerical Methods for Conservation Laws. Second edition*, Lectures in Mathematics ETH Zürich, Birkhäuser Verlag, Basel, 1992.

[38] D. LEVY, G. PUPPO AND G. RUSSO, *Central WENO Schemes for Hyperbolic Systems of Conservation Laws*, Math. Model. Numer. Anal. **33**, 547–571 (1999).

[39] D. LEVY, G. PUPPO AND G. RUSSO, *A Third Order Central WENO Scheme for 2D Conservation Laws*, Appl. Num. Math. **33**, 407–414 (2000).

[40] D. LEVY, G. PUPPO AND G. RUSSO, *Compact Central WENO Schemes for Multidimensional Conservation Laws*, SIAM J. Sci. Comput. **22**, 656–672 (2000).

[41] D. LEVY, G. PUPPO AND G. RUSSO, *Central WENO Schemes for Multidimensional Hyperbolic Systems of Conservation Laws*, submitted.

[42] D. LEVY AND E. TADMOR, *Non-oscillatory central schemes for the incompressible 2-D Euler equations*, Math. Res. Lett. **4**, 321–340 (1997).

[43] K.-A. LIE AND S. NOELLE, *Remarks on high-resolution non-oscillatory central schemes for multi-dimensional systems of conservation laws. Part I: An improved quadrature rule for the flux computation*, submitted.

[44] C.-T. LIN AND E. TADMOR, *High-resolution non-oscillatory central scheme for Hamilton-Jacobi equations*, SIAM J. Sci. Comput. **21**, 2163-2186 (2000).

[45] X.-D. LIU AND E. TADMOR, *Third Order Nonoscillatory Central Scheme for Hyperbolic Conservation Laws*, Numer. Math. **79**, 397-425 (1998).

[46] S. F. LIOTTA, V. ROMANO AND G. RUSSO, *Central schemes for balance laws of relaxation type*, SIAM J. Numer. Anal. **38**, 1337–1356 (2000).

[47] X.-D. LIU, S. OSHER AND T. CHAN, *Weighted Essentially Non-oscillatory Schemes*, J. Comput. Phys. **115**, 200-212 (1994).

[48] X.-D. LIU AND E. TADMOR, *Third Order Nonoscillatory Central Scheme for Hyperbolic Conservation Laws*, Numer. Math. **79**, 397–425 (1998).

[49] I. MÜLLER AND T. RUGGERI,, *Rational extended thermodynamics*, Springer-Verlag, Berlin, 1998.

[50] H. NESSYAHU AND E. TADMOR, *Non-oscillatory Central Differencing for Hyperbolic Conservation Laws*, J. Comput. Phys. **87**, 408–463 (1990).

[51] L. PARESCHI, *Central differencing based numerical schemes for hyperbolic conservation laws with relaxation terms*, preprint (2000).

[52] L.PARESCHI, G.PUPPO AND G.RUSSO, *Improved central WENO schemes for conservation laws*, in preparation.

[53] L. PARESCHI AND G. RUSSO, *Implicit-Explicit Runge-Kutta Schemes for Stiff Systems of Differential Equations*, In Recent Trends in Numerical Analysis; D. Trigiante Ed., Nova Science Publ., 269–288 (2000).

[54] P. L. ROE, *Approximate Riemann Solvers, Parameter Vectors, and Difference Schemes*, J. Comput. Phys. **43**, 357-372 (1981).

[55] V. ROMANO AND G. RUSSO, *Numerical solution for hydrodynamical models of semiconductors*, Math. Models Methods Appl. Sci. **10**, 1099–1120 (2000).

[56] W. ROSENBAUM, M. RUMPF AND S. NOELLE, *An adaptive staggered scheme for conservation laws*, "Hyperbolic Problems: Theory, Numerics, Applications", Proceedings of the 8th international conference held in Magdeburg, Feb. 2000.

[57] R. SANDERS AND W. WEISER, *High resolution staggered mesh approach for nonlinear hyperbolic systems of conservation laws*, J. Comput. Phys. **101**, 314–329 (1992).

[58] SHU C.-W., *Essentially Non-Oscillatory and Weighted Essentially Non-Oscillatory Schemes for Hyperbolic Conservation Laws* in Advanced Numerical Approximation of Nonlinear Hyperbolic Equations, Lecture Notes in Mathematics **1697** (editor: A. Quarteroni), Springer, Berlin, 1998.

[59] J. SHI, C. HU AND C.-W. SHU, *A technique of treating negative weights in WENO schemes*, submitted to Journal of Computational Physics.

[60] SHU C.-W. AND S. OSHER, *Efficient Implementation of Essentially Non-Oscillatory Shock-Capturing Schemes, II*, J. Comput. Phys. **83**, 32-78 (1989).

[61] G. A. SOD, *A Survey of Several Finite Difference Methods for Systems of Nonlinear Hyperbolic Conservation Laws*, J. Comput. Phys. **27**, 1–31 (1978).

[62] J. J. STOKER, *Water waves: The mathematical theory with applications*, Pure and Applied Mathematics, Vol. IV. Interscience Publishers, Inc., New York; Interscience Publishers Ltd., London 1957.

[63] E. TADMOR AND T. TANG, *Pointwise error estimates for scalar conservation laws with piecewise smooth solutions*, SIAM J. Numer. Anal. **36**, 1739-1756 (1999).

[64] E. TADMOR AND T. TANG, *Pointwise convergence rate for nonlinear conservation laws*, Hyperbolic problems: theory, numerics, applications, Vol. II (Zrich, 1998), 925–934, Internat. Ser. Numer. Math., 130, Birkhuser, Basel, 1999.

[65] E. TADMOR AND C. C. WU, *Central scheme for the multidimensional MHD equations*, in preparation.

[66] E. F. TORO, *Riemann solvers and numerical methods for fluid dynamics. A practical introduction.* Second edition. Springer-Verlag, Berlin, 1999.

[67] M. TROVATO AND P. FALSAPERLA, *Full nonlinear closure for a hydrodynamical model of transport in silicon*, Phys. Rev. B-Condensed Matter **57**, 4456-4471 (1998).

[68] W. G. VINCENTI AND C. H. KRUGER, JR., *Introduction to Physical Gas Dynamics*, Krieger Publishing Company, Malabar, Florida, 1982.

[69] G. B. WHITHAM, Linear and Nonlinear Waves, Wiley, 1974.

[70] ZENNARO M., *Natural Continuous Extensions of Runge-Kutta Methods*, Math. Comp. **46**, 119-133 (1986).

3 Methods on unstructured grids, WENO and ENO Recovery techniques

3.1 Introduction to finite volume approximations

Finite volume approximations comprise the most successful class of discretisation techniques for the conservation laws of compressible fluid mechanics. Their success is based not only on their relative simplicity as compared to finite difference and finite element approximations, but also on their flexibility and ability to unite ideas from finite elements with those from finite differences. A first basic finite volume approximation of the Euler equations can be written down for arbitrary grids directly from the conservation laws itself and coded by a novice student within a short amount of time.

However, convergence results are still not known for systems as complicated as Euler's equations. Even simple questions concerning the consistency of finite volume approximations for linear model problems can not be answered satisfactorily. The reader should always be aware that the topics covered in this paper do by no means belong to the save mathematical repertoire.

3.2 Governing equations

Although all of the following considerations are d-dimensional for arbitrary d we shall concentrate on the case $d = 2$. Inviscid, compressible fluid flow is described by the system of Euler equations

$$\partial_t u + \sum_{\ell=1}^{2} \partial_{x_\ell} f_\ell(u) = 0, \qquad (3.1)$$

in which

$$u := \begin{bmatrix} \rho \\ \rho v_1 \\ \rho v_2 \\ \rho E \end{bmatrix}, \quad f_\ell(u) := \begin{bmatrix} \rho v_\ell \\ \rho v_1 v_\ell + p\delta_1^\ell \\ \rho v_2 v_\ell + p\delta_2^\ell \\ \rho H v_\ell \end{bmatrix}, \quad \ell = 1,2$$

denotes the vector of conserved quantities and the flux functions, respectively. If $t \in \mathrm{I\!R}_0^+ := \{t \in \mathrm{I\!R} \mid t \geq 0\}$ denotes time and $x = (x_1, x_2)^T \in \mathrm{I\!R}^2$ space coordinates, the mappings

$$\mathrm{I\!R}^2 \times \mathrm{I\!R}_0^+ \ni (x,t) \overset{\rho, v, p, E, H}{\longmapsto} \left(\rho, v := (v_1, v_2)^T, p, E, H \right)(x,t)$$

denote density, velocity, pressure, total energy and enthalphy, respectively. Enthalphy is defined by

$$H := E + \frac{p}{\rho}$$

and $\delta_i^j = \begin{cases} 1 & ; \quad i = j \\ 0 & ; \quad i \neq j \end{cases}$ denotes the Kronecker symbol. An equation of state is needed to close the system. For ideal gases one uses

$$p = (\kappa - 1)\rho \left(E - \frac{|v|^2}{2} \right),$$

where κ denotes the ratio of specific heats. In the case of dry air one assumes a value of $\kappa = 1.4$.

The Navier-Stokes equations, governing the compressible flow of viscous fluids, are given by

$$\partial_t u + \sum_{\ell=1}^2 \partial_{x_\ell} f_\ell(u) = \frac{1}{\text{Re}} \sum_{\ell=1}^2 \partial_{x_\ell} g_\ell(u), \qquad (3.2)$$

where

$$g_\ell := \begin{bmatrix} 0 \\ \tau_{1,\ell} \\ \tau_{2,\ell} \\ v_1 \tau_{1,\ell} + v_2 \tau_{\ell,2} + \frac{\mu \kappa}{\text{Pr}} \partial_{x_\ell} \epsilon \end{bmatrix}, \quad \ell = 1,2$$

denote the viscous fluxes. Here, $\tau_{i,j} := \mu \left(\partial_{x_j} v_i + \partial_{x_i} v_j \right) + \delta_i^j \lambda \left(\partial_{x_1} v_1 + \partial_{x_2} v_2 \right)$ denote the elements of the stress tensor. The quantity λ is defined by Stokes' hypothesis to be $\lambda = -\frac{2}{3}\mu$ and μ denotes the viscosity coefficient. The quantity ϵ denotes the specific internal energy and is defined by $\epsilon = E - \frac{|v|^2}{2}$. The Reynolds and Prandtl number is denoted by Re and Pr, respectively. Temperature can be computed by means of the formula $T = \kappa(\kappa - 1)\text{Ma}^2 \epsilon$ and the viscosity coefficient is given by the Sutherland law $\mu = T^{1.5}\frac{1+S}{T+S}$, where $S := 110°K/\overline{T}_\infty$ and $\overline{T}_\infty$ denotes the temperature at infinity measured in degree Kelvin.

If we denote by $\nabla_u f_\ell$ the Jacobian matrices of f_ℓ with respect to the conserved quantities the Euler and Navier-Stokes equations can be recast in their quasilinear form

$$\partial_t u + \sum_{\ell=1}^2 \nabla_u f_\ell(u) \partial_{x_\ell} u = 0$$

and

$$\partial_t u + \sum_{\ell=1}^2 \nabla_u f_\ell(u) \partial_{x_\ell} u = \frac{1}{\text{Re}} \sum_{\ell=1}^2 \nabla_u g_\ell(u) \partial_{x_\ell} u,$$

respectively.

We postpone the discussion of the Navier-Stokes equations to a later section and describe some properties of the Euler equations for reference. A system (3.1) consisting of n equations ($n = 4$ for the system of Euler's equations) is called hyperbolic (in time direction), if the matrix

$$A(u,\nu) := \sum_{\ell=1}^{2} \nabla_u f_\ell(u)\nu_\ell \in \mathbb{R}^{n \times n}$$

has exactly n real eigenvalues for all $u \in \mathcal{S} \subset \mathbb{R}^n$ and all $\nu \in \mathbb{R}^2$, where $\mathcal{S}$ denotes the state space of the conserved variables. It is an easy exercise to check that the Euler equations are indeed hyperbolic. Hyperbolic systems describe transport processes.

Although the following properties of the Euler equations can be found in many textbooks (see [7], [8], for example), we list them for easy reference. The Jacobian matrices are given by

$$\nabla_u f_1(u) = \begin{bmatrix} 0 & 1 & 0 & 1 \\ \frac{\kappa-3}{2}v_1^2 + \frac{\kappa-1}{2}v_2^2 & (3-\kappa)v_1 & (1-\kappa)v_2 & \kappa-1 \\ -v_1 v_2 & v_2 & v_1 & 0 \\ (\kappa-1)v_1|v|^2 - \kappa v_1 E & \kappa E - \frac{\kappa-1}{2}(v_2^2 + 3v_1^2) & (1-\kappa)v_1 v_2 & \kappa v_1 \end{bmatrix}$$

and

$$\nabla_u f_2(u) = \begin{bmatrix} 0 & 0 & 1 & 0 \\ -v_1 v_2 & v_2 & v_1 & 0 \\ \frac{\kappa-3}{2}v_2^2 + \frac{\kappa-1}{2}v_1^2 & (1-\kappa)v_1 & (3-\kappa)v_2 & \kappa-1 \\ (\kappa-1)v_2|v|^2 - \kappa v_2 E & (1-\kappa)v_1 v_2 & \kappa E - \frac{\kappa-1}{2}(v_1^2 + 3v_2^2) & \kappa v_2 \end{bmatrix}.$$

The matrix $A(u,\nu)$ is diagonalisable by the product $P^{-1}AP$ with a regular matrix P and their eigenvalues are given by $v \cdot \nu$ (occuring twice), $v \cdot \nu - a|\nu|$ and $v \cdot \nu + a|\nu|$, where $a := \sqrt{\kappa \frac{p}{\rho}}$ denotes the speed of sound.

Another remarkable property of the Euler equations is their rotational invariance. Using the rotation matrix

$$\beta \longmapsto T(\beta) := \begin{bmatrix} 1 & 0 & 0 & 0 \\ 0 & \cos\beta & \sin\beta & 0 \\ 0 & -\sin\beta & \cos\beta & 0 \\ 0 & 0 & 0 & 1 \end{bmatrix}$$

the fluxes of Euler's equations satisfy

$$f_1(u)\cos\beta + f_2(u)\sin\beta = T^{-1}(\beta)f_1(T(\beta)u).$$

Written in terms of the unit vector $n = (n_1, n_2)^T := (\cos\beta, \sin\beta)^T$ the matrix T is given by $T(n) = \begin{bmatrix} 1 & 0 & 0 & 0 \\ 0 & n_1 & n_2 & 0 \\ 0 & -n_2 & n_1 & 0 \\ 0 & 0 & 0 & 1 \end{bmatrix}$ and the rotational invariance is expressed as

$$\sum_{\ell=1}^{2} f_\ell(u)n_\ell = T^{-1}(n)f_1(T(n)u).$$

It is well-known, that smooth solutions of systems (3.1) do exist for small time only, even if the initial data is arbitrarily smooth. Therefore one needs the notion of weak solutions allowing for discontinuities like shock waves or contact discontinuities in compressible flow. In the context of gas dynamics the most natural choice is to base weak solutions on the integral conservation laws from which the Euler and Navier-Stokes equations are derived.

We call a bounded subset $\sigma \subset \mathbb{R}^2$ a control volume, if its boundary $\partial \sigma$ is at least Lipschitz continuous. Then we call u a weak solution of the Navier-Stokes equations, if

$$\frac{d}{dt} \int_\sigma u\, dx + \oint_{\partial \sigma} \{f_1(u)n_1 + f_2(u)n_2\}\, ds = \frac{1}{\text{Re}} \oint_{\partial \sigma} \{g_1(u)n_1 + g_2(u)n_2\}\, ds$$

holds for all control volumes σ with outer unit normal vector $n = (n_1, n_2)^T$.

Note that this formula can be derived from the system (3.2) by integrating over a control volume and applying the Gauss integral theorem. The solution class considered can be described by the function space $BV(\mathbb{R}_0^+; L^1 \cap L^\infty(\mathbb{R}^2; \mathbb{R}^4))$, i.e. the mapping $t \mapsto u(\cdot, t)$ is of bounded variation and the image is an integrable function of the space variables which is almost everywhere bounded.

Due to the rotational invariance of the Euler fluxes weak solutions can also be characterised using the matrix T, i.e.

$$\frac{d}{dt} \int_\sigma u\, dx + \oint_{\partial \sigma} T^{-1}(n) f_1(T(n)u)\, ds = \frac{1}{\text{Re}} \oint_{\partial \sigma} \{g_1(u)n_1 + g_2(u)n_2\}\, ds. \qquad (3.3)$$

The first step towards finite volume approximations is the definition of the cell average operator

$$u(t) \longmapsto \mathfrak{A}(\sigma)u(t) := \frac{1}{|\sigma|} \int_\sigma u\, dx, \qquad (3.4)$$

assigning to each weak solution its integral average on the control volume σ. Here, $|\sigma|$ denotes the area of the control volume. Comparing the definition (3.4) with the weak form (3.3) of the Navier-Stokes equations we end up with the following important observation:

The Navier-Stokes equations are a system of evolution equations for the cell averages of weak solutions, i.e. cell averages satisfy

$$\frac{d}{dt}\mathfrak{A}(\sigma)u(t) = -\frac{1}{|\sigma|} \left(\oint_{\partial \sigma} T^{-1}(n) f_1(T(n)u)\, ds + \frac{1}{\text{Re}} \oint_{\partial \sigma} \{g_1(u)n_1 + g_2(u)n_2\} \right)\, ds. \qquad (3.5)$$

This trivial — but important — observation is the starting point for the development of finite volume approximations.

3.3   Finite volume approximations

The finite volume technique can be described basically as the discretisation of the evolution equation (3.5) for the cell averages of weak solutions. Since control volumes are defined as polygonal subsets of $\mathbb{R}^2$ this technique does not depend on a regular grid and Vankeirsbilck indeed describes finite volume approximations of the Euler equations on polygonal control volumes of varying shapes, see [15]. We want to concentrate on finite volume approximations on conforming (i.e. satisfying some regularity properties) triangulations. Historically two main directions have evolved in this context, namely the primary grid methods using the triangles of the triangulation itself as control volumes, and the secondary grid methods — also known as box methods — using certain polygonal surroundings of each node of the triangulation as control volumes and therebye defining a second tesselation, the secondary grid. We have to point out that any discussion concerning the superiority of a numerical solution computed with finite volume approximations of one category over a numerical solution obtained with a finite volume method of the other class gives no reason to believe that one family of schemes is *better* than the other. Indeed, both approaches follow the same ideas and if two identical discretisations produce different solutions on different types of grid cells we are asked to study the influence of the geometry of the control volumes instead of simply arguing that primary grid methods are *better* (whatever that is!) than box methods or vice versa.

3.3.1 Conforming triangulations

We consider the approximation of the Navier-Stokes equations on a bounded open domain $\Omega \subset \mathbb{R}^2$. For the sake of simplicity we assume that the boundary $\partial\Omega := \overline{\Omega}\backslash\Omega$ is already a polygon. On $\overline{\Omega}$ we establish two types of tesselations.

A triangulation $\mathcal{T}^h$ of $\overline{\Omega}$ is the set of finitely many subsets $T_i \subset \overline{\Omega}, i = 1, \ldots, \#T$, such that the following conditions are satisfied.

- $\overline{\Omega} = \bigcup\limits_{i \in \{1,\ldots,\#T\}} T_i$.

- Every $T_i \in \mathcal{T}^h$ is closed and for the interior $\overset{\circ}{T}_i \neq \emptyset$ holds.

- For two $T_i, T_j \in \mathcal{T}^h$ with $i \neq j$ it holds $\overset{\circ}{T}_i \cap \overset{\circ}{T}_j = \emptyset$.

- The boundary of every $T_i \in \mathcal{T}^h$ is Lipschitz continuos.

A triangulation is called conforming, if the additional condition

- Every one-dimensional edge of any $T_i \in \mathcal{T}^h$ is either subset of $\partial\Omega$ or the edge of another $T_j, j \neq i$

holds.

The parameter h in the notation $\mathcal{T}^h$ corresponds to a typical geometrical length scale of the triangulation which may be represented by the length of the longest edge.

Note that conformity ensures that no hanging nodes, i.e. nodes lying on an edge of another triangle, can occur. Although conformity is not necessary in the context of finite volume approximations it helps to simplify nearly every algorithmic detail, especially in the case of grid adaptivity. The definition of a triangulation given here is identical to that used in finite element methods, see [3].

The definition of a triangulation does not itself include that the T_i have to be triangles. In fact, a very powerful concept in finite volume methods is the use of hybrid grids, i.e. grids consisting of triangular regions and conformly fitted regions of quadrilateral meshes, e.g. for the resolution of boundary layers. However, we concentrate on purely triangular meshes in this formal presentation.

A conforming triangulation $\mathcal{T}^h$ is called primary grid, if all T_i are triangles.

The barycentric subdivision according to [1] can be used to define a secondary grid. Let

$$K_{h,i} := \{T \in \mathcal{T}^h \mid \text{node } i \text{ is vertex of } T\}$$

be the set of all triangles of a primary grid sharing node i. Denote the three edges of triangle T by $e_{T,k}, k = 1,2,3$. The set of all edges sharing node i of the primary grid is then

$$E_{h,i} := \{e_{T,k} \mid T \in \mathcal{T}^h, k \in \{1,2,3\}, \text{node } i \text{ is vertex of } e_{T,k}\}.$$

For each $T \in \mathcal{T}^h$ consider the following barycentric subdivision: Join the barycentre of T with the barycentres (i.e. the midpoints) of its three edges $e_{T,k}, k = 1,2,3$. This divides each triangle into three segments according to figure 3.1. The union of all those segments

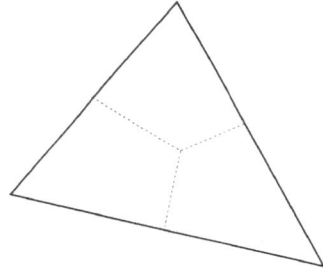

Figure 3.1   Barycentric subdivision of a triangle

of $T \in K_{h,i}$ adjacent to node i is called box B_i around node i. If node i belongs to the boundary $\partial \Omega$ the box is constructed with the two halfs of the boundary edges of the boundary triangles having node i in common. The union $\mathcal{B}^h := \bigcup\limits_{i=1,\ldots,\#B} B_i$ of all boxes is called secondary grid.

The notion of primary and secondary grids follows Heinrich [6]. Primary and secondary grid can be seen in figure 3.2. The situation at the boundary of the domain is shown in figure 3.3. Note that our definition of the secondary grid is consistent at the boundary in the sense that no special treatment is necessary. One could also choose other subdivisions of the primary grid in which the boundary treatment is not straight forward. Consider, for the example, boxes constructed from joining the circumcentres of the triangles sharing a node.

In the following we shall construct finite volume approximations for systems of the type (3.1) on primary and secondary grids. We postpone the discussion of the discretisation of the viscous fluxes to a later section.

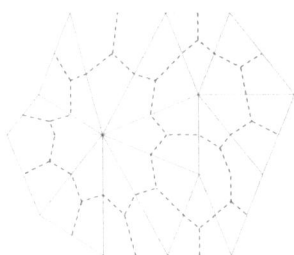

Figure 3.2 Primary and secondary grid

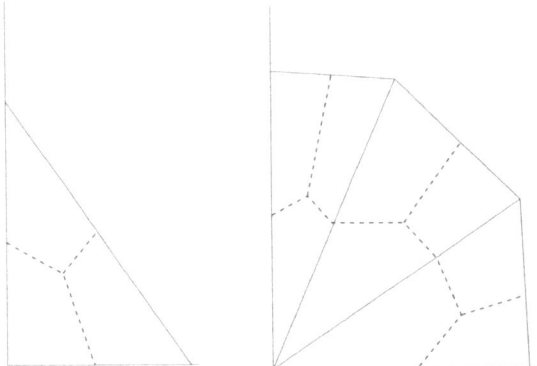

Figure 3.3 Primary and secondary grid at the boundary

3.3.2 Primary grid methods for the Euler equations

On primary grids we need the notion of neighbourhood of triangles.

Let $T_i, i = 1, \ldots, \#T$, denote the triangles of a conforming triangulation $\mathcal{T}^h$. The indexes of the neighbouring triangles are given by

$$N(i) := \{ j \in \mathbb{N} \mid T_i \cap T_j \text{ is an edge of } T_i \}.$$

We now reformulate the evolution equation for cell averages (3.5) in the case of vanishing viscosity on one triangle, i.e. the control volume σ is a triangle T_i. This results in

$$\frac{d}{dt} \mathfrak{A}(T_i) u(t) = -\frac{1}{|T_i|} \sum_{j \in N(i)} \int_{\partial T_j \cap \partial T_i} \sum_{\ell=1}^{2} f_\ell(u) n_{ij,\ell} \, ds,$$

where the outer (with respect to T_i) unit normal vector at the edge $\partial T_i \cap \partial T_j$ is denoted by $n_{ij} = (n_{ij,1}, n_{ij,2})^T$.

If we assume smooth solutions a Gaussian quadrature can be applied to discretise the integral along the edge. If the edge is parametrised by

$$[-1,1] \ni s \xmapsto{x_{ij}} x_{ij}(s) := \frac{1}{2}(x_i + x_j) + \frac{s}{2}(x_j - x_i)$$

where $x_i = (x_{i,1}, x_{i,2})^T$ and $x_j = (x_{j,1}, x_{j,2})^T$ denote the coordinates of the vertex points of $\partial T_i \cap \partial T_j$ then the evolution equation on T_i can be written as

$$\frac{d}{dt}\mathfrak{A}(T_i)u(t) = -\frac{1}{|T_i|}\sum_{j\in N(i)}\frac{|\partial T_i \cap \partial T_j|}{2}\int_{-1}^{1}\sum_{\ell=1}^{2}f_\ell(u(x_{ij}(s),t))n_{ij,\ell}\,ds.$$

The application of quadrature rules of Gaussian type is now straight forward. If we denote the number of Gauss points on $\partial T_i \cap \partial T_j$ by n_G, the Gauss points itself by $x_{ij}(s_\nu), \nu = 1,\ldots,n_G$, and the weights (which do no longer depend on the length of the edge, due to the parametrisation!) by ω_ν, then we get the system

$$\frac{d}{dt}\mathfrak{A}(T_i)u(t)$$

$$= -\frac{1}{|T_i|}\sum_{j\in N(i)}\frac{|\partial T_i \cap \partial T_j|}{2}\left\{\sum_{\nu=1}^{n_G}\sum_{\ell=1}^{2}\omega_\nu f_\ell(u(x_{ij}(s_\nu),t))n_{ij,\ell} + \mathcal{O}\left(h^{2n_G}\right)\right\}. \quad (3.6)$$

Note that this equation does not represent a numerical method! Here u still denotes a weak solution and the derivation of this system has to be understood as being purely formal, since weak solutions of the kind we introduced earlier can not be evaluated at a single point.

The weights of the Gaussian quadrature rule as well as the coordinates of the Gauss points can be looked up in any textbook on numerical analysis, for example in [14]. For the practically most relevant cases $n_G = 1$ and $n_G = 2$ they are listed in the following table.

n_G	ω_ν	s_ν
1	$\omega_1 = 2$	$s_1 = 0$
2	$\omega_1 = \omega_2 = 1$	$s_1 = -s_2 = -1/\sqrt{3}$

In order to transform the evolution equation for triangle T_i into a numerical method the pointwise evaluation of the fluxes must be replaced by a numerical device.

A mapping

$$\mathbb{R}^4 \times \mathbb{R}^4 \times \mathbb{R}^2 \ni (u_L, u_R; n) \overset{H}{\longmapsto} H(u_L, u_R; n) \in \mathbb{R}^4$$

is called numerical flux function or approximate Riemann solver, if it satisfies the consistency condition

$$\forall u \in \mathbb{R}^4 : \quad H(u,u;n) = \sum_{\ell=1}^{2}f_\ell(u)n_\ell.$$

Any of the numerical flux functions used in finite difference methods can also be used in the context of finite volume approximations. Since there are already textbooks available describing dozens of possible numerical fluxes (see [8], for example) we shall not mention any detail of their construction.

If we assume smoothness of the solution the numerical flux function can be introduced into the evolution equation without producing an additional error term. This holds true due to the consistency condition, i.e. we can write

$$\frac{d}{dt}\mathfrak{A}(T_i)u(t) = \tag{3.7}$$

$$-\frac{1}{|T_i|}\sum_{j\in N(i)}\frac{|\partial T_i \cap \partial T_j|}{2}\left\{\sum_{\nu=1}^{n_G}\omega_\nu H\left(u(x_{ij}(s_\nu),t),u(x_{ij}(s_\nu),t);n_{ij}\right) + \mathcal{O}\left(h^{2n_G}\right)\right\}.$$

This is the basic equation from which all primary grid-based finite volume approximations can be derived. The underlying idea is very simple: Given an initial function $u_0(x) := u(x,0)$ the cell average $\mathfrak{A}(T_i)u_0$ can be computed on any triangle. Since an evolution equation (3.7) for cell averages has to be solved it is natural to think of cell averages as arguments in the numerical flux function also. This is the reason for the notion of approximate Riemann solvers: The numerical flux function has to solve the Riemann problem occuring from two constant — but different — states (the cell averages) at the edges of adjacent triangles.

The method: Find $\overline{u}_i(t), i = 1,\ldots,\#T, t \in [0,t^*], t^* > 0$, as a solution of the system of ordinary differential equations

$$\frac{d}{dt}\overline{u}_i(t) = -\frac{1}{|T_i|}\sum_{j\in N(i)}|\partial T_i \cap \partial T_j|H(\overline{u}_i(t),\overline{u}_j(t);n_{ij})$$

$$\overline{u}_i(0) = \mathfrak{A}(T_i)u(0)$$

is called basic finite volume approximation on primary grids.

We shall explain below why the one-point Gaussian quadrature rule was chosen for basic finite volume approximations.

3.3.3   Box methods for the Euler equations

As in the case of primary grid methods we need a notion of neighbouring control volumes. Let $B_i, i = 1,\ldots,\#B$, denote the boxes[1] of a secondary grid $\mathcal{B}^h$. The indexes of the neighbouring boxes are given by

$$N(i) := \{j \in \mathbb{N} \mid B_i \cap B_j \text{ is edge of } B_i\}.$$

In the case of secondary grid methods based on the barycentric subdivision of triangles the boundary between two neighbouring boxes consists of two segments l_{ij}^1 and l_{ij}^2 according to figure 3.4. Therefore the starting point for box methods is the following evolution system for cell averages on box B_i, directly derived from (3.5) for inviscid flow.

$$\frac{d}{dt}\mathfrak{A}(B_i)u(t) = -\frac{1}{|B_i|}\sum_{j\in N(i)}\sum_{k=1}^{2}\int_{l_{ij}^k}\sum_{\ell=1}^{2}f_\ell(u)n_{ij,\ell}^k\,ds. \tag{3.8}$$

---

[1] We shall often denote boxes by σ_i instead of B_i in places where the context is not misleading.

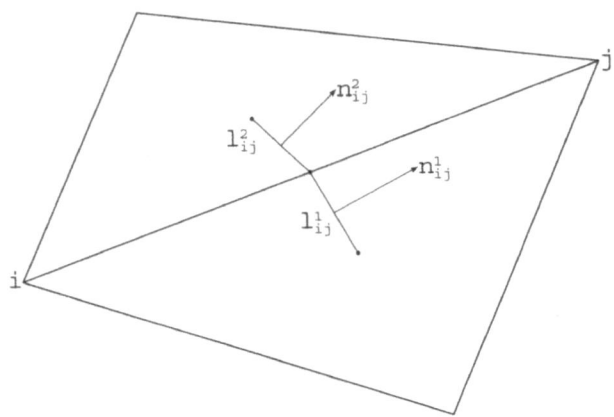

Figure 3.4 Geometry between boxes B_i and B_j

If we denote the barycentres of the two adjacent triangles in figure 3.4 by $c^1 = (c_1^1, c_2^1)^T$ and $c^2 = (c_1^2, c_2^2)^T$, respectively, each segment $l_{ij}^k, k = 1, 2$, can be parameterised by

$$[-1,1] \ni s \xmapsto{x_{ij}^k} x_{ij}^k(s) = \frac{1}{2}\left(c^k + \frac{1}{2}(x_i + x_j) \right) + \frac{s}{2}\left(\frac{1}{2}(x_i + x_j) - c^k \right),$$

where $x_i = (x_{i,1}, x_{i,2})^T$ and $x_j = (x_{j,1}, x_{j,2})^T$ denote the coordinates of nodes i and j. Introducing this parametrisation into (3.8) and using a Gaussian quadrature rule again yields

$$\frac{d}{dt}\mathfrak{A}(B_i)u(t) = -\frac{1}{|B_i|} \sum_{j \in N(i)} \sum_{k=1}^{2} \frac{|l_{ij}^k|}{2} \left\{ \sum_{\nu=1}^{n_G} \sum_{\ell=1}^{2} \omega_\nu^k f_\ell \left(u(x_{ij}^k(s_\nu), t) \right) n_{ij,\ell}^k + \mathcal{O}\left(h^{2n_G}\right) \right\}.$$

As in the case of primary grid methods we introduce a numerical flux function and get the starting point for the development of box methods, i.e.

$$\frac{d}{dt}\mathfrak{A}(B_i)u(t) = \tag{3.9}$$

$$-\frac{1}{|B_i|} \sum_{j \in N(i)} \sum_{k=1}^{2} \frac{|l_{ij}^k|}{2} \left\{ \sum_{\nu=1}^{2} \omega_\nu^k H\left(u(x_{ij}^k(s_\nu), t), u(x_{ij}^k(s_\nu), t); n_{ij}^k \right) + \mathcal{O}\left(h^{2n_G}\right) \right\}.$$

In order to construct a numerical method we proceed as in the case of primary grid methods.

The method: Find $\overline{u}_i(t), i = 1, \ldots, \#B, t \in [0, t^*], t^* > 0$, *as a solution of the system of ordinary differential equations*

$$\frac{d}{dt}\overline{u}_i(t) = -\frac{1}{|B_i|} \sum_{j \in N(i)} \sum_{k=1}^{2} |l_{ij}^k| H(\overline{u}_i(t), \overline{u}_j(t); n_{ij}^k)$$

$$\overline{u}_i(0) = \mathfrak{A}(B_i)u(0)$$

is called basic finite volume approximation on secondary grids or basic box method.

3.3.4 Remarks on the spatial accuracy

In the above definitions of basic finite volume methods one Gauss point was chosen on the edge between control volume i and each of its neighbours. This necessarily drops the order to $\mathcal{O}(h^2)$ but the situation is even worse. If c_i denotes the barycentre of control volume $\sigma_i \in \{T_i, B_i\}$, then a simple Taylor series development shows

$$u(x,t) = \sum_{\mu=0}^{r-1} \frac{1}{\mu!} \sum_{|\alpha|=\mu} (x - c_i)^\alpha \, \partial^\alpha u|_{x=c_i} + \mathcal{O}(|x - c_i|^r)$$

for some $r \geq 1$. Here, $\alpha = (\alpha_1, \alpha_2)$ denotes a multiindex, $x^\alpha := x_1^{\alpha_1} x_2^{\alpha_2}$ and $\partial^\alpha u = \partial_{x_1}^{\alpha_1} \partial_{x_2}^{\alpha_2} u$. Applying the cell average operator yields

$$\mathfrak{A}(\sigma_i)u(t) = \sum_{\mu=0}^{r-1} \frac{1}{\mu!} \sum_{|\alpha|=\mu} \frac{1}{|\sigma_i|} \int_{\sigma_i} (x - c_i)^\alpha \, dx \, \partial^\alpha u|_{x=c_i} + \mathcal{O}(|x - c_i|^r).$$

Choosing $r = 2$ reveals

$$\mathfrak{A}(\sigma_i)u(t) = u(c_i,t) + \frac{1}{|\sigma_i|} \int_{\sigma_i} (x - c_i) \, dx \cdot \nabla_x u(x,t)|_{x=c_i} + \mathcal{O}(|x - c_i|^2).$$

Now the barycentre itself is defined as

$$c_i := \frac{1}{|\sigma_i|} \int_{\sigma_i} x \, dx,$$

i.e. $\frac{1}{|\sigma_i|} \int_{\sigma_i} (x - c_i) = 0$ and thus

$$\mathfrak{A}(\sigma_i)u(t) = u(c_i,t) + \mathcal{O}(h^2).$$

However, this approximation property at the barycentre is not shared by the Gauss points on the edges. Repeating the same Taylor series arguments as above and choosing a Gauss point $x(s_\nu)$ instead of the barycentre yields the weak approximation property

$$\mathfrak{A}(\sigma_i)u(t) = u(x(s_\nu),t) + \mathcal{O}(h).$$

Thus, assuming smoothness of numerical flux and solution, we have

$$H(\mathfrak{A}(\sigma_i)u(t), \mathfrak{A}(\sigma_j)u(t); n_{ij}) = \sum_{\ell=1}^{2} f_\ell(u(x_{ij}(0),t) n_{ij,\ell} + \mathcal{O}(h)$$

and we would like to refer to basic finite volume schemes as being first order accurate in space[2].

---

[2] Note that we do not base our notion of accuracy on the truncation error of the scheme. It is well-known that the notion of truncation error is not well-suited in the case of finite volume methods and that even convergent methods can easily be shown to be inconsistent, i.e. they exhibit a truncation error of order 1, see [5].

The central theme for the construction of higher order finite volume approximations is therefore the recovery of better approximations from given cell average data. This will be described in detail in the second part of this paper but we shall show here the central mechanism. Let time be fixed and let $\bar{u}_i = \mathfrak{A}(\sigma_i)u$ be the given cell average of u on control volume $\sigma_i \in \{T_i, B_i\}$. A function π_i, defined on σ_i, is called recovery function on σ_i, if the conditions

$$\forall x \in \sigma_i : \quad \pi_i(x,t) - u(x,t) = \mathcal{O}(h^r), \quad r > 0$$

and

$$\mathfrak{A}(\sigma_i)\pi_i(t) = \mathfrak{A}(\sigma_i)u(t)$$

are satisfied. The first condition corresponds to an accuracy requirement while the second condition serves to conserve cell averages. Now we replace the values of the weak solution at the Gauss points in (3.7) and (3.9), respectively, by the value of the recovery functions at these points. This leads to the following definition. *The method: Find $\bar{u}_i(t), i = 1, \ldots, \#T, t \in [0, t^*], t^* > 0$, as a solution of the system of ordinary differential equations*

$$\frac{d}{dt}\bar{u}_i(t) = -\frac{1}{|T_i|} \sum_{j \in N(i)} \frac{|\partial T_i \cap \partial T_j|}{2} \sum_{\nu=1}^{n_G} \omega_\nu H\left(\pi_i(x_{ij}(s_\nu),t), \pi_j(x_{ij}(s_\nu),t); n_{ij}\right)$$

$$\bar{u}_i(0) = \mathfrak{A}(T_i)u(0)$$

is called finite volume approximation on primary grids.

The corresponding definition for box methods reads as

The method: Find $\bar{u}_i(t), i = 1, \ldots, \#B, t \in [0, t^*], t^* > 0$, as a solution of the system of ordinary differential equations

$$\frac{d}{dt}\bar{u}_i(t) = -\frac{1}{|B_i|} \sum_{j \in N(i)} \sum_{k=1}^{2} \frac{|l_{ij}^k|}{2} \sum_{\nu=1}^{n_G} \omega_\nu^k H\left(\pi_i(x_{ij}^k(s_\nu),t), \pi_j(x_{ij}^k(s_\nu),t); n_{ij}^k\right)$$

$$\bar{u}_i(0) = \mathfrak{A}(B_i)u(0)$$

is called finite volume approximation on secondary grids or box method.

A primary grid method is q-th order accurate in space, if

$$\sum_{\nu=1}^{n_G} \omega_\nu H\left(\pi_i(x_{ij}(s_\nu),t), \pi_j(x_{ij}(s_\nu),t); n_{ij}\right) = \int_{-1}^{1} \sum_{\ell=1}^{2} f_\ell(u(x_{ij}(s),t))n_{ij,\ell}\, ds + \mathcal{O}(h^q)$$

and a box method is q-th order accurate in space, if

$$\sum_{\nu=1}^{n_G} \omega_\nu^k H\left(\pi_i(x_{ij}^k(s_\nu),t), \pi_j(x_{ij}^k(s_\nu),t); n_{ij}^k\right) = \int_{-1}^{1} \sum_{\ell=1}^{2} f_\ell(u(x_{ij}^k(s),t))n_{ij,\ell}^k\, ds + \mathcal{O}(h^q)$$

holds.

Since the spatial error has two sources, namely the quality of the recovery and the accuracy of the Gaussian quadrature rule, the exponent q satisfies

$$q = \min\{r, 2n_G\},$$

where r is the recovery error defined by

$$H\left(\pi_i(x_{ij}(s_\nu),t),\pi_j(x_{ij}(s_\nu),t); n_{ij}\right) = \sum_{\ell=1}^{2} f_\ell(u(x_{ij}(s_\nu),t))n_{ij,\ell} + \mathcal{O}(h^r)$$

for primary grid methods and by

$$H\left(\pi_i(x_{ij}^k(s_\nu),t),\pi_j(x_{ij}^k(s_\nu),t); n_{ij}^k\right) = \sum_{\ell=1}^{2} f_\ell(u(x_{ij}^k(s_\nu),t))n_{ij,\ell}^k + \mathcal{O}(h^r)$$

for box methods.

3.3.5 Numerical examples

Let us first consider the simple linear model problem

$$\partial_t u + a \cdot \nabla_x u = 0$$
$$u(x,0) = u_0(x), \quad x \in \mathbb{R}^2$$

with

$$a = \begin{pmatrix} a_1 \\ a_2 \end{pmatrix} := \begin{pmatrix} \frac{1}{2} - x_2 \\ x_1 - \frac{1}{2} \end{pmatrix}$$

and

$$u_0(x) = \begin{cases} -\frac{1}{0.0225}((x_1 - \frac{1}{2})^2 + x_2^2) + 1 & ; \quad (x_1 - \frac{1}{2})^2 + x_2^2 \le 0.0225 \\ 0 & ; \quad \text{else.} \end{cases}$$

The initial function is a cone of height one which is centered around $(0.5,0)$. The radius of the cone is 0.15. The linear model equation rotates this cone around $(0,0)$ such that one full rotation corresponds to the time $t = 2\pi$.

We employ a basic finite volume approximation on primary grids and use the numerical flux function of Engquist and Osher (cp. [8]) and a simple forward difference in time is used. We consider two triangulations which we denote by $\mathcal{T}_{h^1}$ and $\mathcal{T}_{h^2}$, respectively. The first contains 400 nodes and 722 triangles while the latter consists of 2500 nodes corresponding to 4802 triangles. The two trianglulations are shown in figure 3.5^3. Figure 3.6 shows the solution of the basic finite volume approximation at time $t = \pi$.

---

[3] All triangulations shown are generated by the optimisation approach [4] of Friedrich.

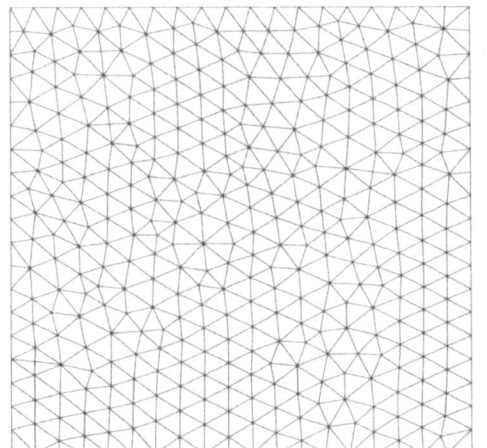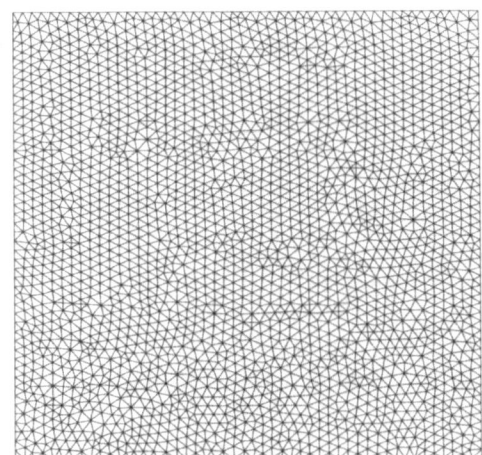

Figure 3.5 Triangulations $\mathcal{T}_{h^1}$ (left) and $\mathcal{T}_{h^2}$

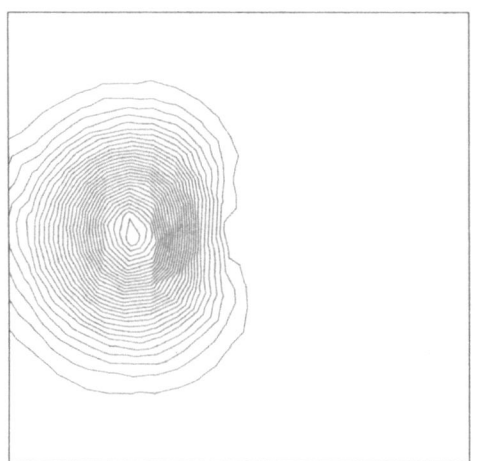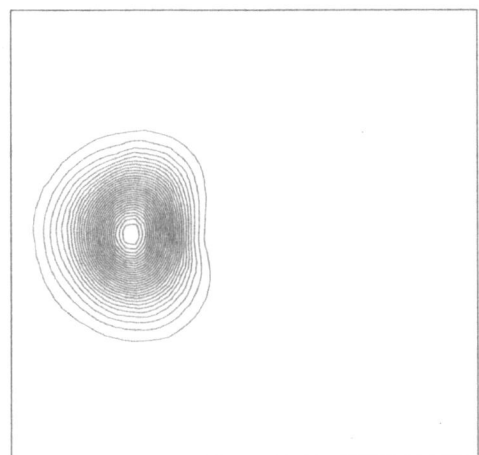

Figure 3.6 Solutions of the basic finite volume approximation

It is now clearly seen that numerical schemes without recovery are too diffusive to be useable in practice. The remaining cone height is shown in the following table.

Triangulation	Typical h	cone height at $t = \pi$
$\mathcal{T}_{h^1}$	0.05	0.176
$\mathcal{T}_{h^2}$	0.02	0.382

As a numerical example let us consider the flow in a channel with forward facing step which is discussed in detail later. Figure 3.7 shows a coarse grid with 2016 triangles and 1089 points. A finer grid consisting of 8064 triangles corresponding to 4139 points is

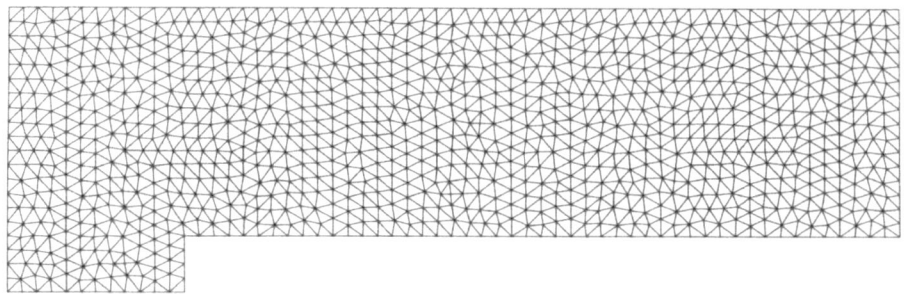

Figure 3.7 Coarse grid

shown in figure 3.8. Applying a primary grid method without any recovery leads to the

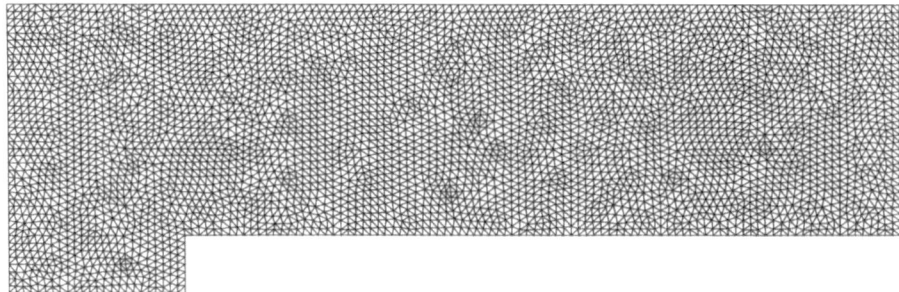

Figure 3.8 Fine grid

solution at the reference time as shown in figure 3.9 where Mach number is presented. For the numerical flux function the flux vector splitting scheme after Steger and Warming is used in comparison with the much more sophisticated scheme of Osher and Solomon (cp. [8]). As can be observed the flow phenomena are not very well represented by both numerical flux functions due to the low accuracy of the basic finite volume method. However, the numerical flux of Osher and Solomon does a better job than the simpler scheme.

On the finer grid we get the corresponding Mach number solutions as shown in figure 3.10. The Osher and Solomon flux is clearly the superior numerical flux function as can be observed in the representation of the contact discontinuity.

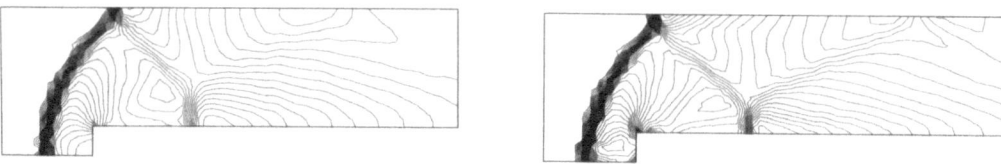

Figure 3.9 Mach number distribution on the coarse mesh. Numerical flux of Steger and Warming (left) and Osher and Solomon

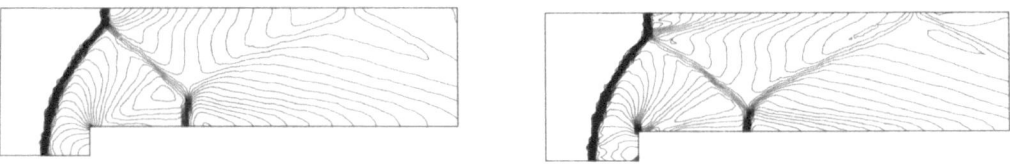

Figure 3.10 Mach number distribution on the fine grid. Numerical flux of Steger and Warming (left) and Osher and Solomon

As a second example we consider transonic flow about a NACA0012 profile. The onflow Mach number is $\mathrm{Ma}_\infty = 0.8$ and the angle of attack $\alpha = 1.25°$. The grid is shown in figure 3.11. As can be observed the cells of the grid on the upper side of the profile

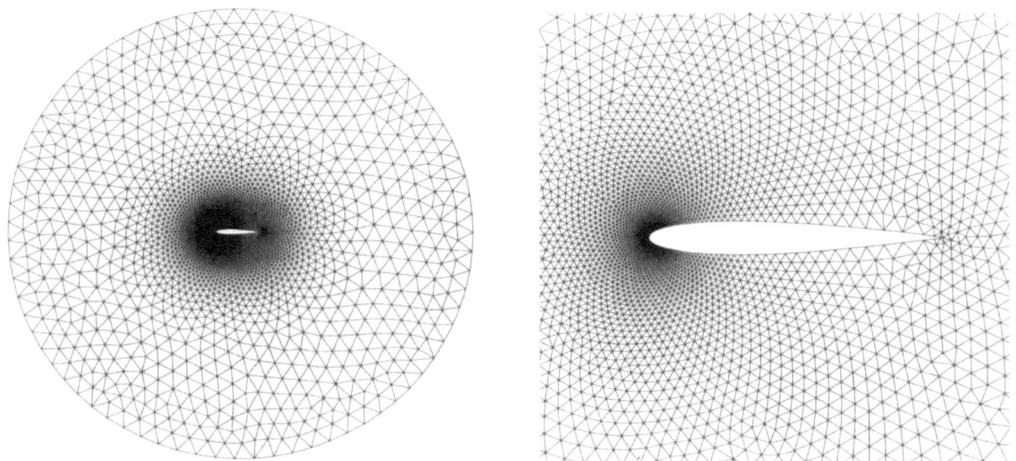

Figure 3.11 A grid for the NACA0012 profile

are very large so that excessive smearing of the upper shock can take place. Applying again the numerical fluxes of Steger and Warming as well as of Osher and Solomon yield the Mach number distribution shown in figure 3.12. The numerical flux of Steger and Warming introduces excessive numerical damping leading to a smeared shock on the upper side while the weak lower side shock is nearly invisible.

Although the topic of recovery is content of the second part of this paper we want to discuss a working second-order scheme on boxes in order to present some numerical results. According to our discussion concerning the order of accuracy above we want to find a linear polynomial π_i on box σ_i so that this polynomial can be evaluated in the

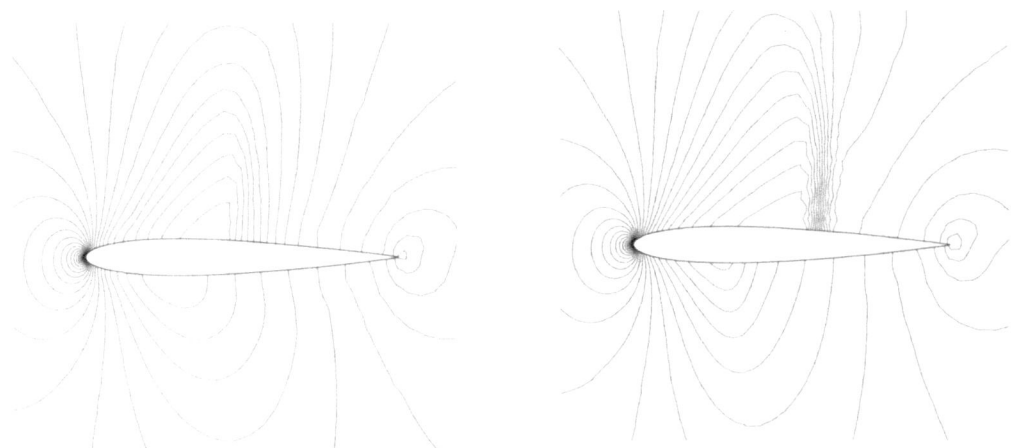

Figure 3.12 Mach number distribution. Numerical flux of Steger and Warming (left) and Osher and Solomon

numerical flux function. We want to reconstruct the primitive variables and therefore denote by u a generic scalar variable. On each of the triangles $T_k \in K_{h,i}$ we can construct a linear polynomial

$$\pi_i^{(k)}(x) = a_{10}^{(k)}(x_1 - c_{i,1}) + a_{01}^{(k)}(x_2 - c_{i,2}) + a_{00}^{(k)}, \quad k = 1, \dots, \#K_{h,i},$$

from the recovery conditions that the cell average of π_i on σ_i as well as on the two other control volumes spanning T should be conserved. The geometrical situation is shown in figure 3.13. ¿From this set of linear polynomials we construct an isotropic gradient $\nabla u|_{\sigma_i}$

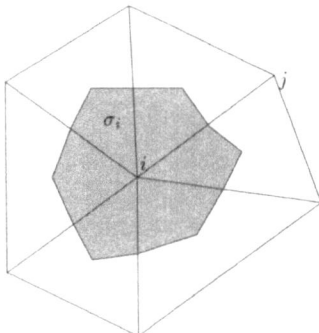

Figure 3.13   Control volume around node i and all $T \in K_{h,i}$

by means of

$$\nabla u|_{\sigma_i} := \frac{1}{|\sigma_i|} \sum_k \int_{\sigma_i \cap T_k} \nabla \pi_i^{(k)} \, dx = \frac{1}{|\sigma_i|} \sum_k \nabla \pi_i^{(k)} |\sigma_i \cap T_k|$$

and define the recovery polynomial π_i on σ_i to be

$$\pi_i(x) = \nabla u|_{\sigma_i}(x - c_i) + \overline{u}_i.$$

Due to the isotropy of the gradient shocks passing the triangles lead to oscillations and an additional limitation is needed. One can use the limiter designed by Barth and Jespersen [2] which was used in [13] to construct a simple finite volume scheme. The limiter is given by

$$\Phi_i = \min_{l_{ij}^k} \overline{\Phi}_{l_{ij}^k}$$

with

$$\overline{\Phi}_{l_{ij}^k} := \begin{cases} \min\left(1, \frac{u_{ij}^{\max} - \overline{u}_i}{\hat{\pi}_i - \overline{u}_i}\right) & , \quad \hat{\pi}_i - \overline{u}_i > 0; \\ \min\left(1, \frac{u_{ij}^{\min} - \overline{u}_i}{\hat{\pi}_i - \overline{u}_i}\right) & , \quad \hat{\pi}_i - \overline{u}_i < 0; \\ 1 & , \quad \hat{\pi}_i - \overline{u}_i = 0 \end{cases} ,$$

where $\hat{\pi}_i$ denotes the value of π_i at the Gauss point of l_{ij}^k and $u_{ij}^{\max} := \max_{j \in N(i)}(\overline{u}_i, \overline{u}_j)$, $u_{ij}^{\min} := \min_{j \in N(i)}(\overline{u}_i, \overline{u}_j)$. From the primitive variables any other flow variable can be computed and used in the numerical flux function.

Applying this scheme to the $\mathrm{Ma}_\infty = 3$ flow around a cyclinder leads to the results as shown in figure 3.14. The Mach number distribution along the stagnation streamline is

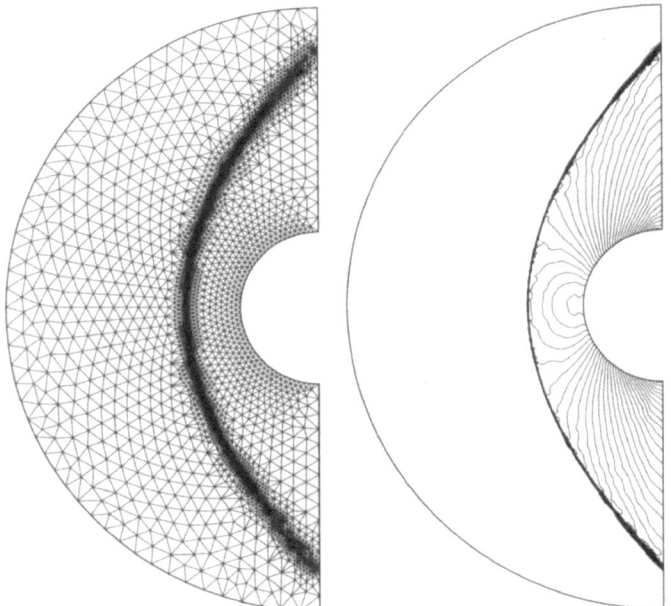

Figure 3.14   Grid and density distribution

shown in figure 3.15. For the sake of comparison two numerical results are presented in this diagram. One computation was done with an explicit time stepping scheme and the other used an implicit discretisation in which the resulting linear systems were solved by the restarted GMRES method with ILU preconditioning (compare the section on time stepping schemes in this paper). The straight vertical curve occuring in the diagram is the location of the shock as predicted by theory. As a transonic test case consider the

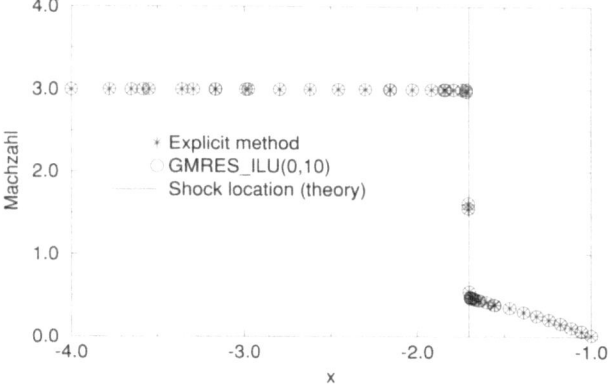

Figure 3.15   Mach number distribution on the stagnation streamline

flow around a Bi-NACA0012 profile as shown in figure 3.16. The onflow Mach number is $Ma_\infty = 0.58$ and the anglke of attack $\alpha = 6°$. The triangulation consists of 26632 triangles corresponding to 13577 points.

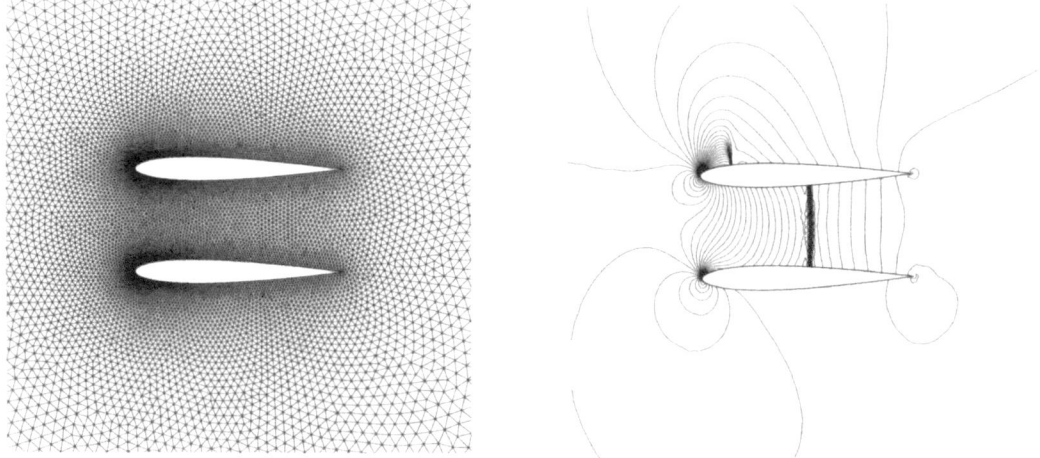

Figure 3.16   Grid and density distribution

3.3.6   Discretisation of the viscous fluxes

The discretisation of the viscous fluxes is certainly problematic if only piecewise constant functions are available on the control volumes. Here recovery is needed to construct functions which possess derivatives at the edges of control volumes.

In the framework of box methods one uses linear interpolants on the primary grid to get linear distributions of the flow variables on the triangles. Evaluating this linear interpolant in the barycentre of each triangle gives us a value of the flow variable under consideration which is used as the value of this variable at the Gauss point of the segment

l_{ij}^k which lies in this triangle. The gradient of the linear interpolant on each triangle is used to get values of the derivatives needed in the diffusive fluxes. This strategy corresponds to an evaluation of the viscous fluxes with a classical central difference or with a standard Galerkin finite element scheme, respectively.

The problem in discretising the compressible Navier-Stokes equations comes with the Reynolds number. It is well known that in the case of very high Reynolds numbers ($\text{Re} \geq \mathcal{O}(10^6)$, say) highly stretched grid cells have to be used in order to take care of the highly anisotropic nature of the flow close to fixed walls. The aspect ratio usually is $1 : 1000$ up to $1 : 10000$. Since all of our methods rely on fluxes in normal direction we then encounter the geometric situation for box schemes as shown in figure 3.17. Note that the normal vectors point now in directions which are nearly perpendicular to the main flow direction and one could argue that problems occur if the physics of the flow is

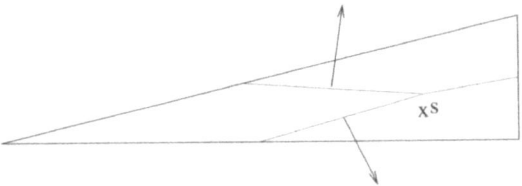

Figure 3.17   Stretched grid cell in Navier Stokes grids

treated so badly (Note that the same situation occurs in a primary grid method!). Indeed, we can construct a simple example where pathologies occur in the numerical solution. We construct a Navier-Stokes grid for a NACA0012 profile as shown in figure 3.18

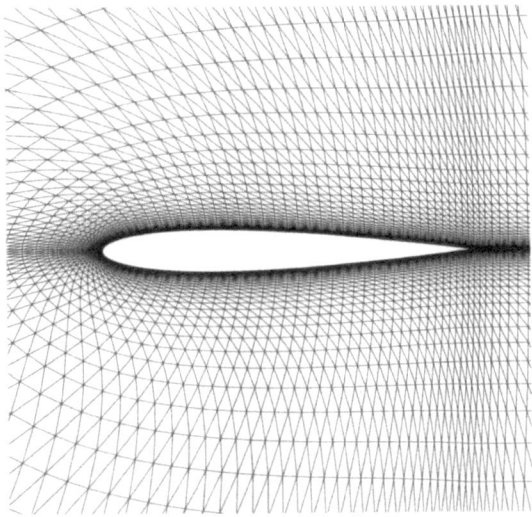

Figure 3.18   Navier-Stokes grid for Euler calculations

Now we compute an inviscid solution of the transonic case $\mathrm{Ma}_\infty = 0.8, \alpha = 1.25°$ with our box method and get the result shown in the left half of figure 3.19. As can be seen

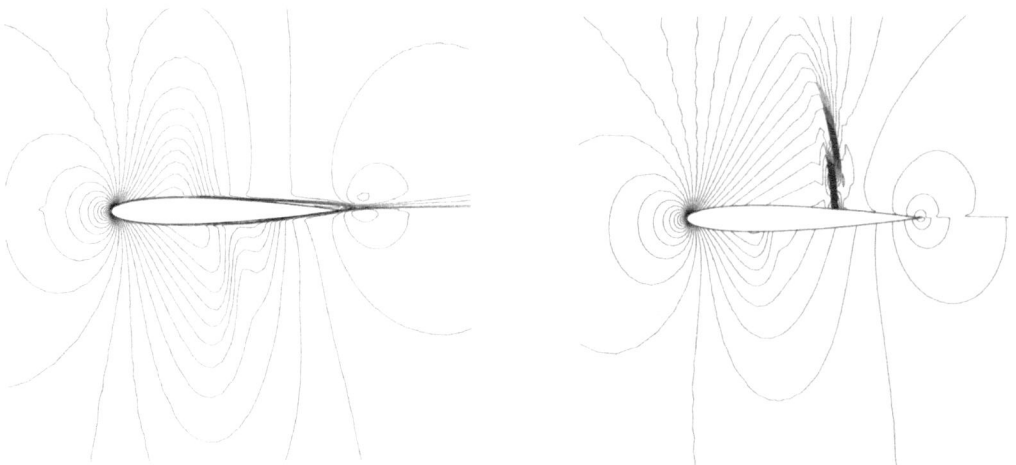

Figure 3.19   Density distribution of inviscid simulation with usual boxes (left) and deformed boxes

the solution is completely wrong since it shows a strong boundary layer which spoils the flow field. We now deform the boxes and move the points which lie in the barycentres of the triangles according to the formula

$$x^s = \sum_{m \in \{i,j,k\}} \alpha^s_m x_m \quad \text{where} \quad \alpha^s_m := \frac{1}{2(|l_i| + |l_j| + |l_k|)} \sum_{\substack{\overline{m} \in \{i,j,k\} \\ \overline{m} \neq m}} |l_{\overline{m}}|$$

which can be seen in figure 3.20.

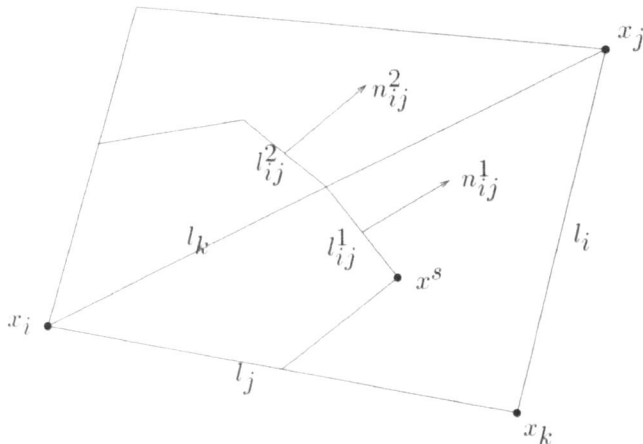

Figure 3.20   Boundary between control volumes σ_i and σ_j

This leads to the geometric situation shown in figure 3.21. Note that we now respect the main flow direction and compute flux balances in this direction. Computing again the inviscid flow field on the same grid as above, but with the deformed boxes, leads to the density distribution shown in the right of figure 3.19. This example shows that com-

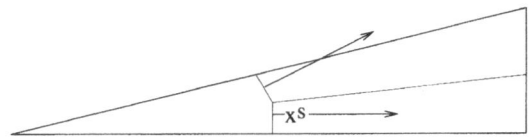

Figure 3.21   Stretched grid cells with deformed boxes

puting viscous solutions in the regime of high Reynolds number may lead to unexpected phenomena. The modification of control volumes as indicated here is a simple way to get rid of some of the problems. Figure 3.22 shows the grid (adapted using the strategies

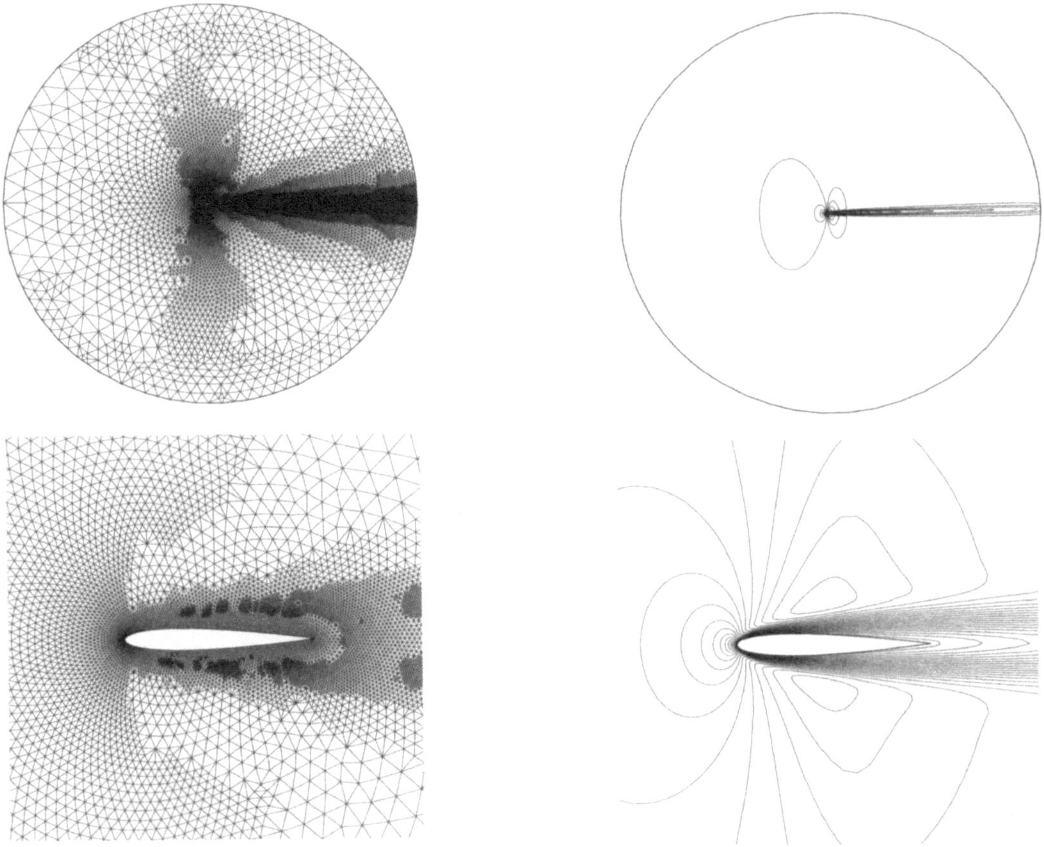

Figure 3.22   Grid and Mach number distribution for viscous transonic flow

described in this paper) for a transonic viscous flow about a NACA0012 profile. The Mach number is $Ma_\infty = 0.85$ and the Reynolds number $Re_\infty = 500$.

As a second example we consider the hypersonic flow around an inlet lip of a hypothetic vehicle. The situation is shown in figure 3.23. A shock hits the bow shock around the inlet

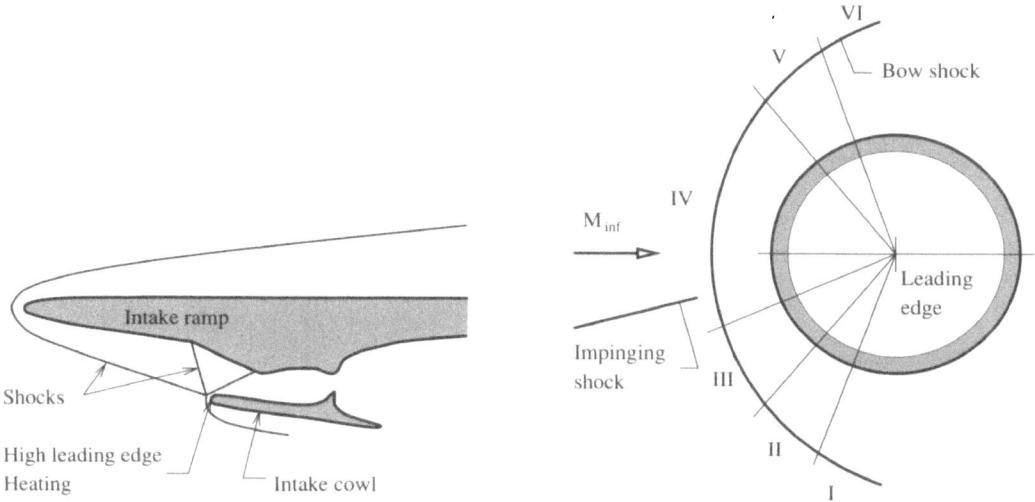

Figure 3.23   Flow around an inlet lip

lip. There are several different flow situations occuring depending on where the impinging shock hits the bow shock. These cases are enumerated in the right part of figure 3.23.

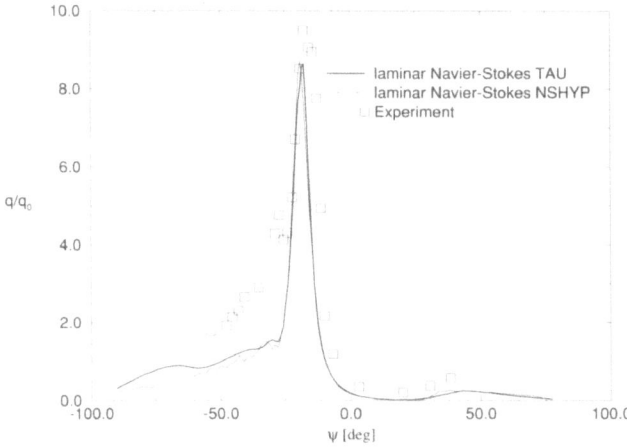

Figure 3.24   Temperature distribution on the lip surface

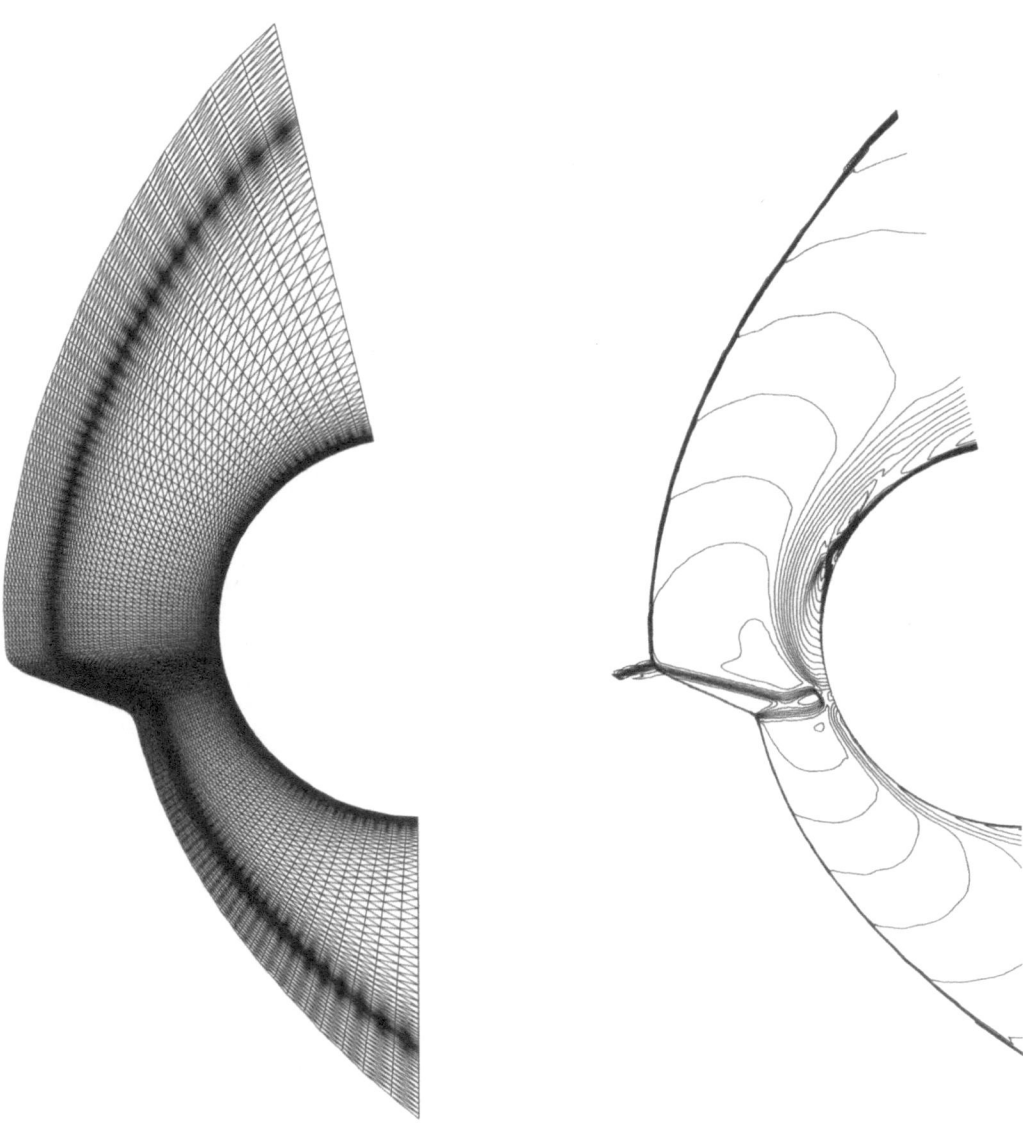

Figure 3.25   Grid and Mach number distribution

The onflow Mach number is $\mathrm{Ma}_\infty = 8.03$, Reynolds number $\mathrm{Re}_\infty = 197500$, temperature $T_\infty = 122.1°K$ and the temperature at the wall is given by $T_{wall} = 294.4°K$ (isothermic wall). Figure 3.25 shows the grid used and the Mach number distribution obtained with the second order box method with Barth-Jespersen limiter as outlined above. An implicit time discretisation was used which is described in more detail below. Figure 3.24 shows the temperature distribution along the surface of the lip. The numerical results of the box method (TAU) are compared to the numerical results of a finite difference scheme (NSHYP) and the measurements done in an experiment.

3.4 Time stepping schemes

Up to this point we have described the basic finite volume idea: transformation of a given partial differential equation in conservation form into a system of ordinary differential equations by means of Gauss' integral theorem and certain discretisation techniques. Although we already presented some numerical results we did not specify algorithms for the discretisation of the time derivative. Let us write the i-th equation of the system of ordinary differential equations in the form

$$\frac{d\overline{u}_i(t)}{dt} = -\mathcal{N}(\overline{u}(t))$$

where we have written $\overline{u}$ for a set of neighbours of cell σ_i in which σ_i is included. The operator $\mathcal{N}$ stands for the full discretisation of the right hand side which includes all operations in space.

3.4.1 The CFL condition

The simplest discretisation in time is the forward Euler difference which reads as

$$\frac{\overline{u}_i(t + \Delta t) - \overline{u}_i(t)}{\Delta t} = -\mathcal{N}(\overline{u}(t)).$$

As is well known explicit discretisations like the forward Euler difference have to satisfy a severe restriction on the time step Δt due to the Courant-Friedrichs-Lewy (CFL) condition. This condition states that during one time step signals starting in one cell are not allowed to interact with signals coming from other cells[4]. Let us treat primary grid methods and look at the situation presented in figure 3.26. We think of our 'information' as living in the barycentre c_i. The distances to the three faces of the triangle are denoted by $k_j, j = 1,2,3$. If ζ_ℓ denotes the eigenvalues of the matrix $\nabla_u f_1(u)n_1 + \nabla_u f_2(u)n_2$ then the CFL condition can be written in the form

$$\frac{\Delta t}{\min_{l=1,2,3} k_l} \max_\ell |\zeta_\ell| \leq \mathrm{CFL}$$

with a constant $\mathrm{CFL} \leq 1$. Conditions for box methods can be derived analogously to this considerations.

---

[4] In fact, this is **not** the classical CFL condition but a simple analogue to imagine what this condition is all about!

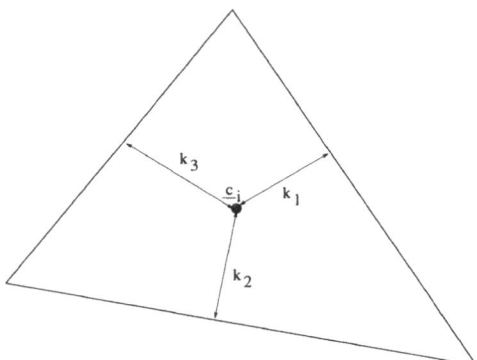

Figure 3.26 Geometry for the CFL condition

3.4.2 Explicit discretisations

One way to weaken the CFL conditon and allow for larger time steps is the use of Runge-Kutta methods (cp. [7]). However, in dealing with ENO-type schemes one has to take care that the chosen time discretisation does not spoil the monotonicity properties of the space discretisation. Shu and Osher observed in [12] that the classical 4-stage Runge-Kutta does not respect the monotonicity properties of the space discretisation and therefore developed a class of so-called TVD-Runge-Kutta schemes. The price to pay is that the CFL condition can not be weakened with this type of discretisations, i.e. the constant CFL has still to be bounded by one from above. The gain of the TVD-Runge-Kutta schemes lies in the accuracy in time as compared to the simple forward Euler difference which results in a $\mathcal{O}(\Delta t)$-approximation in time. The second order (in time) scheme is given by

$$
\begin{aligned}
\overline{u}^{(0)} &:= \overline{u}(t) \\
\overline{u}^{(1)} &= \overline{u}^{(0)} - \Delta t \mathcal{N}(\overline{u}^{(0)}) \\
\overline{u}^{(2)} &= \frac{1}{2}\overline{u}^{(0)} + \frac{1}{2}\overline{u}^{(1)} - \frac{1}{2}\Delta t \mathcal{N}(\overline{u}^{(1)}) \\
\overline{u}(t + \Delta t) &:= \overline{u}^{(2)}
\end{aligned}
$$

while the third order discretisation is given by

$$
\begin{aligned}
\overline{u}^{(0)} &:= \overline{u}(t) \\
\overline{u}^{(1)} &= \overline{u}^{(0)} - \Delta t \mathcal{N}(\overline{u}^{(0)}) \\
\overline{u}^{(2)} &= \frac{3}{4}\overline{u}^{(0)} + \frac{1}{4}\overline{u}^{(1)} - \frac{1}{4}\Delta t \mathcal{N}(\overline{u}^{(1)}) \\
\overline{u}^{(3)} &= \frac{1}{3}\overline{u}^{(0)} + \frac{2}{3}\overline{u}^{(2)} - \frac{2}{3}\Delta t \mathcal{N}(\overline{u}^{(2)}) \\
\overline{u}(t + \Delta t) &:= \overline{u}^{(3)}.
\end{aligned}
$$

The upper index in parathensis denotes the stage number of the scheme. For practical purposes it is convenient to rewrite the Runge-Kutta methods in the form

$$
\begin{aligned}
\overline{u}^{(0)} &:= \overline{u}(t) \\
\overline{u}^{(1)} &= \overline{u}^{(0)} - \Delta t \mathcal{N}(\overline{u}^{(0)}) \\
\overline{u}^{(2)} &= \overline{u}^{(0)} - \frac{1}{2}\mathcal{N}(\overline{u}^{(0)}) - \frac{1}{2}\Delta t \mathcal{N}(\overline{u}^{(1)}) \\
\overline{u}(t + \Delta t) &:= \overline{u}^{(2)}
\end{aligned}
$$

for the second order method and

$$
\begin{aligned}
\overline{u}^{(0)} &:= \overline{u}(t) \\
\overline{u}^{(1)} &= \overline{u}^{(0)} - \Delta t \mathcal{N}(\overline{u}^{(0)}) \\
\overline{u}^{(2)} &= \overline{u}^{(0)} - \frac{1}{4}\mathcal{N}(\overline{u}^{(0)}) - \frac{1}{4}\Delta t \mathcal{N}(\overline{u}^{(1)}) \\
\overline{u}^{(3)} &= \overline{u}^{(0)} - \frac{1}{6}\mathcal{N}(\overline{u}^{(0)}) - \frac{1}{6}\Delta t \mathcal{N}(\overline{u}^{(1)}) - \frac{2}{3}\Delta t \mathcal{N}(\overline{u}^{(2)}) \\
\overline{u}(t + \Delta t) &:= \overline{u}^{(2)}
\end{aligned}
$$

for the third order version. Thus, the second order method is the classical Heun scheme but there is no classical Runge-Kutta scheme which is identical with the third order method.

3.4.3 Implicit time discretisations

In order to overcome the shortcomings of explicit methods due to the severe bound on the time step (especially if Navier Stokes grids are considered) several techniques can be applied. For convergence to steady state multigrid methods are (nearly) standard nowadays. If time accuracy is needed the technology of implicit discretisation still plays the major role in CFD. Meister used a classical Θ-scheme in a box method in [9] to advance in time which can be written as

$$
\underbrace{\overline{u}_i^{n+1} - \overline{u}_i^n}_{\Delta \overline{u}_i^n :=}
$$

$$
= \int_{t^n}^{t^{n+1}} \frac{d}{dt}\overline{u}_i(t)
$$

$$
= -\frac{\Delta t^n}{|\sigma_i|} \sum_{j \in N(i)} \sum_{k=1}^{2} \frac{|l_{ij}^k|}{2} \sum_{\nu=1}^{n_G} \omega_\nu \left\{ (1-\Theta) H \left(\pi_i^{n+1}(x_{ij}^k(s_\nu)), \pi_j^{n+1}(x_{ij}^k(s_\nu)); n_{ij}^k \right) \right.
$$

$$
\left. + \Theta H \left(\pi_i^n(x_{ij}^k(s_\nu)), \pi_j^n(x_{ij}^k(s_\nu)); n_{ij}^k \right) \right\} + \mathcal{O}\left((\Delta t)^{g(\Theta)} \right)
$$

with $\overline{u}_i^n = \overline{u}_i(n\Delta t) = \overline{u}_i(t^n)$, $\pi_i^n(x_{ij}^k(s_\nu)) = \pi_i(x_{ij}^k(s_\nu), n\Delta t)$, and the ordinal function $g : [0,1] \to \mathbb{R}$

$$
g(\Theta) = \begin{cases} 2, & \text{if } \Theta = 1/2 \\ 1, & \text{otherwise.} \end{cases}
$$

For this temporal discretisation, the numerical flux has to be treated at the time level $t^{n+1} = (n+1)\Delta t$, which is accomplished by linearising according to

$$H\left(\pi_i^{n+1}(x_{ij}^k(s_\nu)),\pi_j^{n+1}(x_{ij}^k(s_\nu));n_{ij}^k\right)$$

$$= H\left(\pi_i^n(x_{ij}^k(s_\nu)),\pi_j^n(x_{ij}^k(s_\nu));n_{ij}^k\right) + \frac{\partial H\left(\pi_i^n(x_{ij}^k(s_\nu)),\pi_j^n(x_{ij}^k(s_\nu));n_{ij}^k\right)}{\partial \overline{u}_i}\Delta\overline{u}_i^n$$

$$+\frac{\partial H\left(\pi_i^n(x_{ij}^k(s_\nu)),\pi_j^n(x_{ij}^k(s_\nu));n_{ij}^k\right)}{\partial \overline{u}_j}\Delta\overline{u}_j^n + \mathcal{O}\left(\Delta t^2\right).$$

Thus, we get

$$A_{ii}\Delta\overline{u}_i^n + \sum_{j\in N(i)} A_{ij}\Delta\overline{u}_j^n = b_i$$

with

$$A_{ii} = I + \frac{\Delta t^n}{|\sigma_i|}\sum_{j\in N(i)}\sum_{k=1}^2 \frac{|l_{ij}^k|}{2}\sum_{\nu=1}^{n_G}\omega_\nu\left(1-\Theta\right)\frac{\partial H\left(\pi_i^n(x_{ij}^k(s_\nu)),\pi_j^n(x_{ij}^k(s_\nu));n_{ij}^k\right)}{\partial \overline{u}_i},$$

$$A_{ij} = \frac{\Delta t^n}{|\sigma_i|}\sum_{k=1}^2 \frac{|l_{ij}^k|}{2}\sum_{\nu=1}^{n_G}\omega_\nu\left(1-\Theta\right)\frac{\partial H\left(\pi_i^n(x_{ij}^k(s_\nu)),\pi_j^n(x_{ij}^k(s_\nu));n_{ij}^k\right)}{\partial \overline{u}_j},$$

$$b_i = -\frac{\Delta t^n}{|\sigma_i|}\sum_{j\in N(i)}\sum_{k=1}^2 \frac{|l_{ij}^k|}{2}\sum_{\nu=1}^{n_G}\omega_\nu H\left(\pi_i^n(x_{ij}^k(s_\nu)),\pi_j^n(x_{ij}^k(s_\nu));n_{ij}^k\right)$$

for all inner nodes i. Considering the boundary conditions in the same manner a linear system of equations

$$A\Delta\overline{u} = b$$

with a large sparse non-symmetric matrix $A = (A_{ij})$ has to be solved for each time step.

For the solution of the above system the GMRES algorithm developed by Saad and Schulz [11] is used. Therefore, the system is transformed into an equivalent minimisation problem. First, we define the function $f : \mathbb{R}^n \to \mathbb{R}_0^+$ by

$$f(y) = \|b - Ay\|_2^2$$

and choose an arbitrary initial vector y_0. Starting with $m = 0$ the residual

$$\overline{r}_m = \min_{\substack{y=y_0+z \\ z\in K_m}} f(y)$$

is computed, where

$$K_m(A,r_0) := \text{span}\left\{r_0, Ar_0, A^2r_0, \ldots, A^{m-1}r_0\right\}$$

denotes the m-th Krylov subspace and $r_0 = b - Ay_0$. Now we increase m until $\overline{r}_m$ is below a given tolerance. Then we compute the optimal approximate solution

$$y_m = \arg\min_{\substack{y=y_0+z \\ z\in K_m}} f(y).$$

Considering the fact that the expense to calculate the residual increases with the Krylov subspace dimension it is efficient to limit this dimension. In the case that this limit is reached before that of the tolerance the approximation y_m has to be calculated and used as the initial value during a repetition. This technique is called 'GMRES with restart'.

Since the convergence rate of an iterative method depends on the condition number of the matrix A, an incomplete LU-factorisation is used as a preconditioner in order to decrease the condition number. Hereby the incomplete LU-factorisation is a pair of a lower left (L) and an upper right (U) matrix satisfying the following three conditions:

1. U_{ii} presents the unit matrix for all i,
2. $L_{ij} = U_{ij} = A_{ij}$, if A_{ij} is a null matrix,
3. $(LU)_{ij} = A_{ij}$, if A_{ij} is not a null matrix

and the linear system is transformed into

$$AU^{-1}L^{-1}\tilde{y} = b \quad , \quad y = U^{-1}L^{-1}\tilde{y} \quad .$$

A detailed description of these preconditioned GMRES algorithm in comparison with other implicit and explicit finite volume schemes is presented in [9].

An application in which an implicit time stepping scheme is necessary is the time accurate computation in moving grids as in the case of pitching airfoils. This case is treated in detail in [10]. One can imagine the power of implicit discretisations if one looks back at the flow about a Bi-NACA0012 shown in figure 3.16. In the left part of figure 3.27 the convergence behaviour of the implicit scheme is compared with an explicit second-order TVD-Runge-Kutta scheme. Note that no convergence acceleration was used in the explicit code so that the comparison is a bit unfair. Shown is the discrete L^2-norm of the density variation versus CPU-seconds on a small SUN SPARC-2 work station. In the

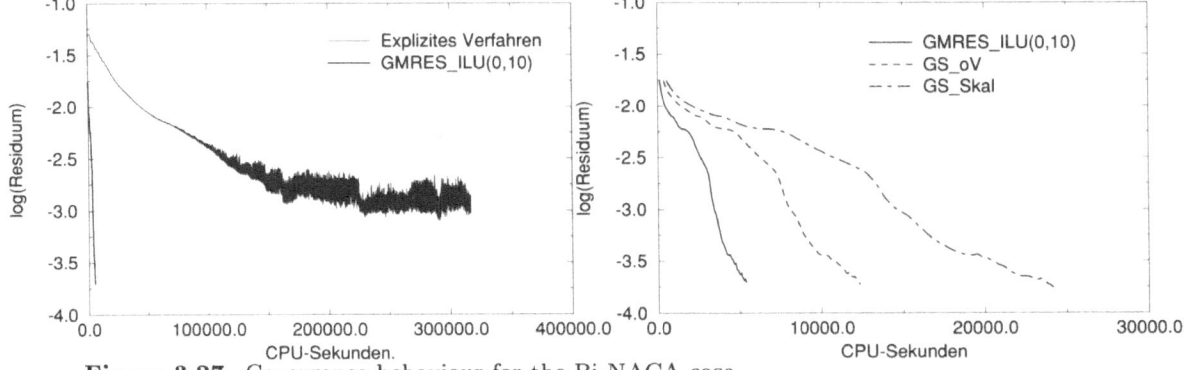

Figure 3.27   Covergence behaviour for the Bi-NACA case

right part of figure 3.27 we see the convergence behaviour of the GMRES method with ILU preconditioning compared with a Gauss-Seidel solver without preconditioning (GS-oV) and using a simple scaling (GS-Skal) as preconditioner. The thesis [10] will contain a thorough comparison of other modern linear system solvers (such as BICGSTAB etc.) applied to realistic flow problems.

3.5 Remarks on the philosophy of ENO schemes

As was outlined above the key to higher order finite volume approximations lies in the computation of recovery functions π_i which are defined locally on each control volume. Since boundedness of total variation of the numerical solution contradicts high order of accuracy Harten et al. [23], [25], [26], [27] designed a class of discretisations which allow spurious oscillations at the order of their truncation error. Due to this property the name *essentially non-oscillatory* (ENO) approximations was coined for these methods. In their approach Harten et al. recovered polynomials of fixed (arbitrary high) degree on each control volume for spatially one-dimensional conservation laws. The ENO algorithm is based on a stencil search in the following sense. Starting with the constant cell average the control volumes to the left and right give rise to two possible linear polynomials preserving the cell average. Since the divided differences reflect the smoothness of the polynomials the least oscillatory is the one with smallest leading divided difference. In the next step three quadratic polynomials are possible which can be constructed from a further enlarged neighbourhood of our control volume. We choose the one which is least oscillating in the sense of smallest leading divided difference and go on to construct a set of cubic polynomials from which we again choose the one which is least oscillating. This procedure goes on until the required degree of the recovery polynomial is reached.

This approach was generalised to multiple space dimensions on structured body-fitted grids by dimensional splitting. Harten and Chakravarthy designed classes of ENO approximations on triangular meshes in [24]. In that case the complexity of the stencil search is much higher than in the one-dimensional case and divided differences can no longer be used to decide whether one polynomial is smoother than another. Additionally, recovering polynomials in multiple space dimensions reduces to the problem of scattered data interpolation. It is well known that polynomials suffer from severe problems if interpolation points are close to being colinear. Meanwhile some progress took place not only in the algorithmic development of multidimensional ENO approximations on unstructured grids (see [15], [34]) but also in the development of numerical analysis for such schemes. We owe Rémi Abgrall the first successful analysis which gives a foundation of polynomial recovery for box methods in the framework of finite element interpolation theory. In [16], [17], [18], [19] he developed a simple stencil selection scheme and the corresponding analysis of choosing the least oscillatory polynomial. Since these methods are very costly in terms of computer time Abgrall and the author developed the technique of generalised Mühlbach expansions in [20]. Using this technique the recovery polynomial can be computed in a step-by-step manner avoiding the solution of ill-conditioned linear systems. The author applied the theory of optimal recovery according to Golomb and Weinberger [22] and Micchelli and Rivlin [29], [30], [31] to the recovery problem in ENO approximations in [33] and [35] and identified the class of radial basis functions as optimal in their native function spaces. The following sections try to summarize these recent developments.

One common plague in all ENO discretizations is the missing convergence to the steady state. Due to the highly nonlinear digital switching of the stencil features in the flow may slightly oscillate and prevent the scheme from convergence. On the other hand oscillatory switching of the stencil may also lead to a breakdown of the accuracy of ENO schemes. Liu, Osher and Chan in [51] therefore designed a weighted ENO scheme (WENO) in which all local recovery polynomials are weighted due to their oscillatory behaviour and summed up to give the resulting recovery function. It was Friedrich in [44] who was first in designing a WENO scheme on triangular meshes. It has turned out in several test cases that the WENO scheme is in general superior to the ENO method. However, all of the basic algorithmic steps of ENO methods are also part of a WENO scheme.

3.6 Polynomial recovery

We remind the reader that a finite volume approximation on primary grids[5] is given by

$$\frac{d}{dt}\overline{u}_i(t) = -\frac{1}{|T_i|}\sum_{j\in N(i)}\frac{|\partial T_i \cap \partial T_j|}{2}\sum_{\nu=1}^{n_G}\omega_\nu H\left(\pi_i(x_{ij}(s_\nu),t),\pi_j(x_{ij}(s_\nu),t);n_{ij}\right)$$

$$\overline{u}_i(0) = \mathfrak{A}(T_i)u(0).$$

Although it is possible to recover the conserved variables directly one often recovers the primitive variables or other combinations of variables. Thus, we describe the recovery process assuming u is a scalar-valued function, i.e. π_i is a scalar-valued recovery polynomial.

Here, π_i is a recovery function (polynomial) on σ_i (either a triangle or a box) satisfying the accuracy requirement $(u - \pi_i)(x,t) = \mathcal{O}(h^r), r \geq 1$ for fixed time t as well as the conservativity property $\mathfrak{A}(\sigma_i)\pi_i(t) = \mathfrak{A}(\sigma_i)u(t)$. We consider now the case if π_i is chosen to be a polynomial of fixed degree $r - 1$, i.e.

$$\pi_i(x,t) = \sum_{\mu=0}^{r-1}\frac{1}{\mu!}\sum_{|\alpha|=\mu}a_\alpha(t)(x - c_i)^\alpha \in \Pi_{r-1}(\sigma_i;\mathbb{R}),$$

where c_i is the barycentre of σ_i. It is then easy to see that π_i is a recovery polynomial, if and only if

$$a_\alpha = \partial^\alpha u|_{x=c_i} + \mathcal{O}\left(h^{r-|\alpha|}\right)$$

holds true, because, dropping dependence on time, Taylor expansion of smooth u

$$u(x) = \sum_{\mu=0}^{r-1}\frac{1}{\mu!}\sum_{|\alpha|=\mu}(x - c_i)^\alpha \partial^\alpha|_{x=c_i} + \mathcal{O}(h^r)$$

yields

$$\pi_i(x) - u(x) = \sum_{\mu=0}^{r-1}\frac{1}{\mu!}\sum_{|\alpha|=\mu}(x - c_i)^\alpha\left(a_\alpha - \partial^\alpha|_{x=c_i}\right) + \mathcal{O}(h^r),$$

i.e.

---

[5] Most of the following results are directly applicable to primary and secondary grid methods.

$$|\pi_i(x) - u(x)| \leq \sum_{\mu=0}^{r-1} \frac{1}{\mu!} \sum_{|\alpha|=\mu} |x - c_i|^{|\alpha|} |a_\alpha - \partial^\alpha u|_{x=c_i}| + \mathcal{O}(h^r)$$

$$\leq \sum_{\mu=0}^{r-1} \frac{1}{\mu!} \sum_{|\alpha|=\mu} h^{|\alpha|} |a_\alpha - \partial^\alpha u|_{x=c_i}| + \mathcal{O}(h^r).$$

Hence, corresponding to $|\alpha|$, the order of $a_\alpha - \partial^\alpha u|_{x=c_i}$ must vary to give the resulting order $\mathcal{O}(h^r)$.

To be able to compute a recovery polynomial of degree at most $r-1$ we need the knowledge of cell averages on dim $\Pi_{r-1} := M$ control volumes. Note that in two dimensions we have

$$M = \binom{r+1}{r-1}.$$

Thus, for the computation of a recovery polynomial $\pi_i \in \Pi_{r-1}(\sigma_i; \mathbb{R})$ of degree at most $r-1$ on control volume σ_i we need the cell average $\mathfrak{A}(\sigma_i)u$ as well as $M-1$ other values of cell averages on neighbouring control volumes. The computation of such neighbourhoods is an art more than a science and will be described in the following section.

3.6.1   Stencil selection algorithms

We know already that we need the value of $\mathfrak{A}(\sigma_i)u$ on σ_i as well as the values $\mathfrak{A}(\sigma_{i_j})u$ on neighbouring control volumes $\sigma_{i_j}, j = 1, \ldots, M-1$. In order to work with consistent notation we define

$$\sigma_{i_0} := \sigma_i.$$

We call every set

$$K(\sigma_i) := \{\sigma_{i_0}, \sigma_{i_1}, \ldots, \sigma_{i_{M-1}}\}$$

of control volumes a node set of σ_i. The intersection of any two members of one node set is assumed to be void. The computation of a number of node sets for each control volume is the task of the stencil selection algorithm. That the construction of such algorithmic ingredients is not a trivial task can be seen from the following simple example. In the grid shown in figure 3.28 we compute the number of different node sets of fixed cardinality. The results for the inner point $p1$ are listed in the following table.

node	length L	# all node sets	# valid node sets	CPU (sec:1/10 sec)
$p1$	3	45	33	0:00
$p1$	6	24710	5232	0:31
$p1$	7	207292	27805	2272:39

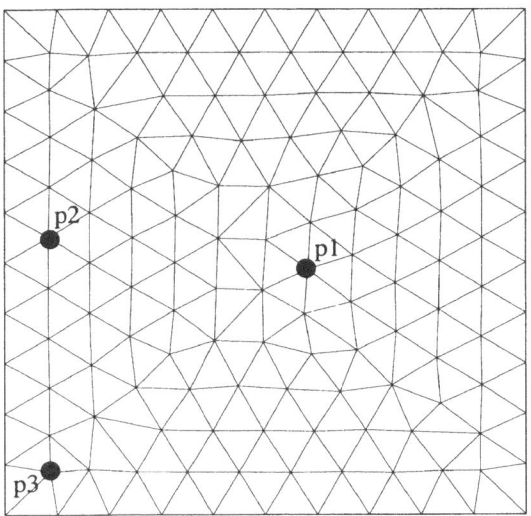

Figure 3.28 A grid for the computation of node sets

We apply a recursive algorithm[6] to detect all graphs in the grid of fixed length passing through $p1$. The number in the column # *all node sets* counts any successful run of the recursion, i.e. a lot of identical graphs are multiply counted. Removing this redundancy leads to the numbers in column # *valid node sets*. CPU time is measured on a SUN 10 workstation using the Unix command `time`. The computation of a quadratic recovery polynomial (6 coefficients in 2-D) corresponds to a graph of length 6. As can be seen the recursive procedure to find all possible node sets would come up with 5232 valid node sets. Basing an ENO algorithm on this stencil selection algorithm would then require to compute 5232 quadratic recovery polynomials for the control volume associated with node p_1 and then choose the one which is least oscillatory. The same numbers of work to do would occur for all other inner points. Computing a cubic polynomial (10 coefficients in 2-D) would already be impossible with this stencil selection. The corresponding numbers for the points $p2$ and $p3$ in the vicinity of the boundary are shown in the following tables.

node	length L	# all node sets	# valid node sets	CPU (sec:1/10 sec)
$p2$	3	41	29	0:00
$p2$	6	14871	2834	0:09

node	length L	# all node sets	# valid node sets	CPU (sec:1/10 sec)
$p3$	3	28	18	0:00
$p3$	6	6121	1030	0:02

As this example shows one needs intelligent algorithms which result in a representative number of node sets. For polynomials of small degree (i.e. linear polynomials) on primary grids this is easily accomplished. For each triangle T_i we call the set

---

[6] I thank Oliver Friedrich for implementing and testing the recursive algorithm.

$$K_{vN}(T_i) := \{T \in \mathcal{T}^h \mid T \cap T_i \text{ is edge of } T_i \text{ and } T \neq T_i\}$$

the von Neumann neighbourhood of T_i. The (at most) three von Neumann neighbours of T_i are denoted by $T_{i_j}, j = 1,2,3$. The set

$$K_M(T_i) := \{T \in \mathcal{T}^h \mid T \cap T_i \text{ is edge of } T_i \text{ or } T \cap T_i \text{ is node of } T_i\}$$

is called Moore neighbourhood of T_i.[7] If the required degree of recovery polynomials

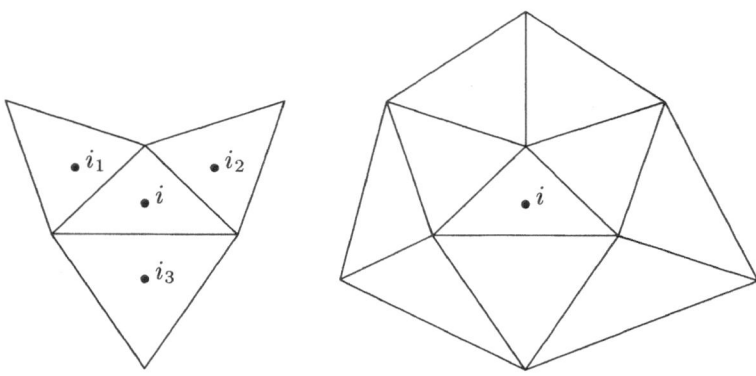

Figure 3.29 Von Neumann (left) and Moore neighbourhood T_i

exceeds one there is a need for node sets out of a larger neighbourhood. Inspired by an idea of Harten and Chakravarthy [24] one could use a sectorial search algorithm. Let T_i be a triangle with nodes x_1, x_2, x_3. The sets

$$S_j := \{x \in \mathbb{R}^2 \mid \text{diag}(d_{jA} \times x) \neq \text{diag}(d_{jB} \times x)\}, \quad j = 1,2,3,$$

defined by the pairs of vectors $(d_{1A} := x_2 - x_1, d_{1B} := x_3 - x_1), (d_{2A} := x_1 - x_2, d_{2B} := x_3 - x_2)$ and $(d_{3A} := x_1 - x_3, d_{3B} := x_2 - x_3)$, are called sectors of T_i. If B_i denotes a box and if $j \in N(i)$, then we define the triangles sharing edge $x_i - x_j$ according to figure 3.30 to be denoted by T_{ij}^1, T_{ij}^2 and their barycentres as c_{ij}^1 and c_{ij}^2, respectively. The sets

$$S_j := \{x \in \mathbb{R}^2 \mid \text{diag}(d_{jA} \times x) \neq \text{diag}(d_{jB} \times x)\}, \quad j = 1,\dots,\text{card } N(i),$$

defined by the pairs of vectors $(d_{jA} := c_{ij}^1 - x_i, d_{jB} := c_{ij}^2 - x_i)$, are called sectors of B_i.

Within each sector a recursive stencil search computing node sets of given cardinality can be employed. Note that sectorial search algorithms are not *a priori* able to compute centered stencils. This may cause trouble in regions of smoothly varying flow, where a central stencil is optimal.

3.6.2   Computation of the recovery polynomial

A node set computed by a stencil selection algorithm is not necessarily usable in an ENO scheme with polynomial recovery. We want to consider $\mathfrak{A}$-unisolvent node sets which are

---

[7] The notation is borrowed from the names of neighbourhoods in the theory of cellular automata.

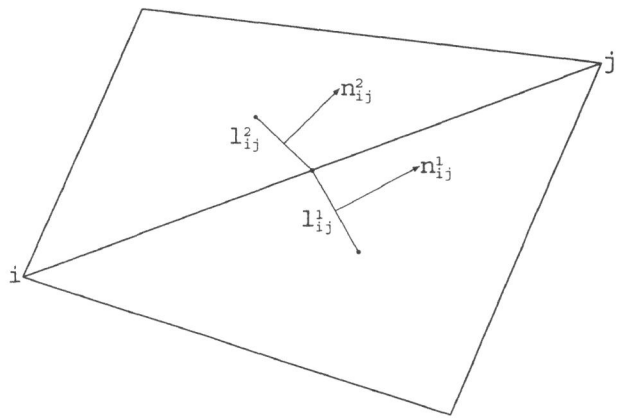

Figure 3.30 Geometry between boxes B_i and B_j

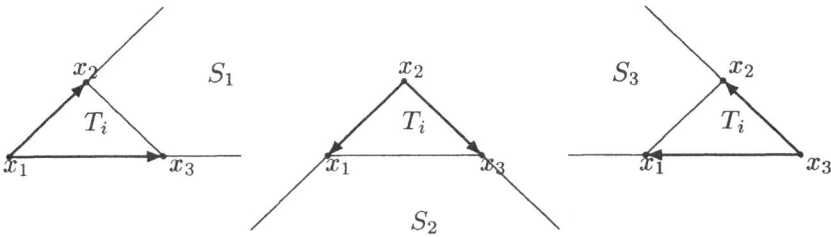

Figure 3.31 Sectors of triangle T_i

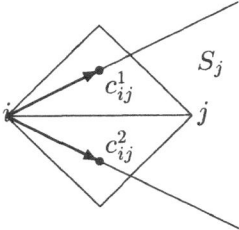

Figure 3.32 Sector S_j of Box B_i

defined as follows. We call a node set $K(\sigma_i)$ of cardinality M $\mathfrak{A}$-unisolvent, if for all $\pi \in \Pi_{r-1}$ and all $\sigma_{i_j} \in K(\sigma_i)$

$$\mathfrak{A}(\sigma_{i_j})\pi = 0 \quad \Rightarrow \quad \pi \equiv 0$$

holds true. This means, that recovery should be uniquely possible on this node set.

Then we can give a recipe concerning the computation of recovery polynomials. Let

$K(\sigma_i) := \{\sigma_{i_0}, \sigma_{i_1}, \dots, \sigma_{i_{M-1}}\}$ be an $\mathfrak{A}$-unisolvent node set with corresponding cell averages $\overline{u}_{i_j} := \mathfrak{A}(\sigma_{i_j})u, j = 0, \dots, M-1$. Then the polynomial

$$\pi_i(x) = \sum_{\mu=0}^{r-1} \frac{1}{\mu!} \sum_{|\alpha|=\mu} (x - c_i)^\alpha a_\alpha$$

defined by the linear system of equations

$$\mathfrak{A}(\sigma_{i_j})\pi_i = \overline{u}_{i_j}, \quad j = 0, \dots, M-1, \tag{3.10}$$

satisfies the condition

$$(u - \pi_i)(x) = \mathcal{O}(h^r).$$

To prove this assertion consider for $\sigma_{i_k} \in K(\sigma_i)$ the equation

$$\mathfrak{A}(\sigma_{i_k})\pi_{i_0} = \overline{u}_{i_k}.$$

Using the abreviation

$$\Gamma_{i_k,\alpha} := \frac{1}{h^{|\alpha|}\mu!|\sigma_{i_k}|} \int_{\sigma_{i_k}} (x - c_i)^\alpha \, dx$$

it follows that

$$\mathfrak{A}(\sigma_{i_k})\pi_{i_0} = \sum_{\mu=0}^{r-1} \sum_{|\alpha|=\mu} \Gamma_{i_k,\alpha} a_\alpha h^{|\alpha|},$$

where we introduced the factor $1 = h^{|\alpha|}/h^{|\alpha|}$ for scaling reasons. Applying the cell average operator to the Taylor expansion

$$u(x) = \sum_{\mu=0}^{r-1} \frac{1}{\mu!} \sum_{|\alpha|=\mu} (a - c_i)^\alpha \, \partial^\alpha u|_{x=c_i} + \mathcal{O}(h^r)$$

yields

$$\mathfrak{A}(\sigma_{i_k})u = \overline{u}_{i_k} = \sum_{\mu=0}^{r-1} \sum_{|\alpha|=\mu} \Gamma_{i_k,\alpha} h^{|\alpha|} \, \partial^\alpha u|_{x=c_i} + \mathcal{O}(h^r).$$

If we now order the coefficients a_α in some fashion resulting in a vector a, then the system (3.10) can be written as a linear system

$$Ga = \vec{\overline{u}}$$

where $\vec{\overline{u}}$ denotes the vector of cell averages. Replacing the vector a by the vector b of corresponding derivatives $\partial^\alpha u|_{x=c_i}$ yields

$$Gb = \vec{\overline{u}} + \mathcal{O}(h^r)$$

and subtraction results in

$$G(b - a) = \mathcal{O}(h^r).$$

Due to the unisolvency of the node set the matrix G is invertible and assuming that the norm of the inverse is of order 1 we end up with

$$|b - a| = \mathcal{O}(h^r)$$

which is the required result.

To see that $\mathfrak{A}$-unisolvency is essential in polynomial recovery consider the situation shown in figure 3.33, where three triangles are shown for which the barycentres are colinear. Computing a linear recovery polynomial $\pi_i(x) = a_{00} + a_{10}(x_1 - c_{i,1}) + a_{01}(x_2 - c_{i,2})$ by solving the linear system

$$\mathfrak{A}(T_j)\pi_i = \overline{u}_j, \quad j \in \{i, i_1, i_2\}$$

results in

$$
\left.
\begin{aligned}
a_{00} + \frac{a_{10}}{|T_i|} \int_{T_i} (x_1 - c_{i,1})\, dx + \frac{a_{10}}{|T_i|} \int_{T_i} (x_2 - c_{i,2})\, dx &= \overline{u}_i \\[2mm]
a_{00} + \frac{a_{10}}{|T_{i_1}|} \int_{T_{i_1}} (x_1 - c_{i,1})\, dx + \frac{a_{10}}{|T_{i_1}|} \int_{T_{i_1}} (x_2 - c_{i,2})\, dx &= \overline{u}_{i_1} \\[2mm]
a_{00} + \frac{a_{10}}{|T_{i_2}|} \int_{T_{i_2}} (x_1 - c_{i,1})\, dx + \frac{a_{10}}{|T_{i_2}|} \int_{T_{i_2}} (x_2 - c_{i,2})\, dx &= \overline{u}_{i_2}
\end{aligned}
\right\}.
$$

Since $c_i = \frac{1}{|T_i|} \int_{T_i} x\, dx$ it follows from this system that the coefficients a_{00}, a_{10}, a_{01} are the solution of the system

$$
\begin{bmatrix} c_{i_1,1} - c_{i,1} & c_{i_1,2} - c_{i,2} \\ c_{i_2,1} - c_{i,1} & c_{i_2,2} - c_{i,2} \end{bmatrix}
\begin{bmatrix} a_{10} \\ a_{01} \end{bmatrix}
=
\begin{bmatrix} \overline{u}_{i_1} - \overline{u}_i \\ \overline{u}_{i_2} - \overline{u}_i \end{bmatrix}
$$

and the constant term is given by $a_{00} = \overline{u}_i$. Thus, linear polynomial recovery on a primary grid is equivalent to linear interpolation on the barycentres of the triangles of the node set. If the barycentres are colinear the above system is not solvable at all.

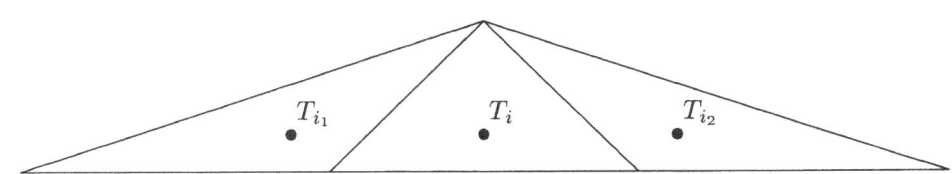

Figure 3.33 Node set which is not $\mathfrak{A}$-unisolvent

3.6.3   Recovery on primary grids

We describe now some basic techniques which result in ENO approximations. A very simple but surprisingly successful technique was described by Durlofsky, Engquist and Osher in [21]. The node sets are taken from the von Neumann neighbourhood and each (inner) triangle is equipped with the three node sets

$$K_1(T_i) := \{T_i, T_{i_1}, T_{i_2}\}, K_2(T_i) := \{T_i, T_{i_1}, T_{i_3}\}, K_3(T_i) := \{T_i, T_{i_2}, T_{i_3}\}.$$

On each node set a linear polynomial

$$\pi_i^{(k)} = a_{00}^{(k)} + a_{10}^{(k)}(x_1 - c_{i,1}) + a_{01}^{(k)}(x_2 - c_{i,2}), \quad k = 1,2,3,$$

is computed by solving the linear systems

$$\mathfrak{A}(T_j)\pi_i^{(k)} = \overline{u}_j, \quad (k,j) \in \{(1,i),(2,i_1),(3,i_2)\}.$$

If there is an extremum at triangle T_i with respect to its von Neumann neighbours, then none of the polynomials is chosen and the value on T_i is not recovered. Otherwise we use the steepest of the three linear polynomials and checkk, wether its use would result in a new extremum. If that is not the case this polynomial is taken to be the recovery on T_i. If a new extremum is created the polynomial with the next steepest gradient is chosen and tested and so on. If all three polynomials would result in a new extremum then no recovery is used on T_i. Note that this procedure corresponds more to the classical TVD-approach than to an ENO approach. The algorithm is shown in the following diagramm.

Algorithm I — Linear polynomial recovery
(after Durlofsky, Engquist and Osher)

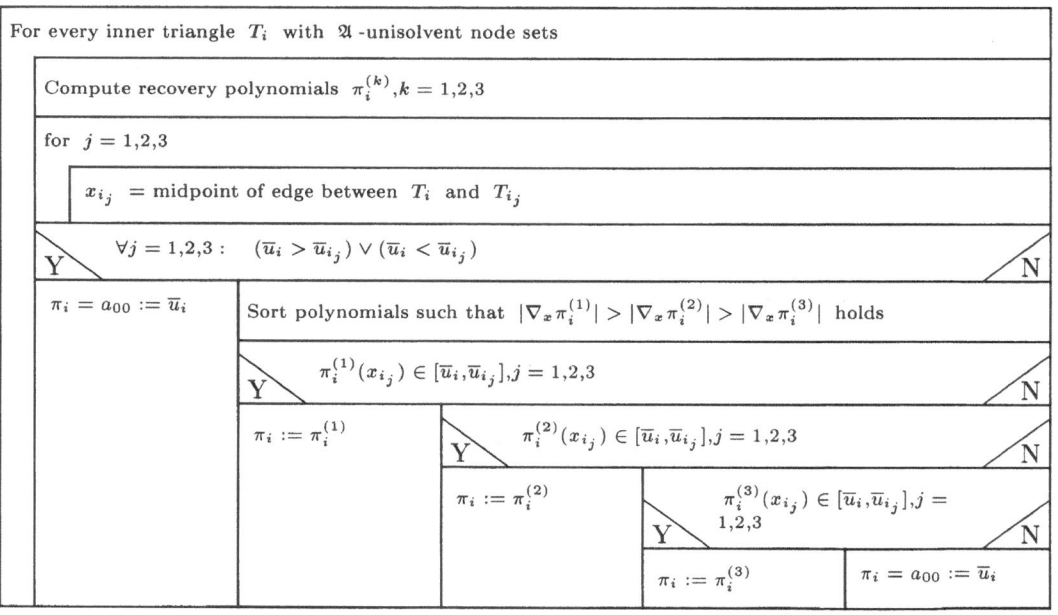

Applying this technique to the linear test problem introduced above results in the cone heights shown in the following table.

Triangulation	Mesh size h	Cone height with/without recovery
$\mathcal{T}_{h^1}$	0.05	0.291 / 0.176
$\mathcal{T}_{h^2}$	0.02	0.635 / 0.382

The corresponding isolines on the grids with 20×20 ($\mathcal{T}_{h^1}$ left) and 50×50 ($\mathcal{T}_{h^2}$) points are shown in figure 3.34. An even simpler linear polynomial recovery algorithm on

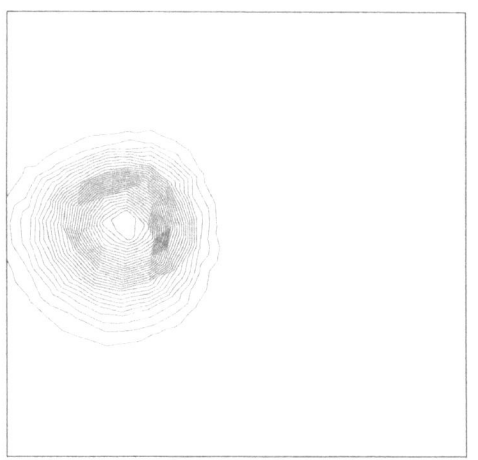
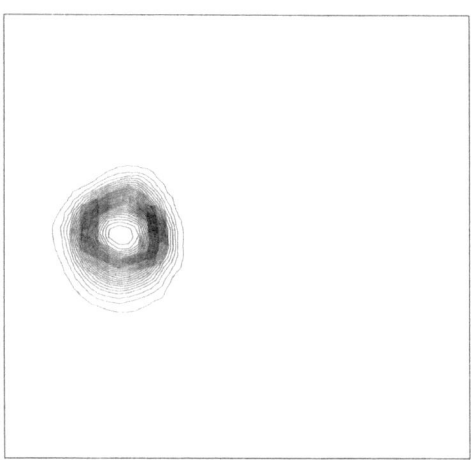

Figure 3.34 Solution of model problem with recovery after Durlofsky, Engquist, Osher

the von Neumann neighbourhood results if the recovery polynomial is chosen to be the one (of the three possible) with smallest gradient.

Algorithm II — Linear polynomial recovery on the von Neumann neighbourhood

For every inner triangle T_i with $\mathfrak{A}$-unisolvent node sets
Compute recovery polynomials $\pi_i^{(k)}, k = 1,2,3$
Choose π_i as the $\pi_i^{(k)}$ with $\|\nabla_x \pi_i\| = \min_{k=1,2,3} \|\nabla_x \pi_i^{(k)}\|$

In contrast to the algorithm described before this one is a true ENO algorithm in that small oscillations are allowed to occur. This can be seen in the fact that the numerical solution takes negative values in contrast to the Durlofsky, Engquist, Osher scheme, namely

$$\mathcal{T}_{h^1} \quad : \quad -0.000013 \leq \overline{u} \leq 0.353$$
$$\mathcal{T}_{h^2} \quad : \quad -0.000014 \leq \overline{u} \leq 0.753.$$

The resulting cone heights are compared with the one obtained by the Durlofsky, Engquist, Osher algorithm (DEO) in the following table.

Triangulation	h	Cone height DEO/Algorithm II
$\mathcal{T}_{h^1}$	0.05	0.291 / 0.353
$\mathcal{T}_{h^2}$	0.02	0.635 / 0.753

The two numerical solutions on $\mathcal{T}_{h^1}$ and $\mathcal{T}_{h^2}$, repectively, are shown in figure 3.35. As

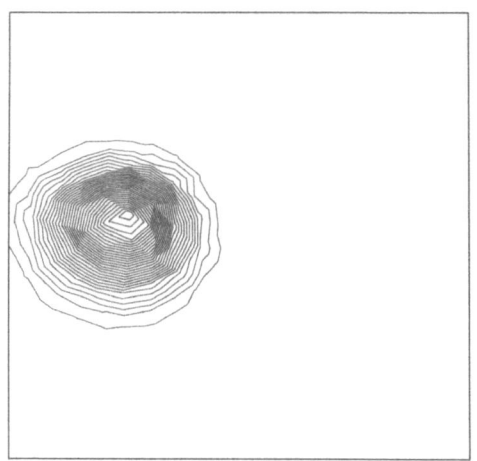 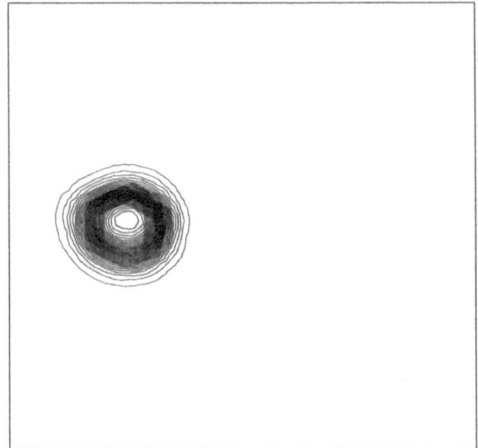

Figure 3.35 Solution with recovery following Algorithm II.

can be observed the result of the ENO scheme preserve the shape of the cone slightly better than the TVD-like algorithm. The cone height is also slightly larger but it has to be noted that an algorithm resulting in negative values of the recovery polynomial would immediately lead into trouble in a code to solve the Euler or Navier Stokes equations.

A simple but costly idea to avoid negative values of the recovery polynomial in the ENO framework is the increase of the number of node sets. This can be accomplished for linear polynomials on the Moore neighbourhood. If $T_{i_j}, j = 1,2,3$ denote the three von Neumann neighbours of T_i we consider node sets out of the set

$$T_i \cup K_{vN}(T_i) \cup \{T \mid T \in K_{vN}(T_{i_j}), j = 1,2,3 \text{ und } T \neq T_i\}.$$

This neighbourhood is shown in figure 3.36. We can now compute six linear polynomials on the node sets

$$
\begin{aligned}
K_1(T_i) &:= T_i \cup T_{i_1} \cup T_{i_{1_1}} \\
K_2(T_i) &:= T_i \cup T_{i_1} \cup T_{i_{1_2}} \\
K_3(T_i) &:= T_i \cup T_{i_2} \cup T_{i_{2_1}} \\
K_4(T_i) &:= T_i \cup T_{i_2} \cup T_{i_{2_2}} \\
K_5(T_i) &:= T_i \cup T_{i_3} \cup T_{i_{3_1}} \\
K_6(T_i) &:= T_i \cup T_{i_3} \cup T_{i_{3_2}}
\end{aligned}
$$

from the six linear systems

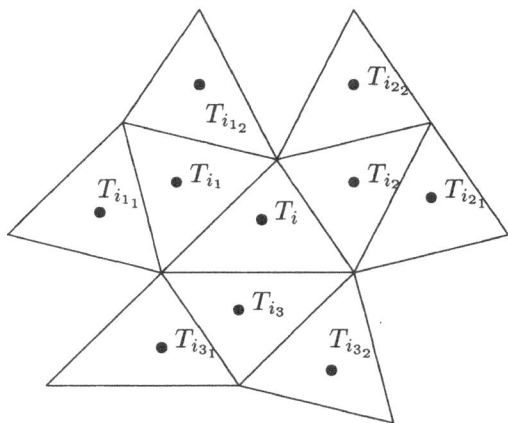

Figure 3.36 Neighbourhood of T_i

$$
\begin{aligned}
\mathfrak{A}(T_i)\pi_i^{(k)} &= \overline{u}_i \\
\mathfrak{A}(T_{i_m})\pi_i^{(k)} &= \overline{u}_{i_m} \\
\mathfrak{A}(T_{i_n})\pi_i^{(k)} &= \overline{u}_{i_n}
\end{aligned}
\qquad
(m,n) =
\begin{cases}
(1,1_1) & ; & k = 1 \\
(1,1_2) & ; & k = 2 \\
(2,2_1) & ; & k = 3 \\
(2,2_2) & ; & k = 4 \\
(3,3_1) & ; & k = 5 \\
(3,3_2) & ; & k = 6
\end{cases}
$$

The final recovery polynomial is again chosen to be the one with least steepest gradient according to Algorithm III.

Algorithm III — Linear polynomial recovery on the Moore neighbourhood

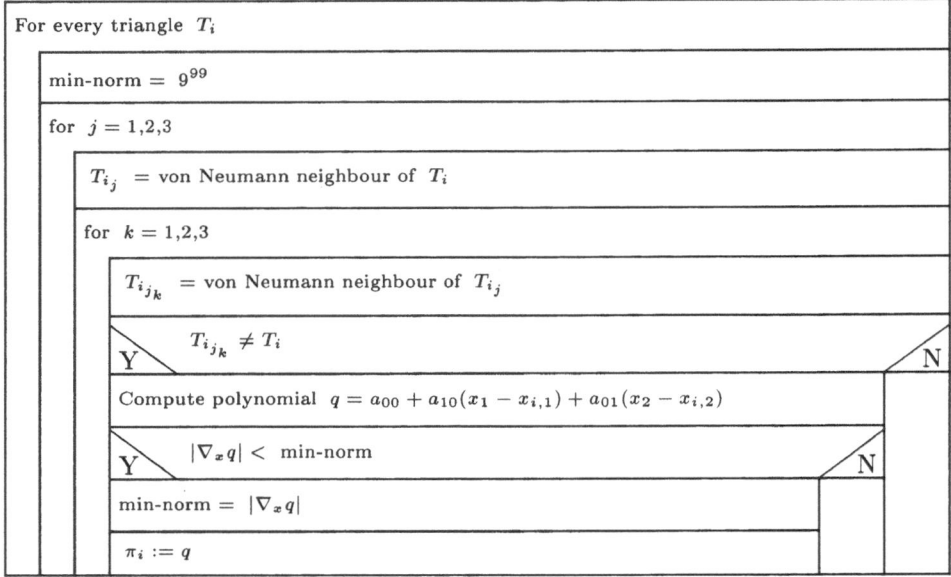

The numerical solution of our modell problem is shown in figure 3.37. The range of values

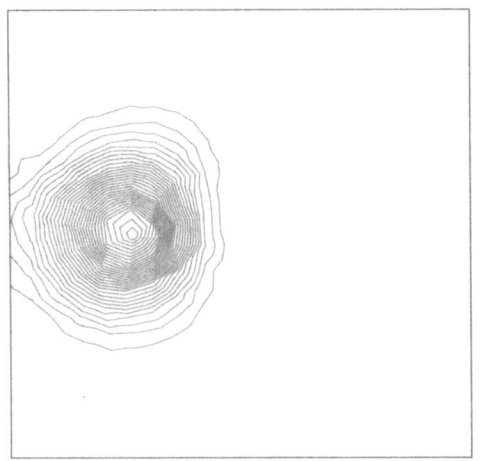
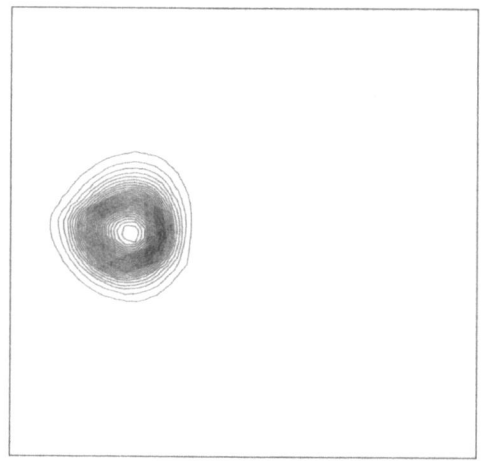

Figure 3.37 Solutions with linear polynomial recovery on the Moore neighbourhood

of the numerical solution is now given by

$$\mathcal{T}_{h^1} \quad : \quad 0.0 \leq \overline{u} \leq 0.279$$
$$\mathcal{T}_{h^2} \quad : \quad 0.0 \leq \overline{u} \leq 0.659.$$

For the computation of quadratic recovery polynomials it is already preferable to use a recursive sectorial stencil search algorithm as described above. Using this stencil search algorithm results in node sets containing six triangles which form a contiguous path in the triangulation and lie entirely in the prescribed sectors. On each node set a quadratic polynomial

$$
\pi_i(x) = \sum_{\mu=0}^{2} \frac{1}{\mu!} \sum_{|\alpha|=\mu} a_\alpha (x - c_i) = a_{00} + a_{10}(x_1 - c_{i,1}) + a_{01}(x_2 - c_{i,2})
$$

$$
+ a_{11}(x_1 - c_{i,1})(x_2 - c_{i,2}) + \frac{1}{2} a_{20}(x_1 - c_{i,1})^2 + \frac{1}{2} a_{02}(x_2 - c_{i,2})^2
$$

can be computed by means of solving

$$
\begin{aligned}
\mathfrak{A}(T_i)\pi_i &= \overline{u}_i \\
\mathfrak{A}(T_{i_k})\pi_i &= \overline{u}_{i_k}, \quad k = 1, \ldots, 5.
\end{aligned}
$$

Although the integrals

$$
\int (x_1 - c_{i,1})^2 \, dx, \quad \int (x_2 - c_{i,2})^2 \, dx, \quad \int (x_1 - c_{i,1})(x_2 - c_{i,2}) \, dx
$$

occuring in the coefficient matrix of this system can be computed exactly it makes more sense to use an appropriate quadrature rule.

If a set of possible quadratic polynomials is computed one has to be selected which is the least oscillatory in a sense to be defined. In order to include information about the behaviour of the higher order coefficients one could use the criterion

$$
W(\pi) := \sqrt{\sum_{\mu=1}^{2} \sum_{|\alpha|=\mu} a_\alpha^2} \tag{3.11}
$$

and choose the polynomial which gives a minimal W. This is used in the following algorithm.

Algorithm IV — Quadratic polynomial recovery with sector search

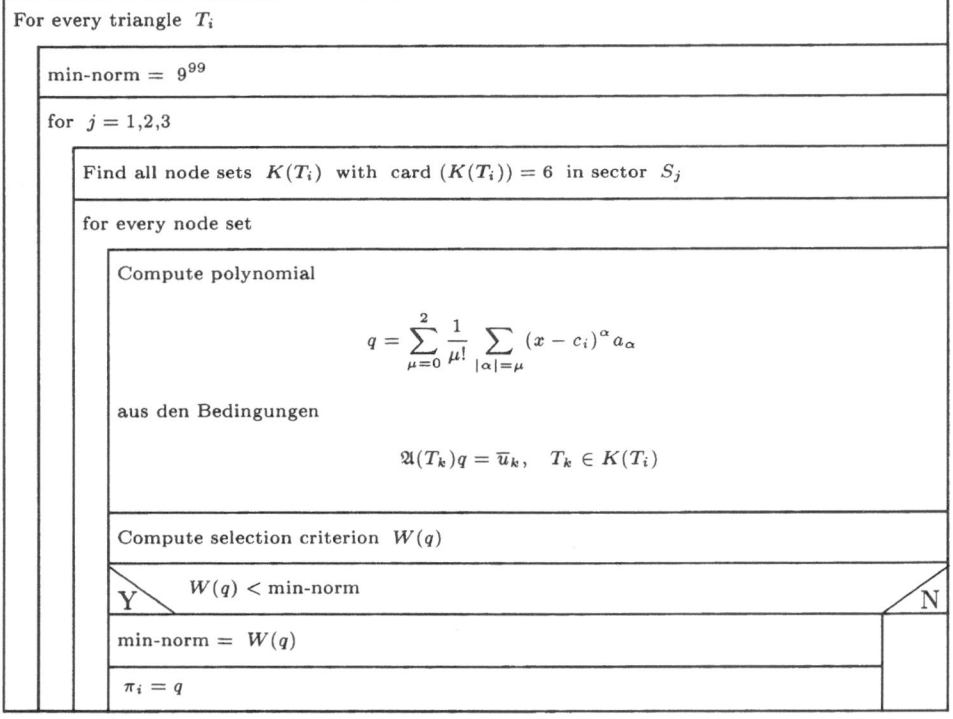

For every triangle T_i

min-norm = 9^{99}

for $j = 1,2,3$

Find all node sets $K(T_i)$ with card $(K(T_i)) = 6$ in sector S_j

for every node set

Compute polynomial

$$q = \sum_{\mu=0}^{2} \frac{1}{\mu!} \sum_{|\alpha|=\mu} (x - c_i)^\alpha a_\alpha$$

aus den Bedingungen

$$\mathfrak{A}(T_k)q = \overline{u}_k, \quad T_k \in K(T_i)$$

Compute selection criterion $W(q)$

Y $W(q)$ < min-norm   N

min-norm = $W(q)$

$\pi_i = q$

The resulting numerical results are shown in figure 3.38. The data range is shown in the

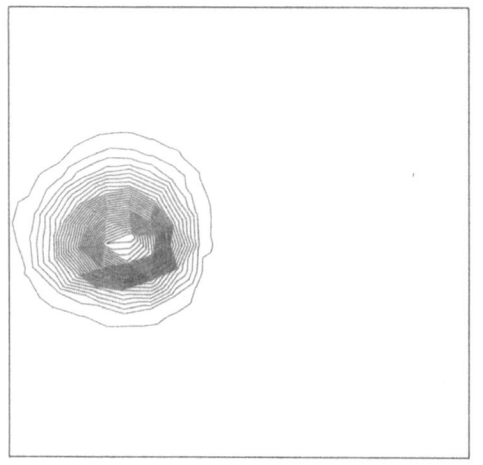
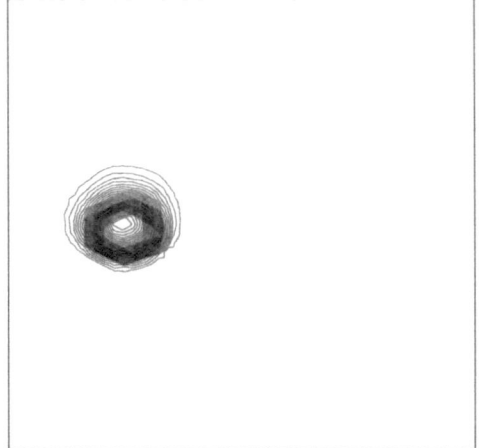

Figure 3.38 Solution with quadratic recovery

following table.

$$\mathcal{T}_{h^1} \quad : \quad -0.004 \le \overline{u} \le 0.432$$
$$\mathcal{T}_{h^2} \quad : \quad -0.000002 \le \overline{u} \le 1.04.$$

As can be seen oscillations have spoiled the solution so that the resulting cone height is even higher than even the initial height. If we use the selection criterion

$$W(\pi) := W_1(\pi) = \sum_{i=1}^{2} \|\partial_{x_i} \pi\|_{L^1(T;\mathbb{R})} \,,$$

instead of (3.11) the situation can be changed. With CFL numbers of 0.5 and 0.9, respectively, we get the numerical solutions on the fine grid $\mathcal{T}_{h^2}$ as shown in figure 3.39. The data range now is

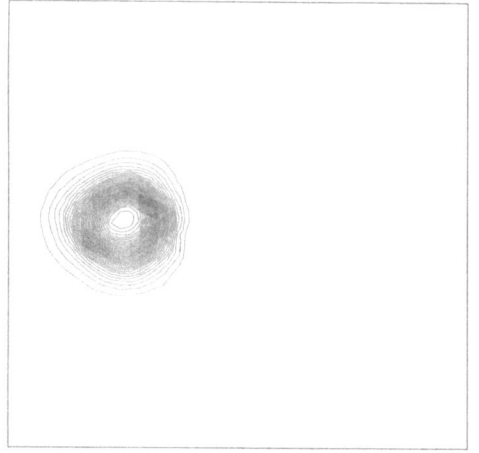 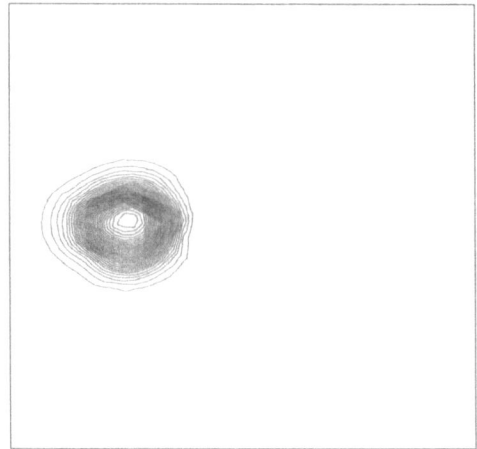

Figure 3.39 Solution with quadratic recovery, CFL=0.5 (left) and CFL=0.9

$$\text{CFL} = 0.5 \quad : \quad -0.000004 \le \overline{u} \le 0.594$$
$$\text{CFL} = 0.9 \quad : \quad 0.000001 \le \overline{u} \le 0.660.$$

Clearly, the small CFL number leads to solutions which are inferior as compared with the higher one. This stems from the fact that more projections onto piecewise constants are necessary to reach a fixed time.

Recovery algorithms for the Euler or Navier-Stokes equations can be constructed along the lines described above. Here we show examples concerning the Euler equations where the primitive variables are recovered. Let us consider the flow in a channel with forward facing step explained in detail later. If we apply the linear recovery after Algorithm II we get the coarse grid solution shown in figure 3.40. To compare with the results of the basic finite volume schemes as described above we again applied the numerical flux functions of Steger and Warming as well as the one of Osher and Solomon. On the fine grid we get the solutions as shown in figure 3.41. If these results are compared with the solutions of the basic finite volume methods above the advantage of this simple linear polynomial recovery can be clearly observed.

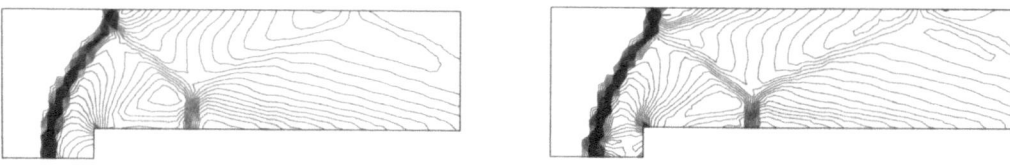

Figure 3.40 Mach number distribution on the coarse grid. Numerical flux function of Steger and Warming (left) and Osher and Solomon.

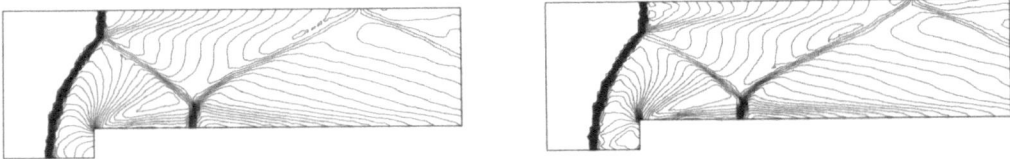

Figure 3.41 Mach number distribution on the fine grid. Numerical flux function of Steger and Warming (left) and Osher and Solomon.

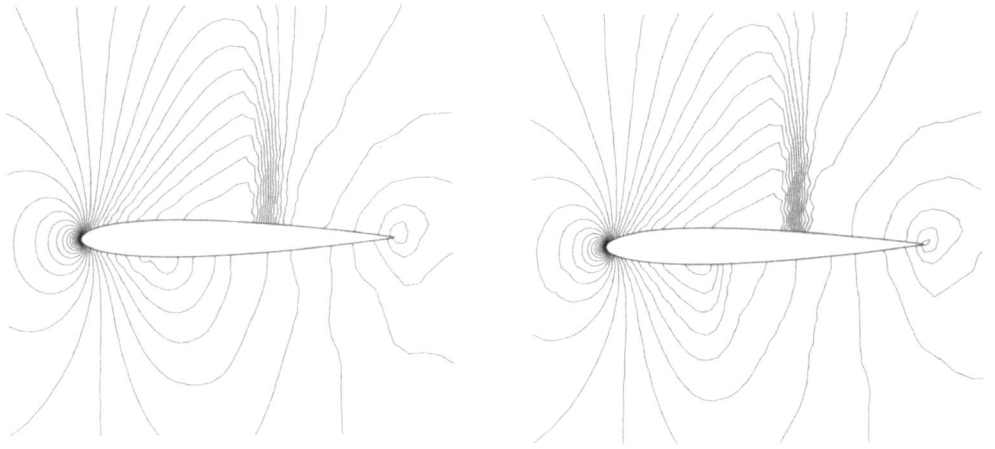

Figure 3.42 Mach number distribution. Numerical flux of Steger and Warming (left) and Osher and Solomon

As a second example we discuss again the transonic flow about an NACA0012 profile as introduced above. Figure 3.42 shows the Mach number distribution for Steger-Warming and Osher-Solomon flux if the simple recovery following Algorithm II is used. Compared to the solutions of the basic finite volume schemes shown in above it is clearly seen that the recovered solutions are superior.

3.6.4   Recovery for box methods

In box methods one can use the duality of primary and secondary grid to construct recovery algorithms of ENO-type. Let us start with the simple example of linear polynomial recovery, i.e. on box σ_i a linear polynomial is sought. We use the notation introduced in above, i.e. $K_{h,i}$ denotes the set of all triangles sharing node i. On each $T \in K_{h,i}$ we can construct a linear polynomial

$$\pi_T(x) = a_{10}^T(x_1 - c_{i,1}) + a_{01}^T(x_2 - c_{i,2}) + a_{00}$$

where c_i again denotes the barycentre of σ_i. The conditions to compute the coefficients are given by

$$\begin{aligned}
\mathfrak{A}(\sigma_i)\pi_T &= \mathfrak{A}(\sigma_i)u \\
\mathfrak{A}(\sigma_i^{T,1})\pi_T &= \mathfrak{A}(\sigma_i^{T,1})u \\
\mathfrak{A}(\sigma_i^{T,2})\pi_T &= \mathfrak{A}(\sigma_i^{T,2})u
\end{aligned}$$

where $\sigma_i^{T,1}, \sigma_i^{T,2}$ denote the other two control volumes which span the triangle T. Out of the set of polynomials π_T we choose the recovery polynomial π_i for σ_i as the one satisfying

$$|\nabla \pi_i| = \min_{T \in K_{h,i}} |\nabla \pi_T|.$$

If this simple ENO scheme is applied to the flow in the channel with forward facing step we get the results as shown in figure 3.43 If a still finer mesh of 14297 grid points is

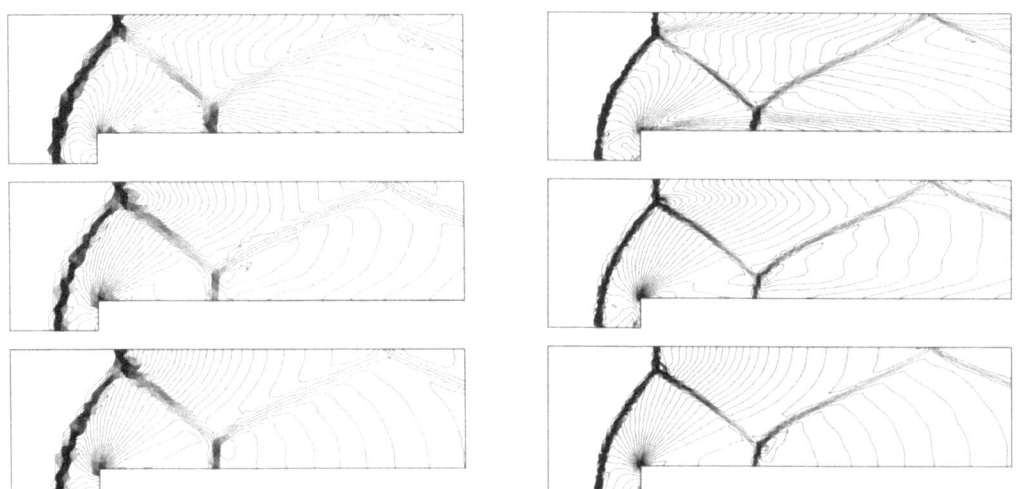

Figure 3.43 Mach number, density and pressure (top to bottom) on the coarse (left) and fine grid (grid shown above)

used we can identify each flow phenomenon in the numerical solution shown in figure 3.44. Applying this scheme to the transonic flow about the NACA0012 (cp. above for the onflow data and the grid) results in the Mach number distribution shown in figure 3.45.

In order to increase the accuracy Abgrall developed in [16] - [19] algorithms based on the approximation theory of finite element methods. We do not want to go into the details of the theory but instead describe the construction of a method of third order which uses quadratic polynomials. We start as in the linear polynomial recovery by constructing a linear polynomial on $K_{h,i}$ which is least oscillating in the sense described above. The

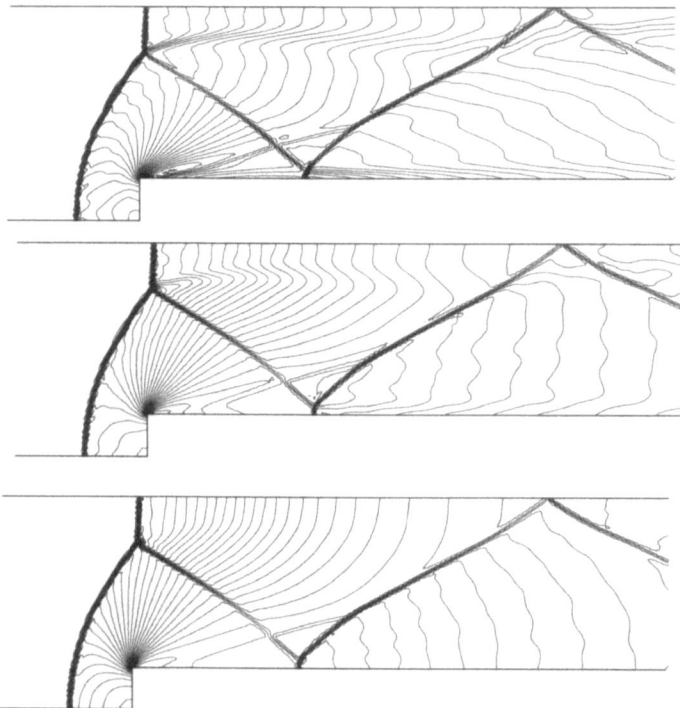

Figure 3.44 Mach number, density and pressure (top to bottom) on a grid with 14297 points

triangle on which this linear polynomial lives is called $T_{\min}$. Abgralls idea is then to construct three node sets for quadratic recovery starting from $T_{\min}$. These three node sets are shown in figure 3.46. Then one constructs a quadratic polynomial on each of the node sets using the conditions that the cell averages have to be conserved on each participating control volume.

It was Abgrall himself who noted that this algorithm is not only very expensive (the linear polynomial is only used to determine $T_{\min}$ and then thrown away) but that the linear systems occuring in the computation of the coefficients of the polynomial are badly conditioned. He moved to the barycentric coordinates of $T_{\min}$ to describe a quadratic polynomial, i.e. he uses the coordinate system

$$\lambda_i(x) = \frac{1}{2|T_{\min}|} \left((x_{k,1} - x_{j,1})(x_2 - x_{j,1}) - (x_1 - x_{j,1})(x_{k,2} - x_{j,2}) \right),$$

$x_\ell = (x_{\ell,1}, x_{\ell,2})$ denoting a node of T_{min} and (i,j,k) cyclic. The k-th quadratic polynomial can then be described as

$$\pi_i^{(k)}(x) = \sum_{m=1}^{3} \left(a_m^{(k)} \lambda_m(x) + \sum_{n>m} A_{mn}^{(k)} \lambda_m(x) \lambda_n(x) \right)$$

and Abgrall could prove that the coefficient matrix of the system

$$\mathfrak{A}(\sigma_j) \pi_i^{(k)} = \overline{u}_j, \quad \sigma_j \in K_k(\sigma_i)$$

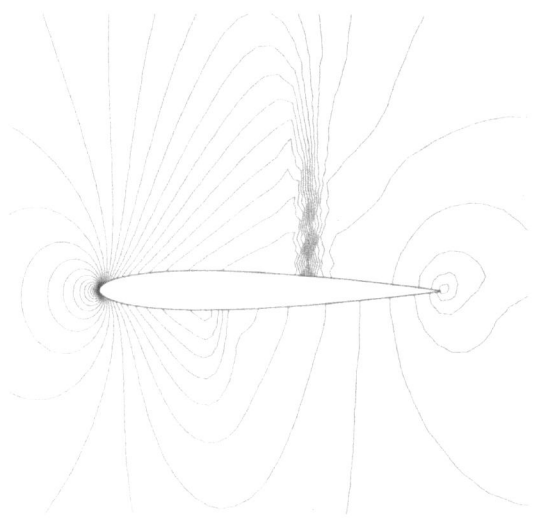

Figure 3.45 Mach number distribution

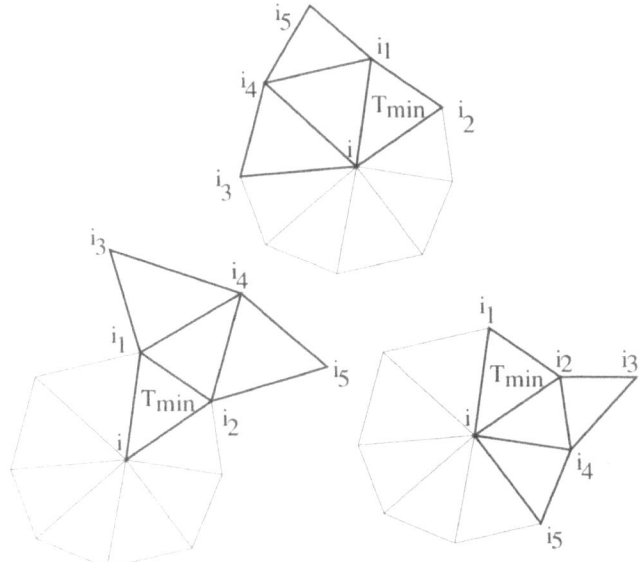

Figure 3.46 Three possible sets $\mathcal{K}(\sigma_i)$ for quadratic recovery

has condition number of order 1. To tackle the problem of economic computation of the quadratic polynomial Abgrall realised that the linear system for the unknown coefficients can be split into two subsystems of size 3×3. To see this we look at the difference

$$\pi_i^{(k),2}(x) - \pi_i^{(k),1} = \sum_{m=1}^{3} \left(a_m^{(k),'} \lambda_m(x) + \sum_{n>M} A_{mn}^{(k),2} \lambda_m(x) \lambda_n(x) \right)$$

between the quadratic polynomial $\pi_i^{(k),2}(x) = \sum_{m=1}^{3} \left(a_m^{(k),2} \lambda_m(x) + \sum_{n>m} A_{mn}^{(k),2} \lambda_m(x) \lambda_n(x) \right)$ and the already known polynomial of degree one $\pi_i^{(k),1}(x) = \sum_{m=1}^{3} a_m^{(k),1} \lambda_m(x)$ on the triangle $T_{\min}$. Thus,

$$a_m^{(k),'} := \left(a_m^{(k),2} - a_m^{(k),1} \right), \quad m = 1,2,3.$$

Taking cell averages of the difference results in the system

$$M \begin{bmatrix} a_1^{(k),'} \\ a_2^{(k),'} \\ a_3^{(k),'} \\ A_{12}^{(k),2} \\ A_{13}^{(k),2} \\ A_{23}^{(k),2} \end{bmatrix} = \begin{bmatrix} \overline{\overline{u}}_i \\ \overline{\overline{u}}_{i_1} \\ \overline{\overline{u}}_{i_2} \\ \overline{\overline{u}}_{i_3} \\ \overline{\overline{u}}_{i_4} \\ \overline{\overline{u}}_{i_5} \end{bmatrix},$$

the matrix M given by

$$M := \begin{bmatrix} \mathfrak{A}(\sigma_i)\lambda_1 & \mathfrak{A}(\sigma_i)\lambda_2 & \mathfrak{A}(\sigma_i)\lambda_3 & \mathfrak{A}(\sigma_i)(\lambda_1\lambda_2) & \mathfrak{A}(\sigma_i)(\lambda_1\lambda_3) & \mathfrak{A}(\sigma_i)(\lambda_2\lambda_3) \\ \mathfrak{A}(\sigma_{i_1})\lambda_1 & \mathfrak{A}(\sigma_{i_1})\lambda_2 & \mathfrak{A}(\sigma_{i_1})\lambda_3 & \mathfrak{A}(\sigma_{i_1})(\lambda_1\lambda_2) & \mathfrak{A}(\sigma_{i_1})(\lambda_1\lambda_3) & \mathfrak{A}(\sigma_{i_1})(\lambda_2\lambda_3) \\ \vdots & \vdots & \vdots & \vdots & \vdots & \vdots \\ \mathfrak{A}(\sigma_{i_5})\lambda_1 & \mathfrak{A}(\sigma_{i_5})\lambda_2 & \mathfrak{A}(\sigma_{i_5})\lambda_3 & \mathfrak{A}(\sigma_{i_5})(\lambda_1\lambda_2) & \mathfrak{A}(\sigma_{i_5})(\lambda_1\lambda_3) & \mathfrak{A}(\sigma_{i_5})(\lambda_2\lambda_3) \end{bmatrix}.$$

Since $\mathfrak{A}(\sigma_{\sigma_j})\pi_i^{(k),2} = \mathfrak{A}(\sigma_{\sigma_j})\pi_i^{(k),1}$ for $j \in \{i,i_1,i_2\}$, it follows that $\overline{\overline{u}}_j = 0$ for $j \in \{i,i_1,i_2\}$. Blocking the matrix M in the form

$$M = \begin{bmatrix} M_{11} & M_{12} \\ M_{21} & M_{22} \end{bmatrix}$$

where

$$M_{11} = \begin{bmatrix} \mathfrak{A}(\sigma_i)\lambda_1 & \mathfrak{A}(\sigma_i)\lambda_2 & \mathfrak{A}(\sigma_i)\lambda_3 \\ \mathfrak{A}(\sigma_{i_1})\lambda_1 & \mathfrak{A}(\sigma_{i_1})\lambda_2 & \mathfrak{A}(\sigma_{i_1})\lambda_3 \\ \mathfrak{A}(\sigma_{i_2})\lambda_1 & \mathfrak{A}(\sigma_{i_2})\lambda_2 & \mathfrak{A}(\sigma_{i_2})\lambda_3 \end{bmatrix},$$

$$M_{12} = \begin{bmatrix} \mathfrak{A}(\sigma_i)(\lambda_1\lambda_2) & \mathfrak{A}(\sigma_i)(\lambda_1\lambda_3) & \mathfrak{A}(\sigma_i)(\lambda_2\lambda_3) \\ \mathfrak{A}(\sigma_{i_1})(\lambda_1\lambda_2) & \mathfrak{A}(\sigma_{i_1})(\lambda_1\lambda_3) & \mathfrak{A}(\sigma_{i_1})(\lambda_2\lambda_3) \\ \mathfrak{A}(\sigma_{i_2})(\lambda_1\lambda_2) & \mathfrak{A}(\sigma_{i_2})(\lambda_1\lambda_3) & \mathfrak{A}(\sigma_{i_2})(\lambda_2\lambda_3) \end{bmatrix},$$

$$M_{21} = \begin{bmatrix} \mathfrak{A}(\sigma_{i_3})\lambda_1 & \mathfrak{A}(\sigma_{i_3})\lambda_2 & \mathfrak{A}(\sigma_{i_3})\lambda_3 \\ \mathfrak{A}(\sigma_{i_4})\lambda_1 & \mathfrak{A}(\sigma_{i_4})\lambda_2 & \mathfrak{A}(\sigma_{i_4})\lambda_3 \\ \mathfrak{A}(\sigma_{i_5})\lambda_1 & \mathfrak{A}(\sigma_{i_5})\lambda_2 & \mathfrak{A}(\sigma_{i_5})\lambda_3 \end{bmatrix},$$

$$M_{22} = \left[\begin{array}{ccc} \mathfrak{A}(\sigma_{i_3})(\lambda_1\lambda_2) & \mathfrak{A}(\sigma_{i_3})(\lambda_1\lambda_3) & \mathfrak{A}(\sigma_{i_3})(\lambda_2\lambda_3) \\ \mathfrak{A}(\sigma_{i_4})(\lambda_1\lambda_2) & \mathfrak{A}(\sigma_{i_4})(\lambda_1\lambda_3) & \mathfrak{A}(\sigma_{i_4})(\lambda_2\lambda_3) \\ \mathfrak{A}(\sigma_{i_5})(\lambda_1\lambda_2) & \mathfrak{A}(\sigma_{i_5})(\lambda_1\lambda_3) & \mathfrak{A}(\sigma_{i_5})(\lambda_2\lambda_3) \end{array} \right],$$

we can conclude that M_{11} is invertible. From

$$\left[\begin{array}{cc} M_{11} & M_{12} \\ M_{21} & M_{22} \end{array} \right] \left[\begin{array}{c} a^{(k),'} \\ A^{(k),2} \end{array} \right] = \left[\begin{array}{c} 0 \\ \bar{\bar{u}} \end{array} \right],$$

where $a^{(k),'} := (a_1^{(k),'}, a_2^{(k),'}, a_3^{(k),'})^T$, $A^{(k),'} := (A_{12}^{(k),'}, A_{13}^{(k),'}, A_{23}^{(k),'})^T$, $\bar{\bar{u}} := (\bar{\bar{u}}_{i_3}, \bar{\bar{u}}_{i_4}, \bar{\bar{u}}_{i_5})^T$ and 0 denotes the null vector in $\mathbb{R}^3$, we further conclude that

$$a^{(k),'} = -M_{11}^{-1} M_{12} A^{(k),2}.$$

Inserting this equation into the second equation of the block results in

$$M_{21} a^{(k),'} + M_{22} A^{(k),2} = \left[-M_{21} M_{11}^{-1} M_{12} + M_{22} \right] A^{(k),2} = \bar{\bar{u}},$$

in which the coefficient matrix is also invertible due to the geometrical properties of the control volumes in $K_k(\sigma_i)$. We arrive at the following algorithm:

1. Compute the three components $A^{(k),2}$ by solving

$$\left[-M_{21} M_{11}^{-1} M_{12} + M_{22} \right] A^{(k),2} = \bar{\bar{u}}.$$

   This requires the inversion of a 3×3 matrix.

2. Compute the first three coefficients of the quadratic recovery polynomial

$$a_m^{(k),2} = a_m^{(k),'} + a_m^{(k),1}, \quad m = 1,2,3.$$

3. Compute the remaining three coefficients by

$$a^{(k),'} = -M_{21} M_{11}^{-1} M_{12} A^{(k),2}.$$

Note that the process described here can be completely generalised by using expansions which mimic the behaviour of one-dimensional Newton polynomials in multiple space dimensions. In [20] these expansions were developed as Mühlbach expansions.

3.7 WENO approximations

On each cell σ_i we have to compute a set of recovery polynomials $\pi_i^{(k)}$ where k denotes the number of the stencil. While in the ENO approximations only one recovery polynomial is chosen we may compute a weighted sum

$$\pi_i := \sum_k \Omega_k \pi_i^{(k)}$$

where the Ω_k are weights with $\sum_k \Omega_k = 1$. An oscillation indicator OI is used to compute the weights in the following manner: If a Sobolev seminorm

$$OI(\pi_i^{(k)}) := \|\nabla \pi_i^{(k)}\|_{L^2(\sigma_i)}$$

is used as an oscillation indicator the weights are computed from

$$\Omega_k := \frac{\Omega(\varepsilon + OI(\pi_i^{(k)}))^{-\beta}}{\sum_j \Omega(\varepsilon + OI(\pi_i^{(j)}))^{-\beta}}$$

Here, Ω is a weight independent of the weights Ω_j which allows a different weighting of different stencils. The parameter ε is chosen to avoid the division by zero and β is a measure of sensitivity of the weights on the oscillation indicator. We set $\varepsilon = 10^{-16}$, $\beta := 8$ and $\Omega = 12$ for a central stencil while $\Omega = 1$ for a one-sided stencil.

In order to document the power of the WENO approach on unstructured grids we consider inviscid flow over a diamond-shaped body shown in figure 3.47 with isolines of density. At time $t = 0$ a shock with Mach number 2.8 starts at the left boundary ($x = -0.3$) of the computational domain. The diamond starts at $x = 0, y = 0$, the top is at $(1/\sqrt{2}, 1/\sqrt{2})$. The air in front of the shock has density $\rho = 1.4$ and pressure $p = 1$. Recovery is done employing polynomials of degree two and an adaptation algorithm was used as described in detail in [45]. The numerical flux function chosen is a Lax-Friedrichs flux.

The numerical flow conditions are most extreme when the shock reaches the top of the diamond and passes it. Experiences with ENO schemes have indicated that quite often a numerical vacuum state occurs which immediately leads to negative density. This is not the case with the WENO method which is much more robust though accurate

A closer look at the solution at time $t = 0.56$ in figure 3.48 reveals that the WENO scheme is in fact able to accurately resolve even weak features in the flow.

3.8   The theory of optimal recovery

While the big advantage of polynomial recovery is its simplicity there are some problems coming naturally with this kind of recovery. If higher degree polynomials are sought the number of coefficients increase dramatically. This corresponds to a recovery algorithm in which data from 'far away' is interpolated to the control volume under consideration and may spoil the accuracy of the recovery instead of increasing the accuracy. In ENO approximations of one-dimensional conservation laws one observes a breakdown of accuracy if the order of the recovery polynomials exceeds six.

These simple observations were the starting point for the application of the theory of optimal recovery to problems of recovery in finite volume approximations. In a sequence of papers [29], [30], [31] Micchelli and Rivlin developed the following framework. Suppose you want to know the value of a linear functional (the feature operator)

$$\mathfrak{F} : \begin{cases} V & \longrightarrow \mathbb{R} \\ u & \longmapsto \mathfrak{F}u \end{cases}$$

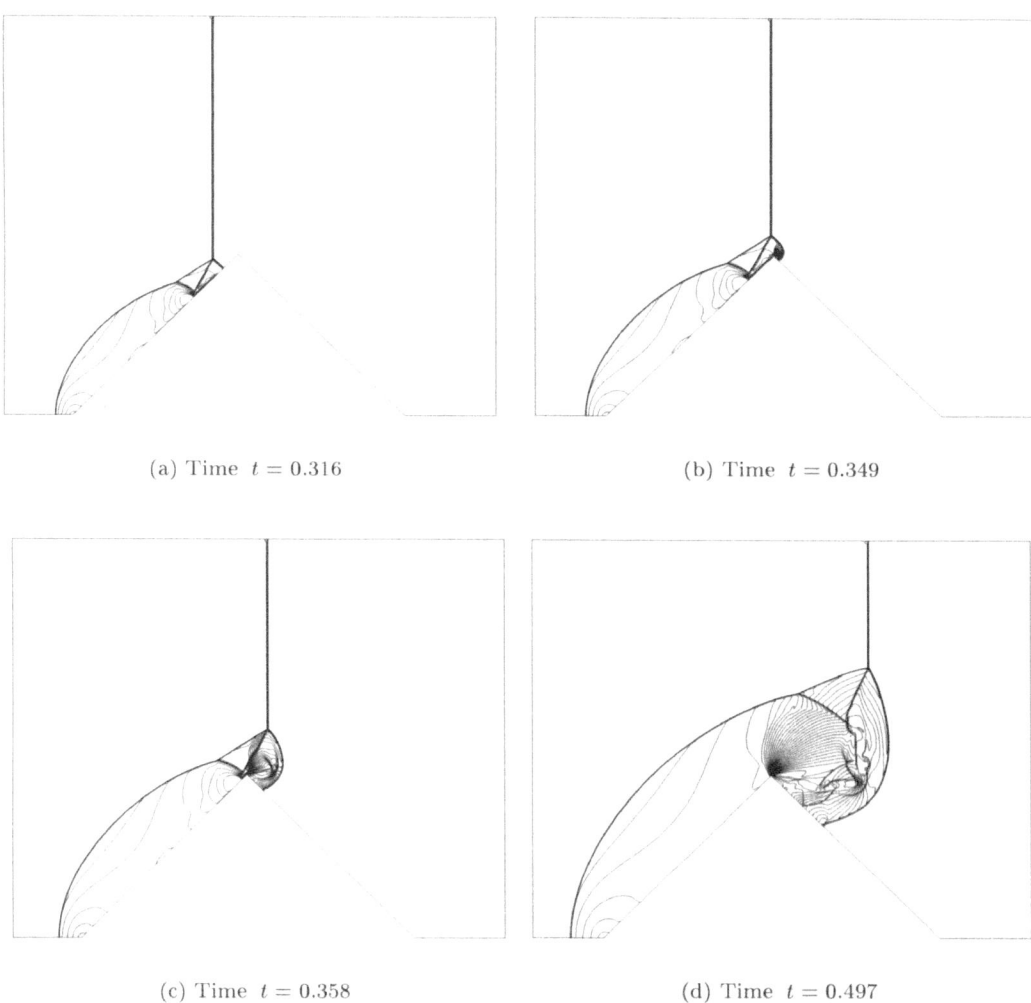

(a) Time $t = 0.316$ (b) Time $t = 0.349$

(c) Time $t = 0.358$ (d) Time $t = 0.497$

Figure 3.47 Shock hitting a diamond-shaped body.

applied to an unkonwn function u living in a function space V. Suppose further that you know M values of linear functionals

$$\lambda_i : \left\{ \begin{array}{ccc} V & \longrightarrow & \mathbb{R} \\ u & \longmapsto & \lambda_i u \end{array} \right. , \quad i = 0, \ldots, M-1,$$

where we assume the linear functionals λ_i and $\mathfrak{F}$ to be pairwise linearly independent. We gather together our linear functionals λ_i in the information operator

$$\mathfrak{I} := (\lambda_0, \ldots, \lambda_{M-1}).$$

Since u is unknown we can only hope to get a good approximation to $\mathfrak{F}u$ by using the given information, i.e. we have to construct an operator

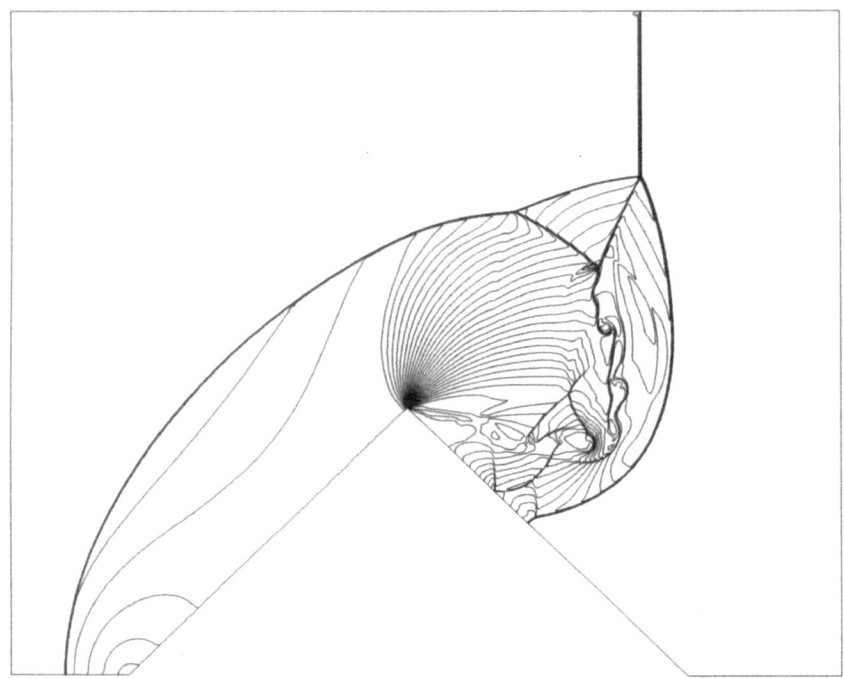

Figure 3.48 Density at time $t = 0.56$.

$$\mathfrak{R} : \begin{cases} \mathbb{R}^M & \longrightarrow & \mathbb{R} \\ \mathfrak{I}u & \longmapsto & \lambda_i \mathfrak{R}\mathfrak{I}u \end{cases}$$

so that the error $|\mathfrak{F}u - \mathfrak{R}\mathfrak{I}u|$ is as small as possible. We would like to call such an operator an optimal recovery operator and the value $\mathfrak{R}\mathfrak{I}u$ is then the optimal recovery of $\mathfrak{F}u$.

If we apply this framework to our specific problem the information operator consists of the cell average operators on the cells in the node set, i.e.

$$\mathfrak{I} := (\mathfrak{A}(\sigma_{i_0}), \mathfrak{A}(\sigma_{i_1}), \dots, \mathfrak{A}(\sigma_{i_{M-1}})),$$

while the feature operator is simply the point evaluation functional

$$\mathfrak{F}u := \delta_x u := u(x)$$

at the Gauss points x of the quadrature rule.

We don't want to tackle the recovery problem in its full generality but instead describe the basic geometric ideas of Golomb and Weinberger ([22]) which clearly indicate the character of optimal recovery operators.

3.8.1   Basic notions

Golomb and Weinberger already noticed in [22] that the knowledge of M linear functionals is not sufficient to solve the recovery problem. This can be seen as follows if we agree on V being a Hilbert space. The equation

$$\mathfrak{I}u = \bar{u}$$

describing the cell averages on all cells in a node set is the equation of a hyperplane of codimension n [8]. Since all linear functionals are supposed to be linearly independent there is a function $w \in V$ such that

$$\lambda_0 w = 0$$
$$\vdots \quad \vdots \quad \vdots$$
$$\lambda_{M-1} w = 0$$
$$\mathfrak{F}w = 1$$

holds. Otherwise the hyperplane given by the equation $\mathfrak{F}w = 1$ would be parallel to one of the hyperplanes defined by the information operator. Now the function $v + \xi w$ is an element of V for all $v \in V$ and all $\xi \in \mathbb{R}$ and for $v = u$ satisfies the equation

$$\mathfrak{I}(u + \xi w) = \mathfrak{I}u + \xi \mathfrak{I}w = \mathfrak{I}u.$$

On the other hand we know that

$$\mathfrak{F}(u + \xi w) = \mathfrak{F}u + \xi \mathfrak{F}w = \mathfrak{F}u + \xi,$$

i.e.

$$\mathfrak{F}u = \mathfrak{F}(u + \xi w) - \xi.$$

Since $\xi \in \mathbb{R}$ was arbitrary we see that unique recovery is not possible at all.

This simple observation led Golomb and Weinberger to the conclusion that a nonlinear constraint is needed which can be found in the condition that only those functions of V are considered which live in the sphere

$$U := \{v \in V \mid \|v\|_V \leq 1\}.$$

In a more general setting the sphere is generated by a restriction operator $\mathfrak{T}$ such that

$$U := \{v \in V \mid \|\mathfrak{T}v\|_V \leq 1\}.$$

One could think of $\mathfrak{T}$ being a differential operator for example. Obviously, we have $U \subset V$ and thus our recovery problem can be cast in the form shown in figure 3.49. The fascinating geometric idea published by Golomb and Weinberger can then be described as follows. We already know that the equations

$$\mathfrak{I}u = \bar{u}$$

define a hyperplane via the set

$$\{v \in V \mid \mathfrak{I}v = \mathfrak{I}u\}.$$

If we further employ the condition

---

[8] Since $\mathfrak{I} = (\lambda_0, \ldots, \lambda_{M-1})$ and each λ_i is a linear functional in V the Fréchet-Riesz theorem guarantees the existence of functions $g_i \in V$ such that $\lambda_i u = (g_i, u)_V$, where $(\cdot, \cdot)_V$ denotes the inner product of V.

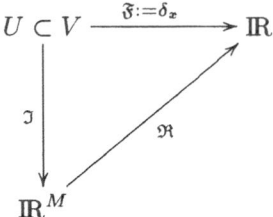

Figure 3.49 Mappings in the problem of optimal recovery

$$u \in U = \{v \in V \mid \|v\|_V \leq 1\}$$

the set U corresponds to a sphere. If our problem has a solution at all, sphere and hyperplane intersect and form a hypercircle

$$B(u) := \{v \in V \mid \Im v = \Im u\} \cap U = \{v \in U \mid \Im v = \Im u\}$$

in the sense of Prager and Synge [36]. It is now easy to see that the centre z_u of the hypercircle must be the optimal recovery of $\Im u$ since it has the smallest distance to all other points in the hypercircle.

Let us discuss the problem of optimal recovery in an abstract sense and discuss all relevant topics in geometric terms using figure 3.50[9]. The kernel of the information operator $\Im$ defined as

$$\ker \Im := \{v \in V \mid \Im v = 0\}$$

is the hyperplane through the origin which is parallel to $\{v \in V \mid \Im u = \Im v\}$. For the centre z_u it holds the characterisation

$$\forall v \in \ker \Im : \quad (z_u, v)_V = 0,$$

i.e. z_u lies in the orthogonal complement of $\ker \Im$, which we write as

$$z_u \in (\ker \Im)^\perp.$$

At the same time z_u is the point of $B(u)$ with smallest distance to the origin, i.e. characterised through

$$\|z_u\|_V^2 = (z_u, z_u)_V = \inf_{\Im v = \Im u} (v, v)_V = \inf_{\Im v = \Im u} \|v\|_V^2.$$

Obviously, the characterisation

$$\forall v \in \ker \Im : \quad (z_u, v)_V \;=\; 0$$
$$\Im z_u \;=\; \Im u$$

is already enough to fix z_u uniquely.

Now the feature operator $\Im$ is assumed to be a bounded linear functional. Let $y \in V$ be that element with unit norm which satisfies

---

[9] I gratefully acknowledge the help of Daniel Hempel who hand-coded (!) this figure in postscript.

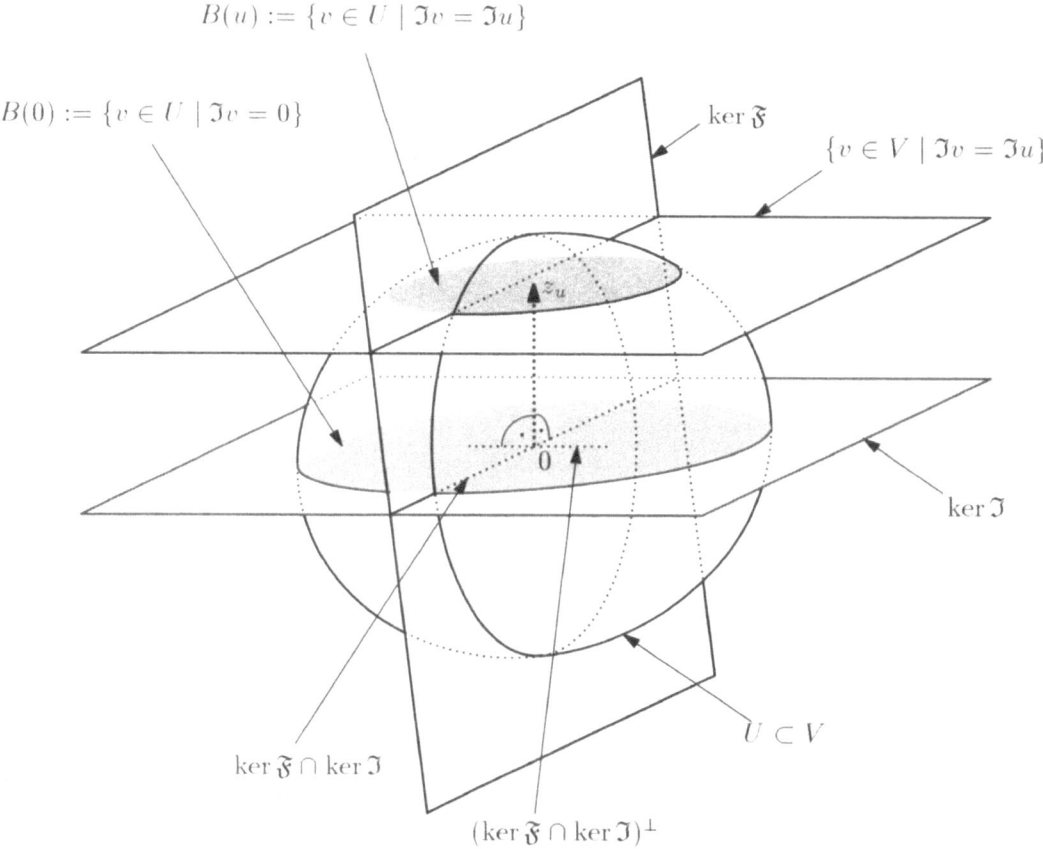

$$B(u) := \{v \in U \mid \Im v = \Im u\}$$

$$B(0) := \{v \in U \mid \Im v = 0\}$$

ker $\mathfrak{F}$

$$\{v \in V \mid \Im v = \Im u\}$$

z_u

0

ker $\Im$

ker $\mathfrak{F} \cap$ ker $\Im$

$U \subset V$

$(\text{ker } \mathfrak{F} \cap \text{ker } \Im)^{\perp}$

Figure 3.50 Sphere, hyperplane and hypercircle in Hilbert space

$$\mathfrak{F}y = \sup_{\substack{v \in \text{ker } \Im \\ \|v\|_V = 1}} |\mathfrak{F}v|_V.$$

The set

$$\text{ker } \mathfrak{F} = \{v \in V \mid \mathfrak{F}v = 0\}$$

is also a hyperplane (of codimension 1) through the origin. The intersection

$$\text{ker } \mathfrak{F} \cap \text{ker } \Im$$

then contains all elements $v \in V$ which annihilate $\mathfrak{F}$ as well as $\Im$ (i.e. $\lambda_0, \ldots, \lambda_{M-1}$). Since $\mathfrak{F}$ is annihilated along the straight line $\text{ker } \mathfrak{F} \cap \text{ker } \Im$ maximal values of $\mathfrak{F}$ can only occur orthogonal to that straight line, i.e. the element y is additionally characterised by

$$y \in \text{ker } \Im \cap (\text{ker } \mathfrak{F} \cap \text{ker } \Im)^{\perp}, \quad (y,y)_V = 1.$$

We have shown that any $v \in B(u)$ is spanned by the linear combination of three vectors, namely

$$v = z_u + \frac{\mathfrak{F}v - \mathfrak{F}z_u}{\mathfrak{F}y}y + w, \quad w \in \ker \mathfrak{F} \cap \ker \mathfrak{I}.$$

The strange-looking coefficients stem from the fact that in case $v = z_u$ we have $z_u = z_u + w$, i.e. $w = 0$, and after application of $\mathfrak{F}$ we need to have

$$
\begin{aligned}
\mathfrak{F}v &= \mathfrak{F}z_u + \frac{\mathfrak{F}v - \mathfrak{F}z_u}{\mathfrak{F}y}\mathfrak{F}y + \underbrace{\mathfrak{F}w}_{=0} \\
&= \mathfrak{F}z_u + \mathfrak{F}v - \mathfrak{F}z_u = \mathfrak{F}v.
\end{aligned}
$$

Due to their construction the three vectors z_u, y and w are pairwise orthogonal. Therefore we have

$$
\begin{aligned}
\forall v \in B(u): \quad (v,v)_V &= \left(z_u + \frac{\mathfrak{F}v - \mathfrak{F}z_u}{\mathfrak{F}y}y + w, z_u + \frac{\mathfrak{F}v - \mathfrak{F}z_u}{\mathfrak{F}y}y + w\right)_V \\
&= (z_u,z_u)_V + \left(\frac{\mathfrak{F}v - \mathfrak{F}z_u}{\mathfrak{F}y}\right)^2 \underbrace{(y,y)_V}_{=1} + (w,w)_V \\
&\geq (z_u,z_u)_V + \left(\frac{\mathfrak{F}v - \mathfrak{F}z_u}{\mathfrak{F}y}\right)^2.
\end{aligned}
$$

In case $v \in (\ker \mathfrak{F} \cap \ker \mathfrak{I})^{\perp}$ no component in the direction of $\ker \mathfrak{F} \cap \ker \mathfrak{I}$ is necessary and equality holds. For the unknown function u we therefore get the result

$$(u,u)_V \geq (z_u,z_u)_V + \left(\frac{\mathfrak{F}u - \mathfrak{F}z_u}{\mathfrak{F}y}\right)^2,$$

which is nothing but

$$(\mathfrak{F}y)^2 \left[(u,u)_V - (z_u,z_u)_V\right] \geq (\mathfrak{F}u - \mathfrak{F}z_u)^2,$$

and, since $(u,u)_V \leq 1$,

$$(\mathfrak{F}u - \mathfrak{F}z_u)^2 \leq (\mathfrak{F}y)^2 \left[1 - (z_u,z_u)_V\right].$$

Taking square roots yields

$$-(\mathfrak{F}y)^2 \left[1 - (z_u,z_u)_V\right]^{1/2} \leq \mathfrak{F}u - \mathfrak{F}z_u \leq (\mathfrak{F}y)^2 \left[1 - (z_u,z_u)_V\right]^{1/2}.$$

This means, that we get the estimate

$$\mathfrak{F}z_u - (\mathfrak{F}y)^2 \left[1 - (z_u,z_u)_V\right]^{1/2} \leq \mathfrak{F}u \leq \mathfrak{F}z_u + (\mathfrak{F}y)^2 \left[1 - (z_u,z_u)_V\right]^{1/2}.$$

Note that the estimate is sharp since the functions

$$u = z_u \pm (1 - (z_u,z_u)_V)^{1/2}$$

lie in $B(u)$ and lead to equality. Thus, the set of possible values of $\mathfrak{F}u$ is an interval of length

$$2(\mathfrak{F}y)(1 - (z_u,z_u)_V)^{1/2}$$

whose centre is z_u. Hence z_u is the optimal recovery of $\mathfrak{F}u$ since it has the smallest distance to all other points in the interval.

The question of explicit formulae for the centre of the hypercircle was also already answered by Golomb and Weinberger. Although the word *Spline* did not occur in their paper (due to its age) all their computations are closely related to Spline algorithm which we want to discuss now.

3.8.2   Splines

We do not want to go into detail in the discussion of the theory of optimal recovery but refer the reader to the papers [29] and [30] as well as to [33], [34], [35]. It turns out that Splines are centres of hypercircles in certain semi-Hilbert spaces V. By a semi-Hilbert we mean a space with seminorm $|\cdot|_V$ for which there may exist nontrivial functions $w \in V$ for which $|w|_V = 0$ holds (i.e. the seminorm has a 'hole', or, mathematically more sophisticated: The kernel $\ker|\cdot|_V = \{v \in V \mid |v|_V = 0\}$ contains more than the null function). A Spline in a semi-Hilbert space is then defined to be a function $\Phi \in V$ which minimises the seminorm, i.e. for which

$$|\Phi|_V = \inf_{\substack{v \in V \\ \Im v = \Im u}} |v|_V$$

holds. We all know that Splines in 1-D are given by piecewise polynomial expressions which satisfy certain compatibility conditions at the nodes where different expressions fit together. The famous cubic Spline, consisting of pieces of cubic polynomials which fit together continuously differentiable, lives in a Sobolew space where the seminorm is given in terms of second derivatives. Here linear polynomials are in the kernel of this seminorm. If one asks for the direct generalisation of the 1-D cubic Spline in multiple space dimensions one gets a surprising answer. The 'cubic Spline' in 2-D is the so-called Thin Plate Spline (because it is the solution of the biharmonic equation). In our setting of recovery from cell average data it is given by

$$\Phi(x) = \sum_{j=0}^{M-1} \alpha_j \mathfrak{A}^y(\sigma_{i_j})\left(|x-y|^2 \log(|x-y|)\right) + v(x),$$

where $\mathfrak{A}^y$ stands for the application of the cell average operator with respect to the variable y and v are functions out of the kernel of the seminorm. As in the case of the 1-D cubic Spline linear polynomials constitute this kernel such that we can write

$$\Phi(x) = \sum_{j=0}^{M-1} \alpha_j \mathfrak{A}^y(\sigma_{i_j})\left(|x-y|^2 \log(|x-y|)\right) + a_{01}x_1 + a_{10}x_2 + a_{00}.$$

We now have to determine $M+3$ coefficients $\alpha_0, \ldots, \alpha_{M-1}, a_{10}, a_{01}, a_{00}$ but we have only M conditions given by the information $\Im u = \overline{u}$ on the node set. If we require the condition

$$\forall q \in \ker|\cdot|_V : \quad \sum_{j=0}^{M-1} \alpha_j \mathfrak{A}(\sigma_{i_j})q = 0$$

we get the remaining 3 conditions needed to determine all coefficients[10]. One can easily prove that the Thin Plate Spline is polynomial reproducing, i.e. recovering from three given cell averages results in the linear polynomial which is constructed to fit the hole in the seminorm. Hence, nontrivial Thin Plate Spline recovery algorithms can only be constructed if more than three cells are in a node set. We use four node sets each containing four triangles which are constructed from the neighbourhood shown in figure 3.36.

---

[10] This condition follows easily from the theory of Splines, see [35] for example. We can think of this condition as 'fixing the hole' in the seminorm.

This figure gives rise to one central node set $K_0(T_i) := T_i \cup T_{i_1} \cup T_{i_2} \cup T_{i_3}$ and the four one-sided node sets $K_1(T_i) := T_i \cup T_{i_1} \cup T_{i_{1_1}} \cup T_{i_{1_2}}$, $K_2(T_i) := T_i \cup T_{i_2} \cup T_{i_{2_1}} \cup T_{i_{2_2}}$, and $K_3(T_i) := T_i \cup T_{i_3} \cup T_{i_{3_1}} \cup T_{i_{3_2}}$. On each of the node sets we solve the linear system

$$\mathfrak{A}(T)\Phi = \mathfrak{A}(T), \quad T \in K_j(T_i), j = 0,1,2,3,$$

together with the conditions

$$\sum_{j=0}^{3} \alpha_j \mathfrak{A}(T) = 0$$

$$\sum_{j=0}^{3} \alpha_j \mathfrak{A}(T) x_1 = 0$$

$$\sum_{j=0}^{3} \alpha_j \mathfrak{A}(T) x_2 = 0,$$

which we rewrite in matrix form as

$$N \begin{bmatrix} \alpha_0 \\ \vdots \\ \alpha_3 \end{bmatrix} = \begin{bmatrix} 0 \\ \vdots \\ 0 \end{bmatrix}.$$

Rewriting also the recovery conditions as

$$\tilde{M} \begin{bmatrix} \alpha_0 \\ \vdots \\ \alpha_3 \\ a_{10} \\ a_{01} \\ a_{00} \end{bmatrix} = \begin{bmatrix} \overline{u}_0 \\ \overline{u}_1 \\ \overline{u}_2 \end{bmatrix}$$

we can reformulate the problem of determining the coefficients of the Thin Plate recovery Spline as

$$\begin{bmatrix} M & N^T \\ N & 0 \end{bmatrix} \begin{bmatrix} \alpha_0 \\ \vdots \\ \alpha_3 \\ a_{10} \\ a_{01} \\ a_{00} \end{bmatrix} = \begin{bmatrix} \overline{u}_0 \\ \vdots \\ \overline{u}_3 \\ 0 \\ 0 \\ 0 \end{bmatrix},$$

where we have denoted the cell averages on the triangles in the node set under consideration with $\overline{u}_0$ to $\overline{u}_3$. It is an easy exercise to get explicit expressions for the matrices M and N, since $\tilde{M} = [M|N^T]$. After computing the four Thin Plate Splines on the four node sets the Spline with smalles total variation over T_i is chosen to be the recovery Spline on T_i.

Applying this simple algorithm to our linear model problem results in the solution as shown in figure 3.51. As can be observed these results are superior in comparison with the linear polynomial (and even quadratic polynomial) recovery. For the coarse grid we get

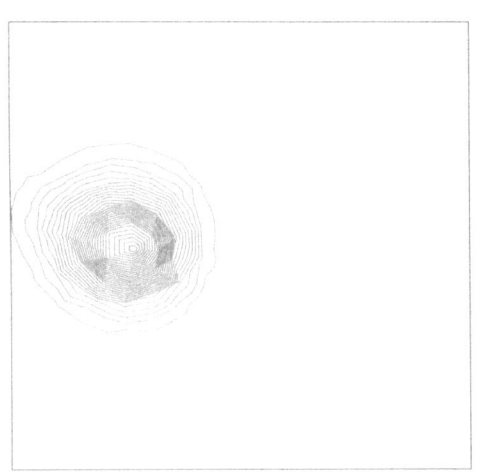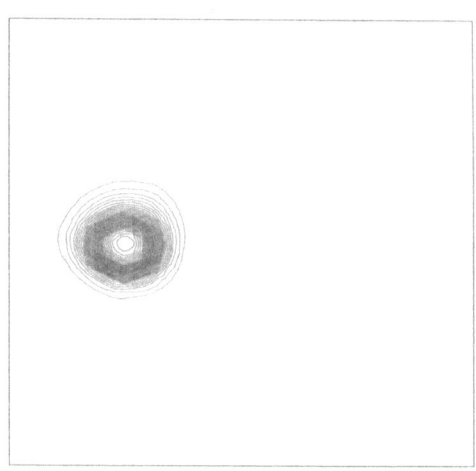

Figure 3.51 Solution with Thin Plate Spline recovery on $\mathcal{T}_{h^1}$ (left) and $\mathcal{T}_{h^2}$.

$$-0.000789 \leq \overline{u} \leq 0.42487$$

while in the case of the finer grid we get

$$0 \leq \overline{u} \leq 0.886.$$

Another advantage is the nearly perfect conservation of shape. In contrast to the polynomial recovery algorithms we can not observe any pollution of the numerical solution.

To finally compare all of the recovery algorithms discussed we choose a very fine grid of 10000 points and 19602 triangles. The algorithms then lead to the results shown in the following table.

$\overline{u}$	basic	DEO	Moore	TPS	quad
min.	0.0	0.0	0.0	0.0	0.0
max.	0.579	0.86	0.878	0.974	0.764

In our notation *basic* denotes the basic finite volume approximation, *DEO* stands for the Durlofsky, Engquist, Osher scheme, *Moore* denotes the linear polynomial recovery on the Moore neighbourhood, *TPS* is the Thin Plate Spline recovery and *quad* is the recovery of quadratic polynomials using a sector search algorithm. Figures 3.52, 3.53 and 3.54 summarize the results.

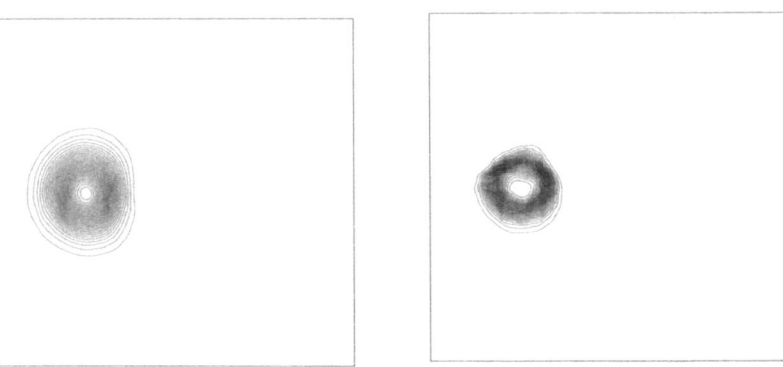

Figure 3.52 Basic FV approximation (left) and linear polynomial recovery after Durlofsky, Engquist, Osher on $\mathcal{T}_{h^3}$.

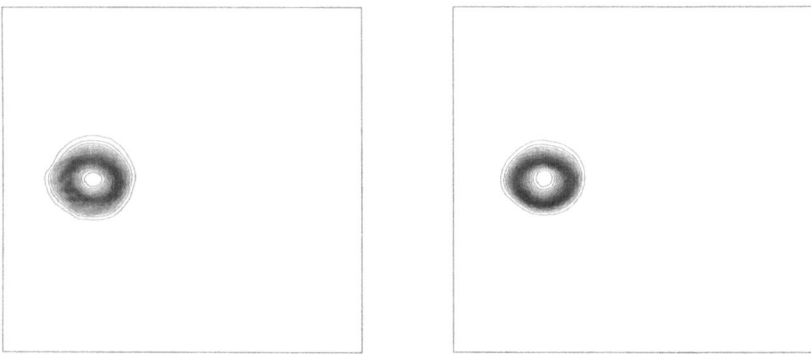

Figure 3.53 Linear polynomial recovery on the Moore neighbourhood on $\mathcal{T}_{h^3}$ (left) and with the Thin Plate Spline.

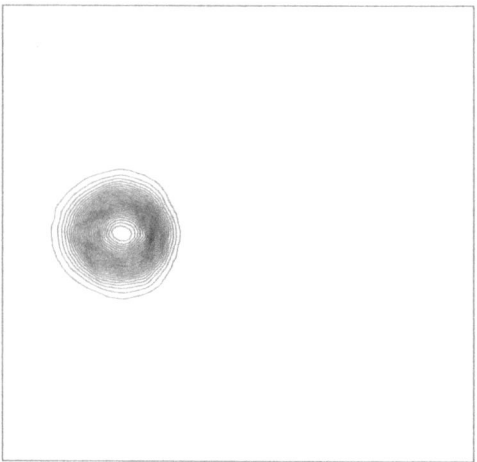

Figure 3.54 Quadratic recovery on $\mathcal{T}_{h^3}$.

The situation in the case of the Euler equations is much less satisfactory but this depends mainly on the fact that the results shown are very preliminary in nature. The Thin Plate Spline tends to oscillate and it was not possible to derive a stable scheme with more than one Gauss point. We hope that other node sets or additional limitation of the values of the Splines will result in better schemes as we know today.

To demonstrate the dependence on the CFL number we did two computations in the channel with forward facing step (for a discussion of this flow field see the section on adaptation). Figure 3.55 compares the results of linear polynomial recovery with Thin Plate Spline recovery in the case of CFL=0.9. The main drawback of the Thin Plate

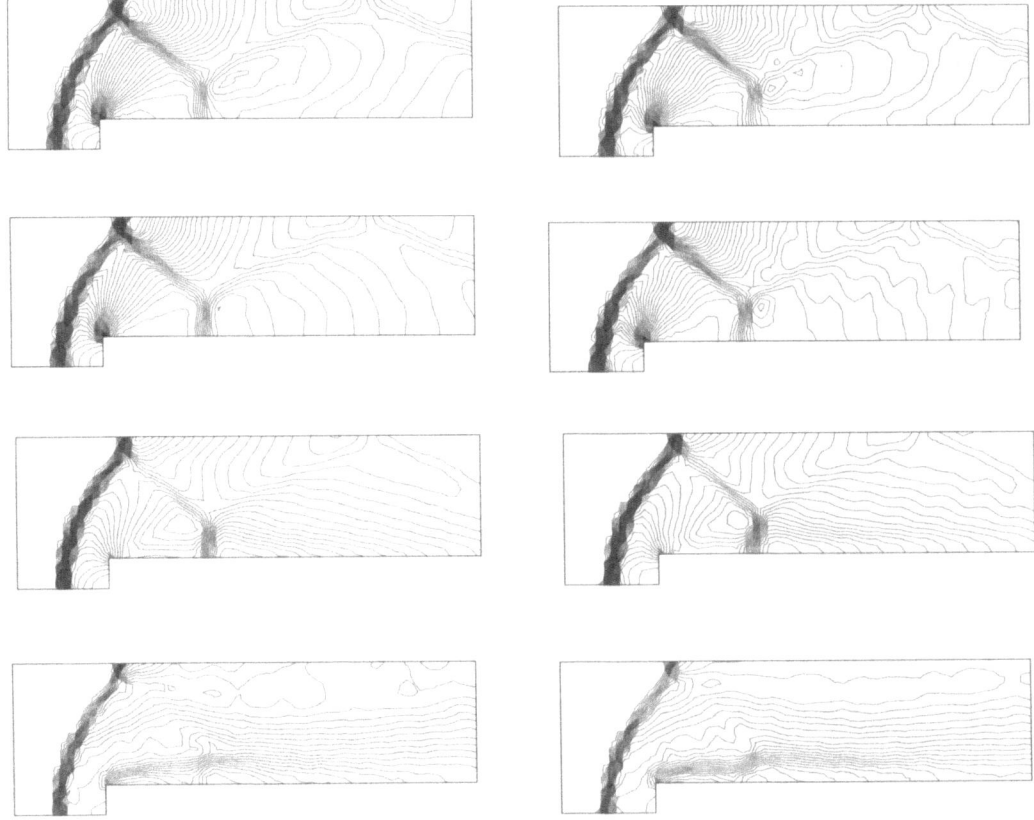

Figure 3.55 Solution with linear polynomial recovery (left) and Thin Plate Spline recovery. From above: density, pressure, Mach number, entropy. CFL=0.9.

Spline — excessive oscillations — can be clearly observed in this figure. Using a CFL number of 0.5 results in the solutions as shown in figure 3.56. Now the solution based on the Thin Plate Spline recovery algorithm looks much better as compared to the CFL number 0.5.

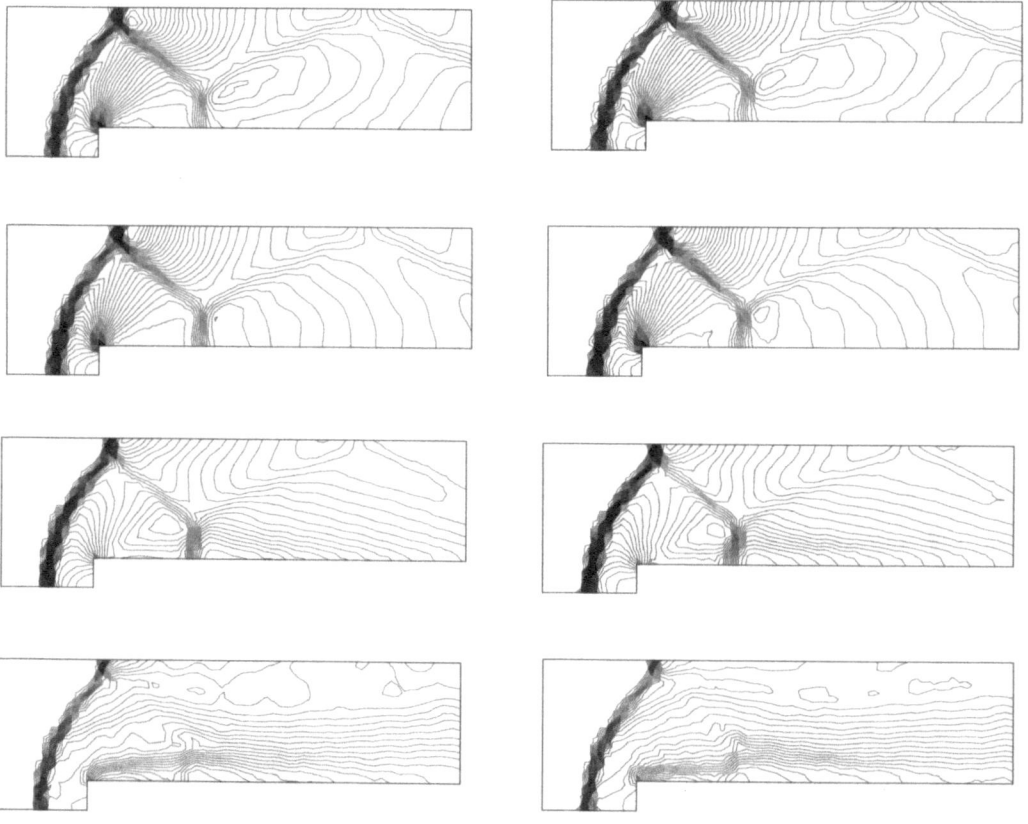

Figure 3.56 Solution with linear polynomial recovery (left) and Thin Plate Spline recovery. From above: density, pressure, Mach number, entropy. CFL=0.5.

One can now ask if there are other Splines in other function spaces which may result in better numerical solutions than the Thin Plate Spline. Indeed, families of Splines and associated semi-Hilbert spaces can be found in the class of radial basis functions. We do not want to go into the details but refer the reader to [33], [34], [35] for application of radial basis functions in the recovery problem of ENO methods. The paper [28] contains a proof that the recovery problem with radial basis functions is uniquely possible under mild assumptions. The family of radial basis functions contains Splines like the Gaussian, the multiquadric and inverse multiquadic and other functions which were in use by practicians long before the theory was developed.

For practical purposes it would be desirable to have radial basis functions at hand which do not need an additional polynomial and which are local in the sense that they have compact support. Classes of such functions were constructed by Schaback and his collaborateurs in Göttingen, see [32], [37] and [38]. Preliminary results of ENO approximations using these functions are very promising, see [35], but research is at the very beginning in this area.

3.9   Grid adaptivity for box methods

We describe adaptivity in the context of finite volume box methods as described in before. These methods work with piecewise constant cell average data on boxes and therefore define a continuous, piecewise linear interpolant of the cell averages on the primary grid. It is this local information that we use to refine triangles.

We begin with considering the technique of red-green-refinement and coarsening of a given triangulation. The following part of this lecture is essentially taken from Hempel's report [48] who developed and implemented this particular refinement strategy.

3.9.1   A refinement procedure

In this subsection we will concentrate on the geometrical problem of how to refine or coarsen a grid in the case of a conforming triangulation in the euclidean plane.

Definition 3.9.1 A finite set of nondegenerated triangles in the euclidean plane is called triangulation. Triangulations are conforming if the intersection of different triangles is empty or consists either of a common vertex or a common edge. In the last case the triangles are called neighbours. An edge of a triangle is called a boundary edge of the triangulation if no further triangle contains this edge; otherwise we call it an inner edge. Furthermore we call a vertex of a triangle a nonconforming node or a hanging node if there is another triangle which contains this point but the point is not a vertex of this triangle.

Our aim is to refine or recoarsen conforming triangulations. While adapting a triangulation we divide triangles into subtriangles, called *children* or *daughters* of a common *mother*, or we unify daughters of a common mother, called *sisters*, and restore their mother. For that purpose we sometimes destroy the conformity but the result of the presented adaption procedure will always be a conforming triangulation.

The adaption procedure gets three input parameters. The first parameter is a conforming triangulation containing a set of triangles and a set of vertices. The second parameter is an array of integers, which specifies for each triangle in the triangulation whether it should be refined, coarsened or if it has a suitable size. The third parameter is an array of so called *histories*, which describes for each triangle in the triangulation the process of refinements which led to this triangle. This data structure can be implemented very efficiently by using just a few bytes per triangle.

The procedure first refines all triangles which are specified for a refinement. After this further refinements will be done to eliminate nonconforming nodes. Then the procedure looks for locally bounded regions in the triangulation, which are specified for recoarsening. If the triangles in this regions are previously refined and if this regions meet some further conditions the mothers of the triangles in this domains are restored. The result is again a conforming triangulation.

The recoarsening routine we are using here has two advantages compared with the two-dimensional one described in [39] and [47]: It is always possible to recoarsen a refined triangulation up to its initial state and the routine which determines regions which can be recoarsened depends only on a few number of triangles around a vertex. Furthermore it is possible to extend the recoarsening technique to grids of tetrahedrae, which are refined by the isotropic refinement algorithm described in [40].

It is often desirable to transfer data of a given numerical solution on the input grid to the output triangulation. The necessary interpolation rules depend on how the data is represented in the grid. Therefore this part cannot be implemented independently of the problem to be solved. Nevertheless we will give some examples in the last section for interpolation rules used in a finite volume application.

3.9.2   Red-Green Refinement

The refinement procedure is based on subdividing triangles in two different ways as shown in the following figure. The *red refinement* inserts three points, the midpoints of the

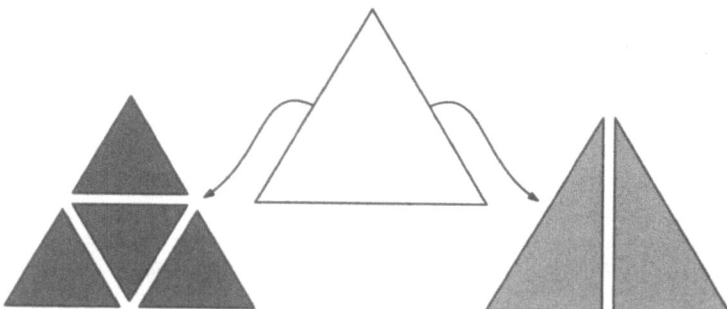

Figure 3.57   Red and green refinement of a triangle

triangle's edges as new points of the triangulation and divides the triangle into four similar subtriangles. To avoid triangles with nonconforming nodes we use the *green refinement*: If a triangle has exactly one nonconforming node we divide it in two subtriangles along the median of the edge with the nonconforming node. The resulting children of a green refinement are called *green* triangles.

These methods to divide single triangles can be used to define a refinement procedure for conforming triangulations $\mathcal{T}$ where a subset of triangles are specified for a refinement:

Algorithm 1

1. Eliminate all green refinements in $\mathcal{T}$ by restoring the mothers of green triangles. If a green triangle was specified for a refinement the restored mother is also marked for a refinement.

2. Refine all triangles red which are specified for a refinement.

3. While there exist triangles in $\mathcal{T}$ with more than one nonconforming node, they are refined red.

4. Apply the green refinement for all triangles which have exactly one nonconforming node.

The algorithm terminates with a conforming triangulation. All children of red refinements are similar to their mothers and green triangles will be removed in the next refinement step. Thus the refinement procedure is *stable*, which means that the triangles' inner angles are limited from below ($\geq C > 0$) for each sequence of grid refinements.

3.9.3   History

If we want to restore mothers of triangles to eliminate green refinements or to recoarsen the grid we need the information which triangles are children of the mother we want to restore. A solution to this problem may be to store the mother of a triangle (and the mother of the mother etc.) and to have references between them but this will produce a huge amount of memory overhead. In this section we will explain a suitable data structure for book-keeping of triangle refinements.

With each triangle of a given triangulation we associate a data structure called `History`:

History	
Bool is_green_result	If the triangle is a result of a green refinement.
Integer green_sister	If is_green_result is TRUE, green_sister specifies the number of the sister of this triangle.
Integer red_refinements	The number of red refinements, which led to this triangle. This also is the actual size of the stack below.
Stack of 2-bit-numbers red_hist[11]	Each child resulted from a red refinement has a number 0, 1, 2 or 3. Each daughter inherits this stack from her mother and pushs her own number on top of this stack.

The first member is_green_result specifies, if a triangle is a result of a green refinement. In this case, the second member green_sister specifies the location of the sister.[12] These two data member are sufficient for book-keeping of green refinements, since a green refinement will be eliminated before further refinements will be done.

The third data-member red_refinements specifies the number of red refinements which led to a triangle. This also is the actual size of the stack red_hist.

For a red refinement we enumerate the children of a red refined triangle: child 0 is located at point 0, child 1 is located at point 1 and child 2 is located at point 2 of the mother. Furthermore point i of child i is identical with point i of the mother triangle ($i = 0,1,2$). Child 3 is the inner triangle of a red refinement. The stack red_hist contains for each red refinement the number of the child. With these informations we are able to reconstruct all ancestors of triangles and thus we do not have to store triangles which have been refined.

3.9.4   Recoarsening

In this chapter we will present a new algorithm to restore triangles which have been refined by the refinement algorithm described in the last chapter. This recoarsening procedure will produce a triangulation, which also can be obtained by a sequence of refinements of the initial mesh. Furthermore this algorithm is able to recoarsen each triangle which has been refined by the refinement algorithm.

The algorithm is based on the following main loop:

---

[11] In the authors application the stack red_hist has actually a fixed maximal size of 8 bytes used for 32 refinements of an initial triangle. This will be enough for all realistic applications because after n refinements the edges of a triangle are shortened by a factor of 2^{-n}.

[12] It is sufficient to store a number 0, 1 or 2, to determine which of the three neighbours the sister is.

Algorithm 2

> for all triangles $T \in \mathcal{T}$ which are specified for recoarsening
> > for all three vertices P of T
> > > if the pair (P,T) spans a *resolvable patch* $\mathcal{P}$
> > > > recoarse this resolvable patch $\mathcal{P}$.

To explain, what a *resolvable patch* is, we need the following definitions:

Definition 3.9.2 Let P be a grid point of the triangulation $\mathcal{T}$. A triangle T located at P is called simply red, if it is a result of a red refinement, its sisters have not been red or green refined and the sisters are not located at the point P; see figure 3.58 for an example.

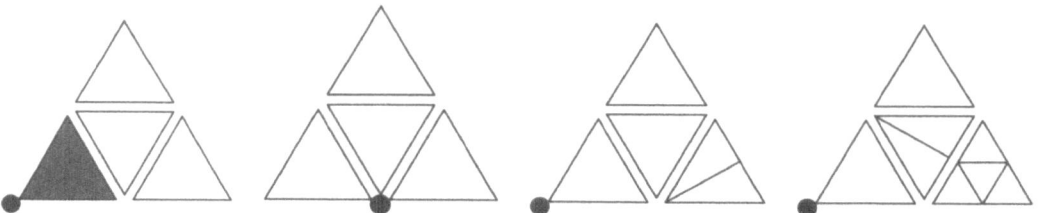

Figure 3.58   The red triangle at the left-hand side shows a simply red triangle. The point in the second figure is located at several sisters and therefore the triangles are not simply red. In both figures at the right-hand side there are already red and green refined sisters and thus none of the triangles are called simply red.

Definition 3.9.3 A finite sequence $\{T_i\}_{i=0...k}$ ($k \in \mathbf{N}_0$) of pairwise different triangles in $\mathcal{T}$ which are located at a common vertex P is called a thread of P, if all $T_i (i \in \{0...k\})$ are simply red and T_i and T_{i-1} are neighbours ($\forall i \in \{1...k\}$), see figure 3.59. A triangle $S \in \{T \in \mathcal{T} : T \neq T_i, i = 0...k\}$ is called successor of the thread $\{T_i\}_{i=0...k}$, if P is a vertex of S and S is a neighbour of at least one $T_j \in \{T_i : i = 0...k\}$. A thread $\{T_i\}_{i=0...k}$ is called a maximal thread of P if all successors of this thread have a lower number `red_refinements` than the triangles in $\{T_i\}_{i=0...k}$.

Obviously triangles in a thread have the same number `red_refinements`. With these definitions we will now describe the resolvable patches:

Definition 3.9.4 Let $T \in \mathcal{T}$ be a triangle and P one of its vertices. Assume there is a maximal thread $\{T_i\}_{i=0...k}$ of P with $T \in \{T_i\}_{i=0...k}$. Let $\mathcal{P}$ be the set of triangles in the thread together with their sisters. Then $\mathcal{P}$ is called a resolvable patch spanned by the triangle T and its vertex P, if all triangles in $\mathcal{P}$ are specified for recoarsening.

Figure 3.60 illustrates on the left-hand side some configurations for resolvable patches (red triangles) and on the right-hand side possibilities to recoarsen the patch without producing hanging nodes.

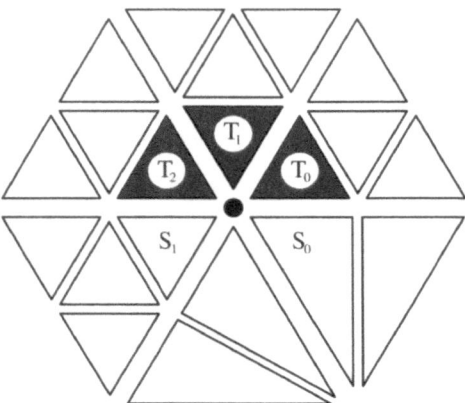

Figure 3.59   A thread $\{T_i\}_{i=0...2}$ around a point and two successors S_0 and S_1 .

The upper part of figure 3.60 shows a maximal thread without any successors. This leads to a resolvable patch and the triangles in this patch can be recoarsened as shown on the right-hand side.

In the middle of figure 3.60 a maximal thread with two green triangles as successors is shown. As can easily be seen here, successors of a maximal thread are always green. On the right-hand side the red triangles are transformed into green triangles and the green triangles can be resolved.

The lower part of figure 3.60 illustrates a situation at the boundary of a triangulation. The maximal thread has no successor and the resolvable patch can be recoarsened as shown on the right-hand side.

We will now explain how to resolve the triangles in a resolvable patch $\mathcal{P}$ spanned by a triangle $T \in \mathcal{T}$ and one of its vertices P .

Algorithm 3

1. Remove all triangles in $\mathcal{P}$ and restore their mothers.

2. Remove all green triangles which produce hanging nodes in the mothers of $\mathcal{P}$, and restore the mothers of this green triangles.

3. Refine all triangles green which have one nonconforming node.

The only difficulty in this algorithm is to see that successors of a maximal thread spanned by T and P are removed in the second step of the algorithm. This is valid because they are green and they produce hanging nodes in the mothers of the triangles in $\mathcal{P}$. Furthermore this algorithm avoids in the second step that there are two pairs of green triangles with the same refined edge.

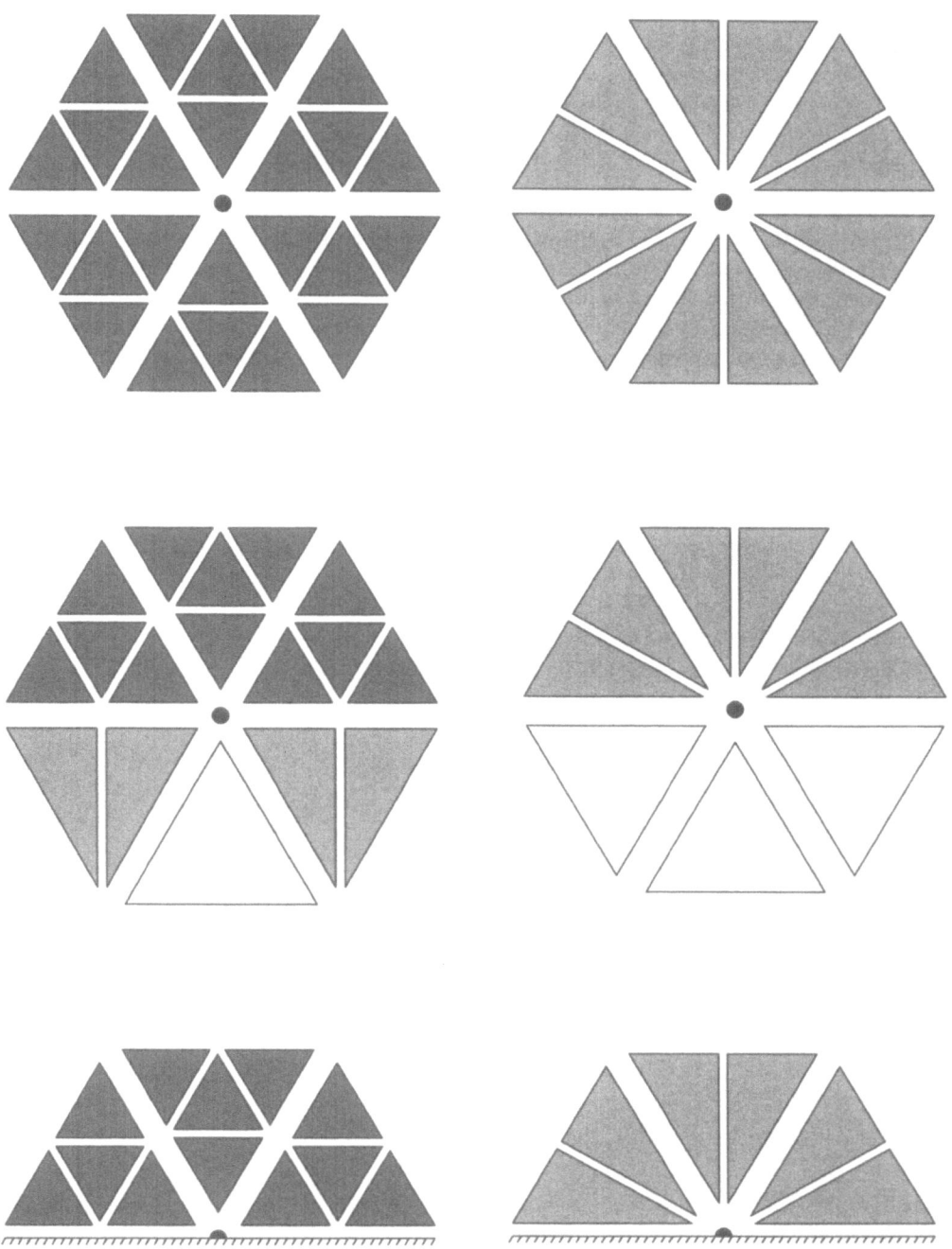

Figure 3.60 Resolvable patches on the left-hand side and recoarsened triangles at the right-hand side

After recoarsening a resolvable patch the triangulation is conforming and as can easily be seen the resulting triangulation also is obtainable by a sequence of refinements of the initial mesh. The amount of work which has to be done to determine a resolvable patch and to resolve the triangles can be estimated from above by $\mathcal{O}(N_p)$, where N_p is the number of triangles around a point. Because this number is typically small ($N_p \approx 6$ for isotropic triangulations) the procedure for recoarsening a grid is very cheap ($\mathcal{O}(N_p \cdot n_c) = \mathcal{O}(n_c)$ with n_c denoting the number of triangles, which have been specified for recoarsening).

With a sequence of recoarsening steps it is possible to recoarsen each refined triangulation up to its initial state. To see that, one has to look at the triangles with the largest number `red_refinements` in the triangulation. We will assume that they are specified for recoarsening. These triangles form at least one resolvable patch and thus the triangulation contains resolvable patches as long as there are triangles with a positive number `red_refinements`. This implies the possibility to recoarsen a grid up to its initial state.

3.9.5   Conservativity

Physical data (for example the density ϱ of the fluid) in finite volume box approximations is represented by piecewise constant functions on control volumes. Each control volume is unambiguously associated with a grid point of the triangulation, see figure 3.61. The piecewise constant data on the control volumes corresponds to integral averages of the solutions of the PDEs. This property should be respected in an adaptive algorithm in which interpolation of cell average values is required.

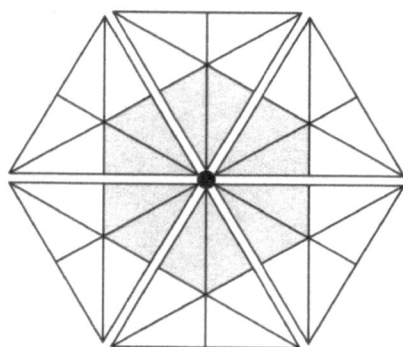

Figure 3.61 Control volume associated with a grid point.

We will explain here the case of interpolating piecewise constant functions on the control volumes: Let $\varrho(P)$ be the value of the function on the control volume associated with the grid point P. If we insert a new point P at the barycenter between two vertices P_0 and P_1, we take the arithmetical mean of $\varrho(P_0)$ and $\varrho(P_1)$ as the value of $\varrho(P)$ and the values $\varrho(P_0)$ and $\varrho(P_1)$ remain unchanged.

$$\varrho(P) \;=\; \tfrac{1}{2}(\varrho(P_0) + \varrho(P_1))$$

Otherwise if we delete a grid point P which is the midpoint of two points P_1 and P_2 of an edge, we distribute the value $\varrho(P)$ to the new values $\tilde{\varrho}(P_0)$ and $\tilde{\varrho}(P_1)$. Let $|P|$, $|P_0|$ and $|P_1|$ be the areas of the control volumes before deleting the point P. Then we use the formula

$$\tilde{\varrho}(P_0) = \frac{|P_0| \cdot \varrho(P_0) + \frac{1}{2}|P| \cdot \varrho(P)}{|P_0| + \frac{1}{2}|P|},$$

$$\tilde{\varrho}(P_1) = \frac{|P_1| \cdot \varrho(P_1) + \frac{1}{2}|P| \cdot \varrho(P)}{|P_1| + \frac{1}{2}|P|}.$$

Both, the formula for inserting and the one for removing grid points, are *conservative* in the following sense: Let Ω be the domain of control volumes which have been changed while inserting and deleting grid points. If $\mathcal{P}$ is the set of grid points in Ω before adapting the grid and $\widetilde{\mathcal{P}}$ the set afterwards, then the following relationship is valid

$$\int_\Omega \varrho(x,y)\,\mathrm{d}(x,y) = \sum_{P \in \mathcal{P}} |P| \cdot \varrho(P) \;\equiv\; \sum_{\tilde{P} \in \widetilde{\mathcal{P}}} |\tilde{P}| \cdot \tilde{\varrho}(\tilde{P}) = \int_\Omega \tilde{\varrho}(x,y)\,\mathrm{d}(x,y)$$

The following examples are computed with this formula, except that Hempel has implemented the refinement and recoarsening algorithm in a way that no unnecessary operations[13] will be done. For this purpose one needs two further interpolation rules: One for tranforming a green triangle directly to a red triangle and one for the inverse process.

3.10   Error and residual

The basic ideas behind residual based error indicators are easily explained. To start with, consider a heuristic approach in which certain features of the flow field — shocks, for example — should be better represented in a numerical solution and therefore the grid should be finer at locations where those phenomena occur. In the case of inviscid compressible flow one could choose the modulus of the Mach number, density or pressure to obtain sensors which are able to locate shocks. However, if the value of one such sensor on one triangle is high, it will be even higher after refining the triangle, since the gradient of flow variables is growing in magnitude at a shock the finer the grid is. Thus, this approach does not lead to a desirable automatic algorithm in which the computation stops after a certain tolerance is reached. Besides these deficiencies no flow variable can capture all of the phenomena of compressible flow fields as can be seen in the case of pressure gradients which are not able to detect contact discontinuities.

In order to overcome the aforementioned difficulties Johnson and his collaborators [46] were the first to introduce the idea of residual based error indication into CFD. Although their ideas are based on properties of their finite element method (like Galerkin orthogonality, streamline diffusion and artificial viscosity) and therefore not applicable to finite volume the success of their analysis was convincing and it seemed worth to follow the general philosophy. Consider an abstract partial differential equation with an operator $\mathcal{L}$ of first order

---

[13] For example the green coarsening in the first step of the refinement algorithm. This operation is obsolete for domains which remain unchanged.

$$\mathcal{L}u = 0$$

in which we assume boundary conditions already being included. If u^h denotes the solution of a numerical method which is at least differentiable on the triangles of a primary grid then

$$e^h := u^h - u$$

denotes the error of the approximate solution. If we apply the operator $\mathcal{L}$ to the numerical solution u^h we can clearly not expect to get $\mathcal{L}u^h = 0$ but a remaining quantity which we call residual, i.e.

$$r^h := \mathcal{L}u^h.$$

We shall always look for local error control, i.e. we consider the residual computed triangle-wise. If now $\mathcal{L}$ were a linear differential operator possessing a bounded inverse operator $\mathcal{L}^{-1}$ then application of $\mathcal{L}$ to the error leads to

$$\mathcal{L}e^h = \mathcal{L}u^h - \mathcal{L}u = r^h.$$

Due to the invertibility of the operator we finally would get error control in the form

$$\|e^h\|_X \le \|\|\mathcal{L}^{-1}\|\|_{Y \to X} \|r^h\|_Y \tag{3.12}$$

for some spaces X and Y. Thus, error control is possible through a computable quantity — the residual — if an upper bound on the norm of the inverse operator is at hand. Then the above inequality can be used for reliable *a posteriori* error estimation and adaptive error control.

However, it may happen that the error is very much overestimated and that grids are generated which are much too fine. This would degenerate the efficiency of the adaptive method. One can assure efficiency of the adaptive process if, in addition to (3.12), an inequality of the form

$$C\|r^h\|_Y \le \|e^h\|_X$$

is available with a computable constant C.

If $\mathcal{L}$ is nonlinear and most of its properties are unknown (as is the case for the Euler equations), it is unlikely that a relation of the form

$$C_1\|r^h\|_Y \le \|e^h\|_X \le C_2\|r^h\|_Y$$

will be available with computable constants C_1, C_2. We shall describe in the following the use of heuristically motivated *ad hoc* indicators which stood at the beginning of our efforts, the theory of Mackenzie, Süli and Warnecke concerning *a posteriori* error control of Friedrichs systems, the application of this theory to the field of numerical gas dynamics, and a result due to Süli which satisfactorily links the developments in the theory of Friedrichs systems with our initial heuristic approaches.

Unless otherwise stated we consider the steady case, i.e. $\partial_t u = 0$, and generalise afterwards to the time-dependent case.

3.11 Experience with L^2

Keeping an eye on the development of *a posteriori* error control in finite element methods for linear elliptic equations shows that the space L^2 of square integrable functions is the right space to work with. Although there is no evidence to assume that this space is the right one for hyperbolic conservation laws we started to work using

$$\|r^h\|_{L^2(T)} := \sum_{i=1}^{4} \|r_i^h\|_{L^2(T)}$$

on the triangles of the primary grid, where the numerical solution u^h was the Lagrange interpolant of the cell averages on the triangle T. Here, r_i^h denote the $i-$ th component of the residual corresponding to the continuity equation, momentum and energy equation, respectively.

3.11.1 On the necessity of proper weighting

Although preliminary tests with $\|r^h\|_{L^2(T)}$ showed the ability of this indicator to capture the phenomena of compressible flow, a simple counter-example due to Süli showed the general inability of $\|r^h\|_{L^2(T)}$ to be used as a building block in an automatic algorithm.

Consider a grid $\{x_1,\dots,x_{i-1},x_i,x_{i+1},\dots,x_N\}, x_1 = 0, x_N = 1$, in one space dimension with uniform spacing $h := x_{i+1} - x_i$ and a function $x \to u(x)$ defined by

$$u(x) = \begin{cases} 0 & ; & 0 \le x < \frac{x_i+x_{i+1}}{2} \\ 1 & ; & \frac{x_i+x_{i+1}}{2} < x \le 1 \end{cases}$$

and a continuous numerical solution based on piecewise linear polynomials on the intervalls $[x_i, x_{i+1}]$, i.e.

$$u^h(x) = \begin{cases} 0 & ; & 0 \le x < x_i \\ \frac{1}{h}(x - x_i) & ; & x_i \le x < x_{i+1} \\ 1 & ; & x_{i+1} \le x \le 1 \end{cases} .$$

The situation can be seen in figure 3.62. Since we concentrate on first order differential

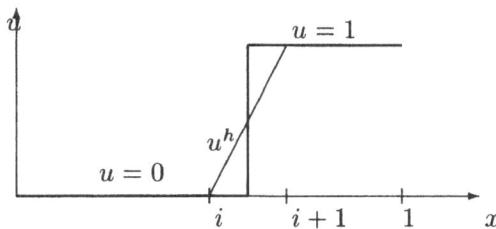

Figure 3.62 Model example

operators $\mathcal{L}$ the residual is essentially eqivalent to the x -derivative of u^h , i.e.

$$r^h(x) \sim \partial_x u^h(x) = \begin{cases} 0 & ; & 0 \le x < x_i \\ \frac{1}{h} & ; & x_i \le x < x_{i+1} \\ 0 & ; & x_{i+1} \le x \le 1 \end{cases}$$

The L^2-norm on the intervall $[x_i, x_{i+1}]$ is easily calculated to give

$$\|r^h\|_{L^2([x_i, x_{i+1}])} = \sqrt{\int_{x_i}^{x_{i+1}} |r^h|^2 \, dx} \sim \sqrt{\int_{x_i}^{x_{i+1}} \frac{1}{h^2} \, dx} = \frac{1}{\sqrt{h}}$$

and this quantity blows up at discontinuities as the grid is refined.

Note that the quantity $\|r^h\|_{L^2(T)}$ on a single element T (intervall, triangle, etc.) scales with space dimension. Consider a square control volume of side length h in two dimensions and a function defined on it which is 0 in the left half of the element and 1 in the right (i.e. a discontinuity which is parallel to the x_2-axis) as in figure 3.63. The residual

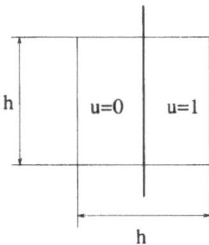

Figure 3.63 Square control volume in 2-D

is essentially the gradient of u^h which is given by $\nabla u^h = (1/h, 0)^T$ if we choose linear interpolation on the square. Thus,

$$\|r^h\|_{L^2(T)} = \sqrt{\int_0^h \int_0^h \frac{1}{h^2} \, dx_1 \, dx_2} = 1.$$

This shows $\|r^h\|_{L^2(T)} = \mathcal{O}(1)$ in two space dimensions. In general, if d denotes the space dimension, the local quantity scales like

$$\|r^h\|_{L^2(T)} = \mathcal{O}\left(h^{\frac{d-2}{2}}\right).$$

However, the global quantity, extending over a domain Ω, always behave like

$$\|r^h\|_{L^2(\Omega)} = \mathcal{O}\left(h^{-\frac{1}{2}}\right).$$

This is easily seen in 2-D on the unit square which is divided in $N \times N$ square elements with side lengths h, i.e. $Nh = 1 \Rightarrow N = 1/h$, according to figure 3.64. If we denote the square volumes by T_{ij} then

$$\|r^h\|_{L^2(\Omega)} = \sqrt{\int_0^1 \int_0^1 |r^h|^2 \, dx_1 \, dx_2} = \sqrt{\sum_{i=1}^N \sum_{j=1}^N \int\int_{T_{ij}} |r^h|^2 \, dx_1 \, dx_2}.$$

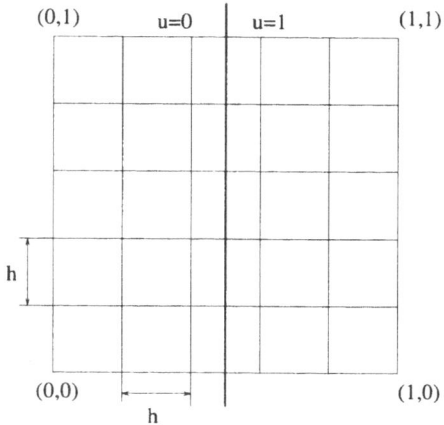

Figure 3.64 Domain in $\mathbb{R}^2$

Now $r^h \neq 0$ only on N conrol volumes, namely the row of volumes through which the discontinuity passes, say row with index $i = i^*$. On each of these volumes the gradient of u^h — still assuming linear interpolation — is $\nabla u^h = (1/h, 0)^T$. Thus, it remains to compute

$$\|r^h\|_{L^2(\Omega)} = \sqrt{\sum_{j=1}^{N} \int_{T_{i^*j}} \frac{1}{h^2} \, dx_1 \, dx_2} = \sqrt{\sum_{j=1}^{N} 1} = \sqrt{N} = \sqrt{\frac{1}{h}}.$$

Obviously this result does not depend on dimension. However, the behaviour of the unweighted L^2-norm is unsatisfactory for the use in an automatic adaptive code.

Consider the transonic flow about a NACA0012 airfoil as a simple steady test case. The Mach number at infinity is $\mathrm{Ma} = 0.8$ and the angle of attack is $\alpha = 1.25°$. We expect a strong shock on the upper side and a weak one on the lower side of the profile. The grid in the vicinity of the profile as well as the whole grid can be seen in figure 3.65. Note that the leading edge region of the profile is already overly refined in this initial grid.

Using a second-order box method as described above results in the numerical solution as shown in figure 3.66, where the distribution of pressure can be seen. Note that although both shocks are visible in the numerical solution the weak lower side shock is quite badly resolved.

Figure 3.67 shows isolines of different classical refinement indicators used in CFD, namely the modulus of gradients of density, pressure, Mach number and thermodynamical entropy. Note that in the scale chosen entropy is not detecting the weak shock on the lower side. From our point of view the leading edge region should not be defined too much since it is already very well resolved. Note that all classical indicators would indeed refine this region. This behaviour is shared by the unweighted L^2-norm indicator as can be seen in figure 3.68, where the grid after three refinement cycles is shown. The corresponding pressure distribution is shown in figure 3.69. In order to show that the classical refinement indicators would still further refine their isolines after three refinement cycles is shown in figure 3.70.

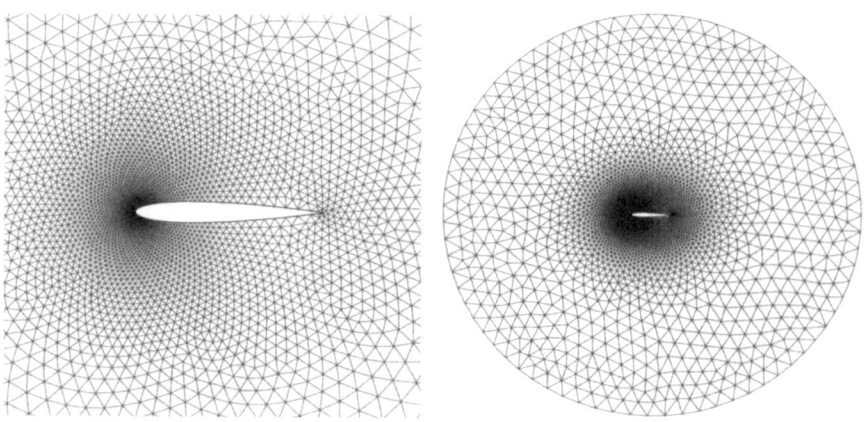

Figure 3.65 Grid for NACA0012

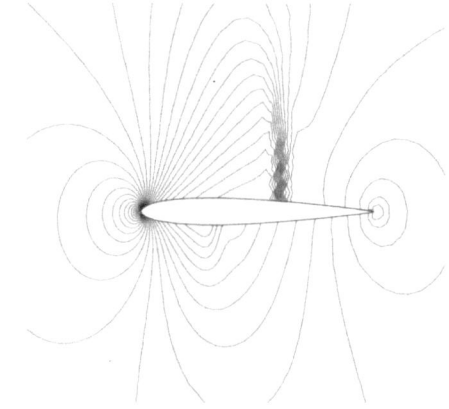

Figure 3.66 Pressure distribution on the initial grid

In order to keep the desirable features of the unweighted L^2-indicator, namely the good capturing of the flow phenomena, a proper weighting of the L^2-norm was sought. Inspired by the results of Johnson and his group [46] the weighting with the local triangle diameter h_T, taken as the length of the longest side of T, was further exploited.

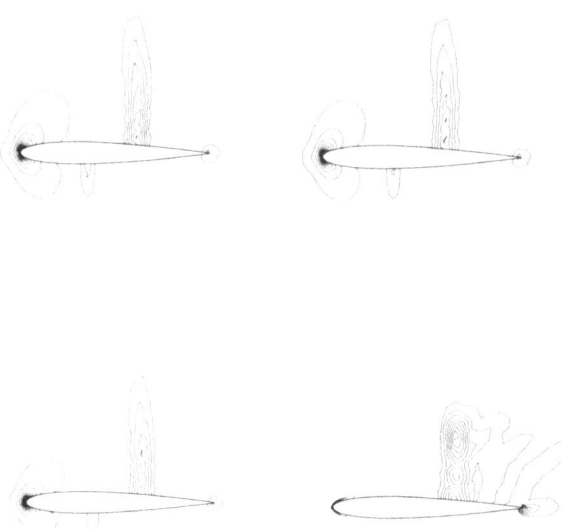

Figure 3.67 Modulus of gradients of density, pressure, Mach number, and entropy (left to right).

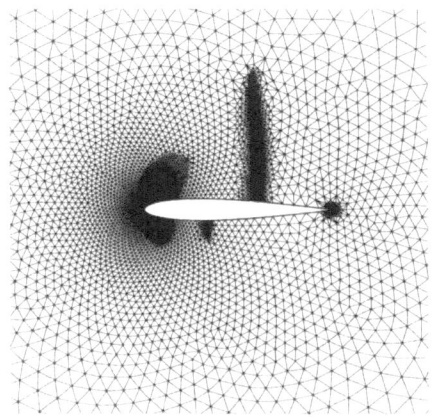

Figure 3.68 Grid after three refinement cycles

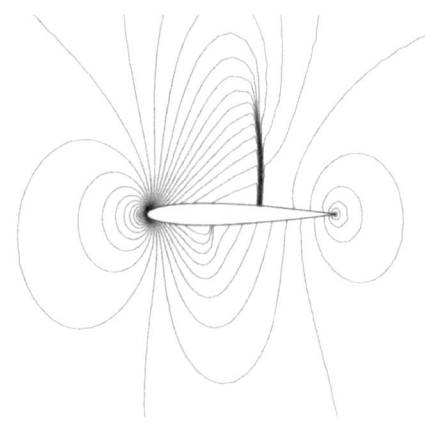

Figure 3.69 Pressure distribution on the refined grid

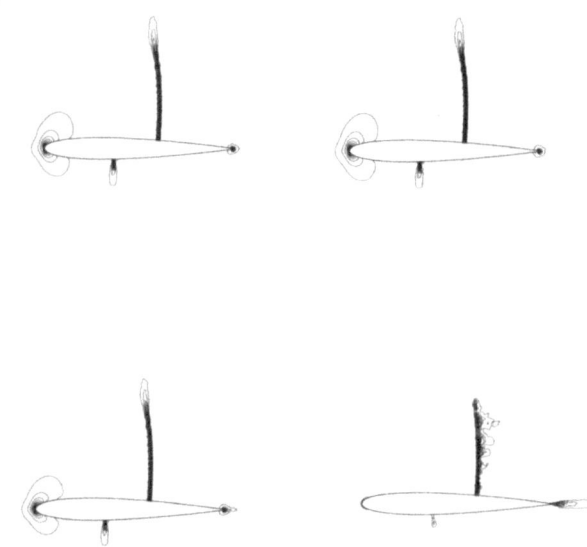

Figure 3.70 Modulus of gradients of density, pressure, Mach number, and entropy after adaption (left to right).

3.11.2 The weighted L^2-norm

¿From a purely heuristic standpoint the order of $1/\sqrt{h}$-behaviour of the L^2-norm can be removed by the weighted norm

$$\|r^h\|_{L^2_h(T)} := h_T \|r^h\|_{L^2(T)}$$

where h_T denotes the length of the longest side of T. Although the use of this type of weighting was purely heuristical in the context of finite volume approximations, indicators were used successfully in the Streamline Diffusion Finite Element Method of Johnson et al. [46]. Additionally, numerical tests confirmed that h_T gave proper weighting compared to weights $h_T^q, q > 1$.

Consider the transonic flow about a NACA0012 airfoil as discussed in the previous subsection. After three refinement cycles with the weighted L^2-norm indicator the grid and Mach number distribution shown in figure 3.71 are obtained.

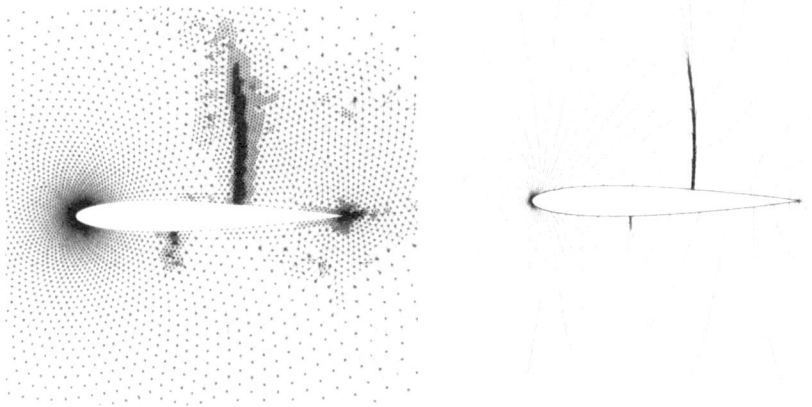

Figure 3.71 Grid after three refinement cycles and Mach number distribution

A much more complicated flow field is provided by the flow through a supersonic combustion chamber as shown in figure 3.72. The flow enters the slightly diverging chamber from the left with an inflow Mach number of Ma = 2. At the back side of the wedge hydrogen is injected. As can be seen from the Mach number distribution shown in figure 3.72 the flow field is dominated by strong shock-shock interactions. Note that the due to the non-symmetric channel the flow field is also non-symmetric.

The zoom of the back side of the wedge shown in figure 3.72 shows the Kelvin-Helmholtz-type instability caused by the injected hydrogen. Note that all of the relevant flow phenomena are well resolved and captured by the weighted L^2-norm indicator.

However satisfactory these results may seem it was not possible to carry the finite element analysis of Johnson and his co-workers directly over to the case of our finite volume method. In order to gain flexibility Süli therefore proposed to depart from using the weighted L^2-norm and proposed the use of a weak norm.

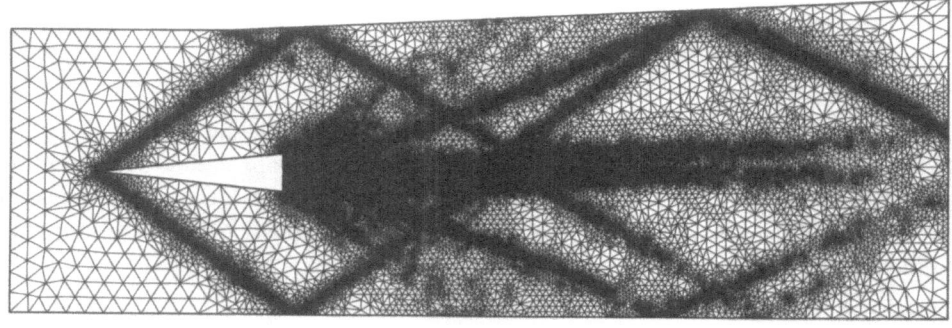

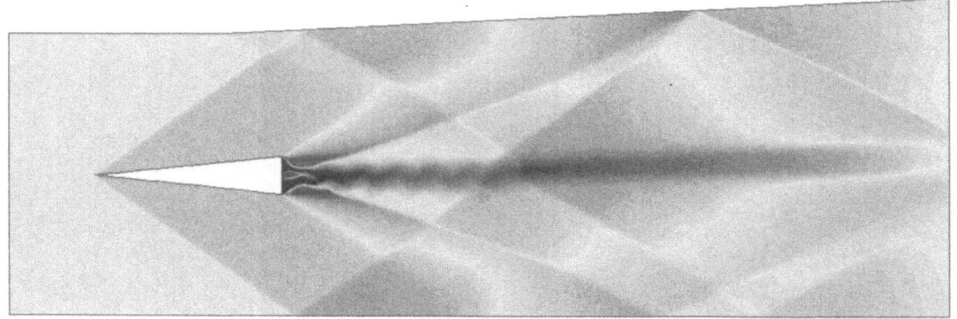

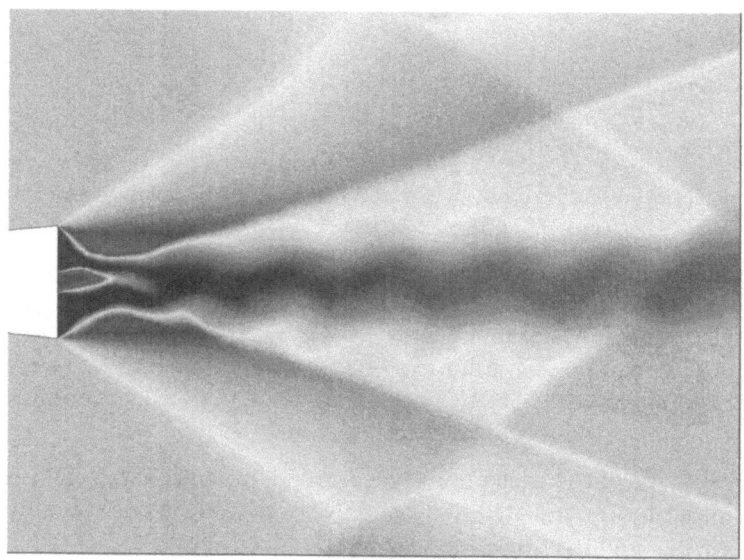

Figure 3.72 Flow through a supersonic combustion chamber

If we consider u to be an element of the space L^2 then an application of a first order differential operator would lead us in the negative Sobolev space H^{-1}. In this space we find ordinary functions as well as certain measures and distributions. If we denote by ∂_x^α the operators $\partial_{x_1}^{\alpha_1} \partial_{x_2}^{\alpha_2}$ for $\alpha = (\alpha_1, \alpha_2)^T \in \mathbb{N}^2$, then H^{-1} is defined as the dual of

$$H_0^1(\Omega) := \{v \in L^2(\Omega) \mid \partial_x^\alpha v \in L^2(\Omega) \text{ for } |\alpha| = 1 \text{ and } v \text{ has compact support on } \Omega\},$$

i.e. functions which are non-zero only in the interior of Ω. As a dual space it carries a weak norm given by

$$\|u\|_{H^{-1}(\Omega)} := \sup_{\Phi \in H_0^1} \frac{|(u,\Phi)_\Omega|}{\|\Phi\|_{H_0^1(\Omega)}}$$

where $(u,\Phi)_\Omega := \int_\Omega u(x)\Phi(x)\,dx$ denotes the L^2-inner product. Thus, measuring the residual in this norm amounts to computing the supremum

$$\|r^h\|_{H^{-1}(T)} := \sup_{\Phi \in H_0^1} \frac{|(r^h,\Phi)_T|}{\|\Phi\|_{H_0^1(T)}}$$

over all H_0^1-functions on triangle T. Obviously this can not be done on a computer and one has to think about approximative algorithms.

In [54] we used a simple subdivision algorithm. On each $T \in \mathcal{T}^h$ a subdivision according to figure 3.73 was established. This subdivision gives rise to the definition of three linear

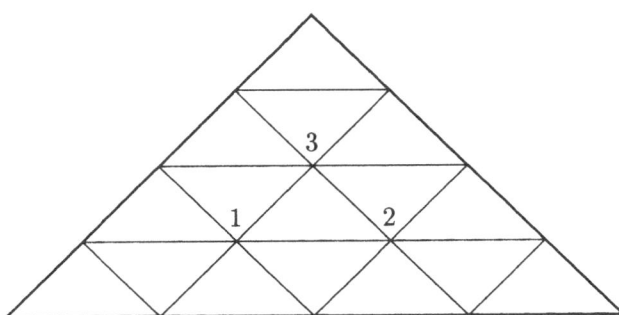

Figure 3.73 Subdivision for the computation of $\|\phi\|_{H_0^1(T)}$

hat functions $\Phi_i, i = 1,2,3$, which take the value 1 at the subdivison node i and 0 everywhere. On the subtriangles sharing subdivision node i the function Φ_i is a linear polynomial. Clearly, these functions belong to $H_0^1(T)$. We now replace the supremum over an infinity of possible functions Φ by the maximum over these three functions defined on the subdivision. This gives

$$\|r^h\|_{\tilde{H}^{-1}} := \max_{i=1,2,3} \frac{|(r^h,\Phi_i)_T|}{\|\Phi_i\|_{H_0^1(T)}}$$

as an approximation to the true H^{-1}-norm of the residual. Note that it is by no means obvious that this approximation process leads to a reliable estimate of the true H^{-1}-norm. The only reason to believe that the approximation is not too bad stems from the fact that numerical experiments with a further subdivision, leading to 21 test functions Φ_i, showed the same results up to plot accuracy.

Applied to the test case of steady transonic flow about the NACA0012 profile the weak norm indicator was able to detect the flow phenomena close to the obstacle (i.e. both shocks) as good as the weighted L^2-norm indicator. However, certain instabilities occured far away from the profile. Figure 3.74 shows the resulting grid and the correspond-

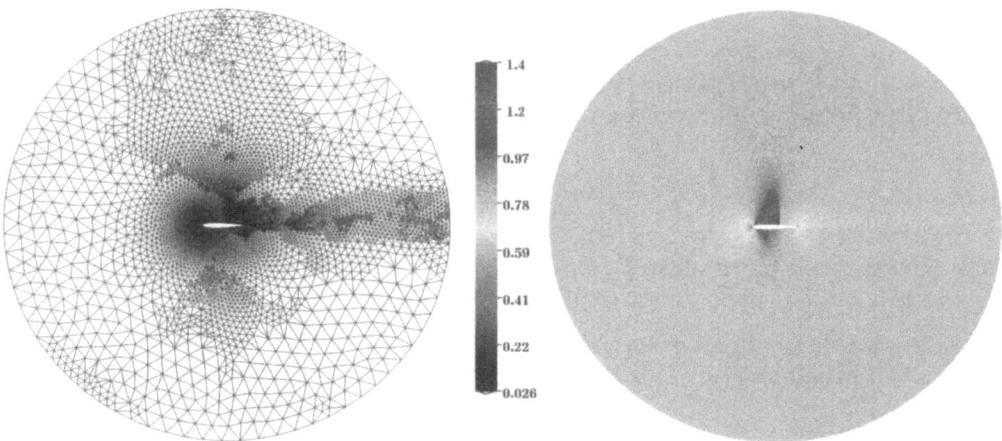

Figure 3.74   Transonic flow after adaption

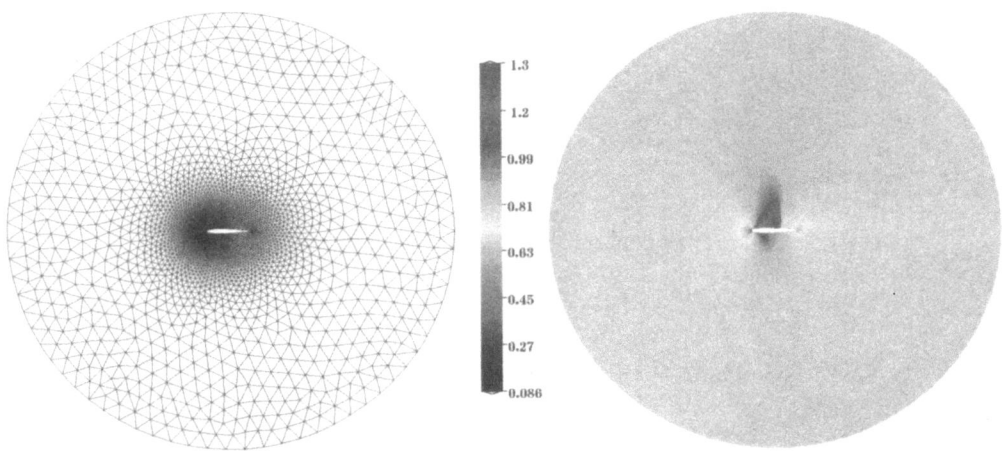

Figure 3.75   For comparison: Transonic flow before adaption

ing Mach number distribution after three adaption cycles. Figure 3.75 shows for the sake of comparison the initial grid and the corresponding Mach number distribution. As can be seen there is no obvious reason to adapt the grid on the upper half up to the farfield since the solution does not change much there. It turns out that the implementation of the weak norm is extremely sensitive to small perturbations and the 'right' tolerance to keep the balance between overadaption and leaving the grid as it was before adaption is hard to achieve. We shall come back to this point in the discussion of the dual graph-norm. The use of the weak norm resulted in none of the cases in which we tested it to superior results as compared to the weighted L^2-norm.

3.11.3 Unsteady Problems

One of the conceptual advances of the streamline diffusion finite element methods analysed by Johnson et al. is the consequent use of finite elements in space and time, i.e. tru space-time elements are used such that the temporal residual $\partial_t u^h$ is computable without difficulty. In our case the solution is known at certain points $n\Delta t, (n+1)\Delta t, \ldots$ in time which corresponds to a finite difference method in time. Writing the unsteady residual as

$$r^h := \partial_t u^h + \mathcal{L}u^h$$

amounts to approximate the temporal part $\partial_t u^h$ on each triangle $T \in \mathcal{T}^h$. Instead of utilising space-time elements and thus computing space-time residuals this can be done by a simple finite difference technique and the use of linear interpolation on T:

Compute

$$\partial_t u_i^h := \frac{u^h(x_i, (n+1)\Delta t) - u^h(x_i, n\Delta t)}{\Delta t}$$

for every node point $x_i, i = 1,2,3$, of triangle T and interpolate these three values by means of a linear polynomial on T

$$\partial_t u^h := \sum_{i=1}^{3} \partial_t u_i^h \Phi_i(x),$$

where Φ_i again denotes the linear polynomial basis function on T uniquely defined by $\Phi_i(x_j) = \delta_i^j$. Then the (spatial!) weighted L^2-norm can be computed as $r^h = \partial_t u^h + \mathcal{L}u^h$.

This procedure is applied to the test case of flow in a channel with forward facing step according to Woodward and Colella [59]. At time $t = 0$ a step is introduced spontaneously in a $\text{Ma}_\infty = 3$ flow. Complex flow patterns evolve and at time $t = 8$ a complicated shock system has developed in the channel as shown in figure 3.76. Note that the time $t = 8$ corresponds to the size of the channel as well as to the quantities used for non-dimensionalisation. Our time corresponds to $t = 4$ in the paper of Woodward and Colella [59]. The corresponding density distributions are shown in figure 3.78.

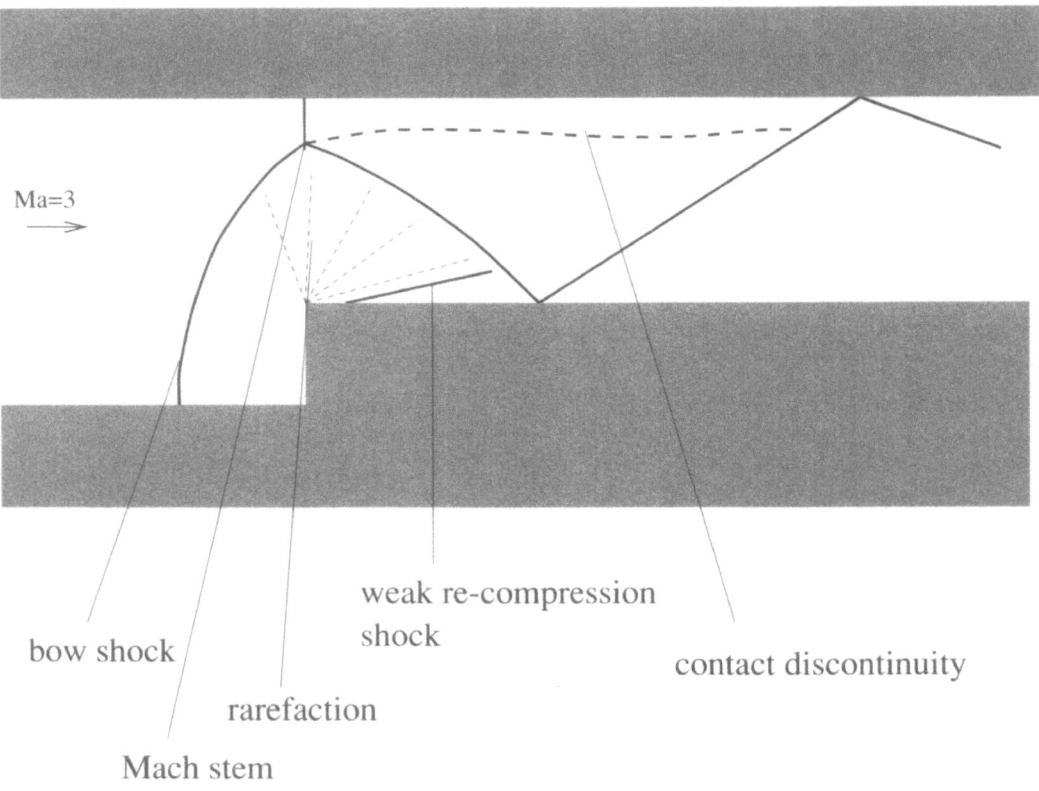

Figure 3.76 Flow phenomena in the channel with forward facing step

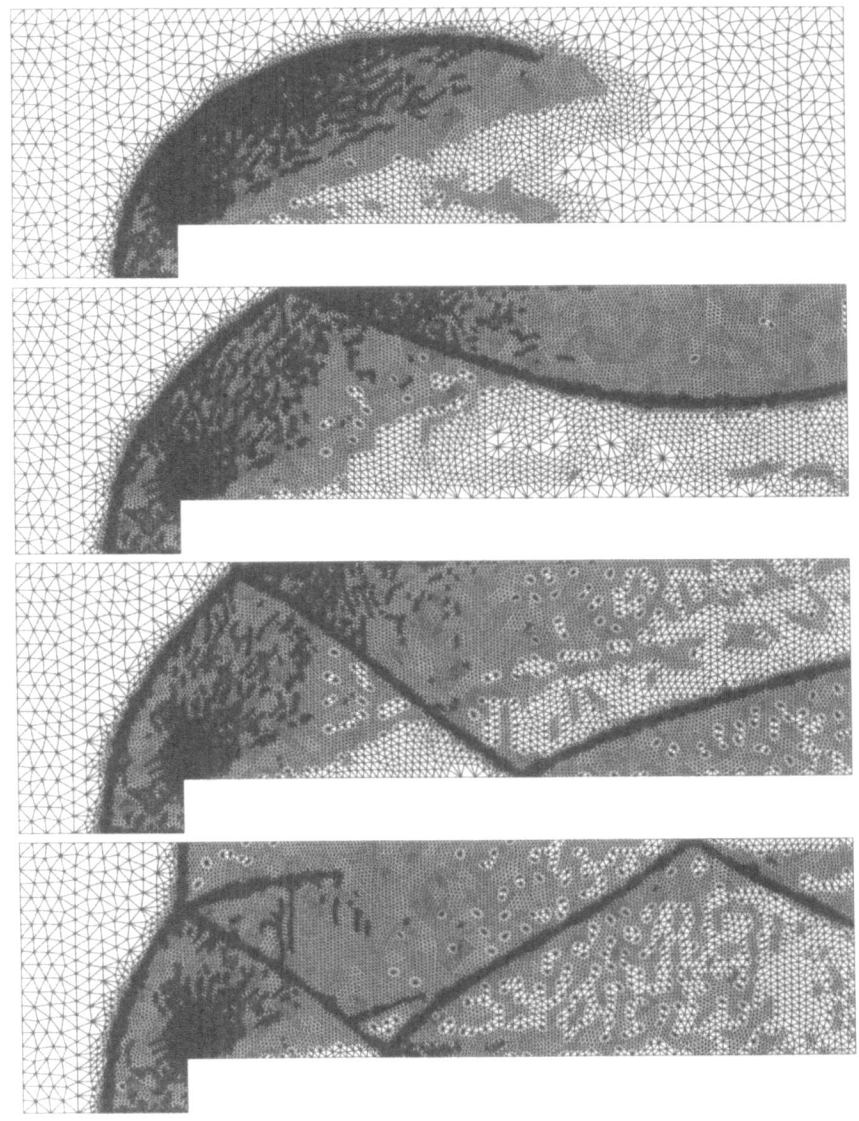

Figure 3.77 Adapted grids at different times

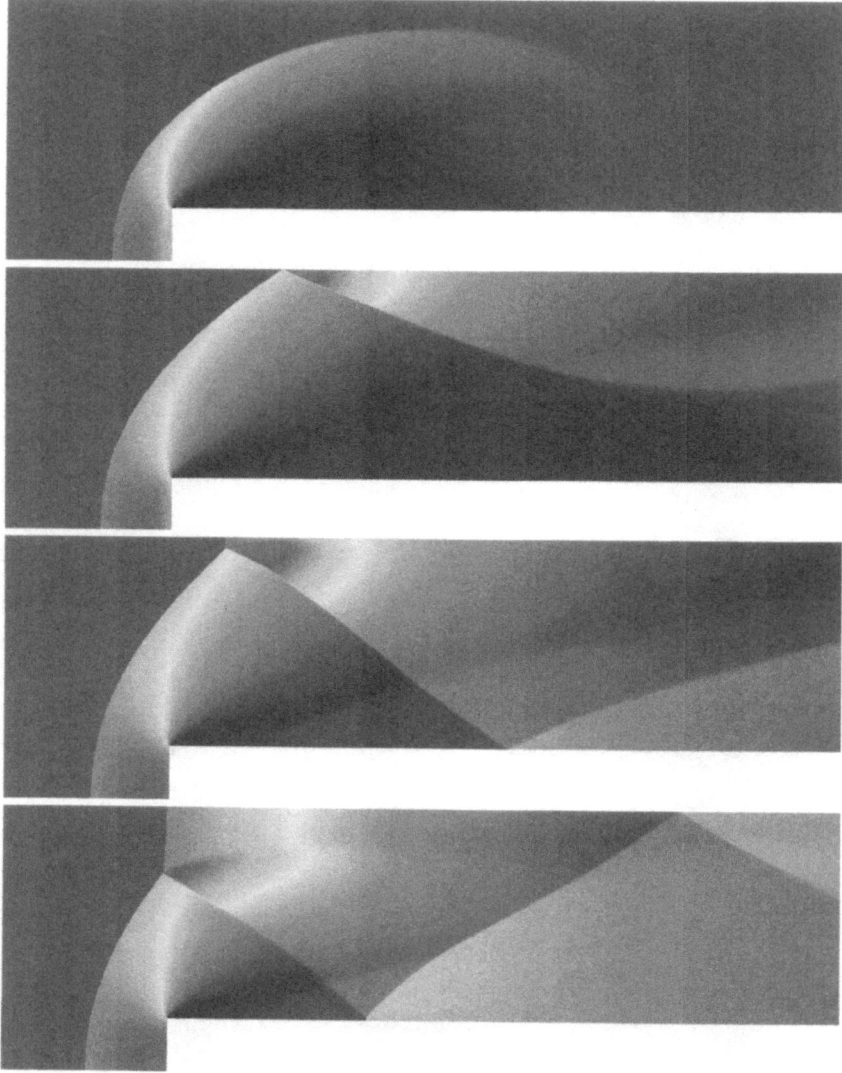

Figure 3.78 Density distributions corresponding to the grids in figure 3.77

As can be seen the weighted L^2-norm indicator has detected all relevant flow phenomena. The shock system is clearly visible as is the contact discontinuity starting at the Mach stem. Note that the corner point of the step is a true corner singularity since it corresponds to the footpoint of a rarefaction wave. Thus, a rarefaction shock occurs at the corner point. The weighted L^2-norm indicator has also detected phenomena associated with this special point and refined the region in its vicinity.

However, at the time these numerical results were presented in [55] there was no analysis available to prove that the weighted L^2-norm indicator is a true error indicator. This proof was made possible by the analysis of Mackenzie, Süli and Warnecke in [52] and [53] in which linear symmetric systems were considered and the residual was taken in the so-called dual graph-norm.

3.12 The dual graph-norm

The results presented in this section are taken from [57].

3.12.1 Friedrichs systems

In [52] and [53] Mackenzie, Süli and Warnecke developed a theory of *a posteriori* error analysis for general Petrov-Galerkin discretisations of so-called Friedrichs systems, i.e. linear systems

$$Lu := \sum_{i=0}^{n-1} A_i \partial_{x_i} u + Cu = f,$$

with matrices A_i and C depending on x. Suppose $u : \mathbb{R}^n \supset \Omega \to \mathbb{R}^m$, then the matrices belong to $\mathbb{R}^{m \times m}$. The formal adjoint operator L^* is given by

$$L^* u := -\sum_{i=0}^{n-1} \partial_{x_j}(A_j^* u) + C^* u$$

where A_i^* and C^* denote the transposed matrices. We want to specialize on the case of time-dependent hyperbolic problems, i.e. the case in which one of the x_i, say x_0 is time t. Note that Mackenzie, Süli and Warnecke considered a much more general class of systems to which their analaysis applies, see [52].

Restricting to the spatial two-dimensional case, we consider systems of the form

$$Lu := \sum_{j=0}^{2} A_j(x) \partial_{x_j} u + C(x)u = 0, \tag{3.13}$$

where now $x = (x_0, x_1, x_2)^T := (t, x_1, x_2)^T$ is a space-time coordinate. According to Friedrichs [43] we call the system *symmetric positiv*, if the following two conditions are satisfied:

1. for each space-time prism P_{in} there exists a weight function $w_{in} \in C^1(\overline{P}_{in})$, positive on $\overline{P}_{in}$, and a positive constant c_{in} such that, for all $x \in \overline{P}_{in}$, in a component-wise sense,

$$\frac{1}{2}(K(x) + K^*(x)) \geq c_{in}I,$$

where

$$K := C + \sum_{j=0}^{2} \left(\partial_{x_j} \log w_{in}\right) A_j - \frac{1}{2}\sum_{j=0}^{2} \partial_{x_j} A_j;$$

2. $A_j(x) = A_j^*(x)$ for all $x \in \overline{P}_{in}$, $j = 0,1,2$.

Note that this conditions are satisfied if the system is *symmetric hyperbolic*, i.e. if the matrices $A_j, j = 0,1,2$, are symmetric and A_0 is positive definite. The weight function can then be chosen as

$$w_{in} := \exp(\xi_{in}(x - x_{in}^c)),$$

where ξ_{in} is a local time-like direction on P_{in} (suitably scaled so as to satisfy hypothesis (1) above), and x_{in}^c is the centroid of the space time prism P_{in}. Obviously, time itself is a time-like direction so that $\xi_{in} = (1,0,0)^T$ in our context.

If the weight function looks a bit weired one can replace condition (1) by the requirement that there may exist a vector $\xi \in \mathbb{R}^3$ such that the symmetric part of the matrix

$$K_\xi := C + \frac{1}{2}\sum_{j=0}^{2} \partial_{x_j} A_j(x) + \sum_{j=0}^{2} \xi_i A_i(x)$$

is positive definite, uniformly on the space-time domain, i.e. there exists $c_0 > 0$ such that

$$\frac{1}{2}\left(K_\xi(x) + K_\xi^*(x)\right) \geq c_0 I, \quad x \in \overline{P}_{in}.$$

The hyperbolicity in the sense of Lax:

$$\exists c_0 > 0 \, \exists \xi \in \mathbb{R}^3 \backslash \{0\}: \quad \sum_{j=0}^{2} \xi_i A_i(x) \geq c_0 I, \qquad (3.14)$$

is a sufficient condition for the above requirement.

We are now going to show how the system of Euler equations can be transformed, at least in the sense of local linearisation, into a Friedrichs system.

3.12.2   Symmetrising the Euler equations

The system of Euler equations governing compressible, inviscid flow was introduced above and we shall use the notation introduced there. Note that although the Jacobi matrices $\nabla_u f_i, i = 1,2$, can be diagonalised individually, they can not be diagonalised simultaneously. To see this, note that the product of the Jacobian matrices is not symmetric, i.e.

$$(\nabla_u f_1(u) \nabla_u f_2(u))^* \neq \nabla_u f_1(u) \nabla_u f_2(u).$$

If both Jacobian matrices could be diagonalised simultaneously there would exist an orthogonal transformation matrix $D(u) \in \mathbb{R}^{4 \times 4}$ such that

$$D^{-1}(u) \nabla_u f_i(u) D(u) = \Lambda_i, \quad i = 1,2,$$

where Λ_i is the required diagonal matrix. If we assume the existence of such a matrix the calculation

$$
\begin{aligned}
D^{-1}(u) \nabla_u f_1(u) \nabla_u f_2(u) D(u) &= D^{-1}(u) \nabla_u f_1(u) D(u) D^{-1}(u) \nabla_u f_2(u) D(u) \\
&= \lambda_1(u) \lambda_2(u) =: \Lambda_3(u) = \Lambda_3^*(u) \\
&= \left(D^{-1}(u) \nabla_u f_1(u) \nabla_u f_2(u) D(u) \right)^* \\
&= D^{-1}(u) \left(\nabla_u f_1(u) \nabla_u f_2(u) \right)^* D(u)
\end{aligned}
$$

shows that the product $\nabla_u f_1(u) \nabla_u f_2(u)$ is then necessarily symmetric, which leads to a contradiction.

Nevertheless, the system of Euler equations can be symmetrised by means of entropy variables. The entropy density is given by

$$\eta(u) := -\rho s,$$

where $s := \log p \rho^{-\kappa}$ denotes the thermodynamical entropy. We can now introduce new variables, the so-called entropy variables, into the Euler equations by means of the transformation

$$u \longmapsto U(u) := \nabla_u \eta(u).$$

It is well-known (cp. [49]) that this change of variables symmetrises the system of Euler equations. Applying this transformation leads to

$$\partial_t u(U) + \sum_{i=1}^2 \partial_{x_i} f_i(u(U)) = 0 \tag{3.15}$$

which is still in conservation form. Using the chain rule, yields

$$(\nabla_U u) \partial_t U + \sum_{i=1}^2 (\nabla_u f_i(u(U))) (\nabla_U u) \partial_{x_i} U = 0,$$

or, denoting $A^0(U) := \nabla_U u$, we can write this as

$$A^0(U) \partial_t U + \sum_{i=1}^2 (\nabla_u f_i(u(U))) A^0(u) \partial_{x_i} U = 0. \tag{3.16}$$

The mapping $u \longmapsto U$ is given by

$$U(u) = \frac{\kappa - 1}{p} \begin{bmatrix} \frac{p}{\kappa - 1}(\kappa + 1 - s) - \rho E \\ \rho v_1 \\ \rho v_2 \\ -\rho \end{bmatrix} =: \begin{bmatrix} U_1 \\ U_2 \\ U_3 \\ U_4 \end{bmatrix}, \tag{3.17}$$

while the inverse $U \longmapsto u$ is given by

$$u(U) = \frac{p}{\kappa - 1} \begin{bmatrix} -U_4 \\ U_2 \\ U_3 \\ 1 - \frac{1}{2}\frac{U_2^2 + U_3^2}{U_4} \end{bmatrix},$$

see [49], for example. Thus,

$$A^0 = \frac{1}{\kappa - 1} \begin{bmatrix} \rho & \rho v_1 & \rho v_2 & \frac{p}{\kappa-1} + \frac{1}{2}\rho|v|^2 \\ & p + \rho v_1^2 & \rho v_1 v_2 & \frac{1}{2}\rho v_1|v|^2 + \frac{\kappa v_1 p}{\kappa-1} \\ & & p + \rho v_2^2 & \frac{1}{2}\rho v_2|v|^2 + \frac{\kappa v_2 p}{\kappa-1} \\ & \text{symm} & & \frac{1}{4}\rho|v|^2 - \frac{\kappa|v|^2 p^2}{(\kappa-1)^2} + \frac{\kappa p^2}{\rho(\kappa-1)^2} \end{bmatrix},$$

while its inverse $A^{0^{-1}}$ is given by

$$A^{0^{-1}} = \frac{\kappa - 1}{p^2} \begin{bmatrix} \frac{1}{4}\rho|v|^2 + \frac{\kappa p^2}{\kappa-1} & \frac{1}{2}\rho v_1|v|^2 & \frac{1}{2}\rho v_2|v|^2 & \frac{p}{\kappa-1} - \frac{1}{2}\rho|v|^2 \\ & \rho v_1^2 + \frac{p}{\kappa-1} & \rho v_1 v_2 & -\rho v_1 \\ & & \rho v_2^2 + \frac{p}{\kappa-1} & -\rho v_2 \\ & \text{symm} & & \rho \end{bmatrix}.$$

Thus, we can rewrite the system as

$$\sum_{i=0}^{2} A_i(U)\partial_{x_i} U = 0$$

where $A_0(U) := A^0(U)$ and $A_i(U) := \nabla_u f_i(u(U))A^0(U), i = 1,2$, are symmetric 4×4 matrices. A direct proof of the symmetry can be found in [57].

Applying $A^{0^{-1}}$ to (3.16) from the left, we arrive at the system

$$\partial_t U + \sum_{i=1}^{2} A^{0^{-1}}(U)\nabla_u f_i(u(U))A^0(U)\partial_{x_i} U = 0.$$

We can rewrite this as

$$\sum_{i=0}^{2} \tilde{A}_i(U)\partial_{x_i} U = 0$$

where the $\tilde{A}_i$ are non-symmetric 4×4 matrices given by

$$\tilde{A}_0(U) \quad := \quad I,$$
$$\tilde{A}_i(U) \quad := \quad A^{0^{-1}}(U)A_i(U), \quad i = 1,2.$$

Equivalently, if we define the matrix $A_0 := I$, all $\tilde{A}_i$ are given by

$$\tilde{A}_i(U) := A^{0^{-1}}(U)\nabla_u f_i(u(U))A^0(U), \quad i = 0,1,2.$$

Next we perform a local linearisation of (3.16) about a constant mean state in each cell, thus obtaining a linear symmetric positive system in entropy variables for which an *a posteriori* error analysis, similar to the one proposed in [53], may be carried out.

We consider the non-linear system (3.15) and assume the existence of a mean constant state $U_c \in \mathbb{R}^4$ such that the decomposition

$$U = U_c + V$$

holds for a small non-constant perturbation function V. It follows that

$$
\begin{aligned}
u(U) &= u(U_c + V) = u(U_c) + \nabla_U u(U_c)V + \mathcal{O}(|V|^2) \\
&= u(U_c) + A^0(U_c)V + \mathcal{O}(|V|)^2
\end{aligned}
$$

and

$$
\begin{aligned}
f_i(u(U)) =: F_i(U) &= F_i(U_c + V) = F_i(U_c) + \nabla_U F_i(U_c)V + \mathcal{O}(|V|^2) \\
&= F_i(U_c) + \nabla_U f_i(u(U_c))V + \mathcal{O}(|V|^2) \\
&= F_i(U_c) + \nabla_u f_i(u(U_c))(\nabla_U u)V + \mathcal{O}(|V|^2) \\
&= F_i(U_c) + \nabla_u f_i(u(U_c))A^0(U_c)V + \mathcal{O}(|V|^2).
\end{aligned}
$$

Writing $u_c := u(U_c)$ and dropping the $\mathcal{O}(|V|^2)$ terms, we get the symmetric system

$$A^0(U_c)\partial_t V + \sum_{i=1}^{2} \nabla_u f_i(u_c)A^0(U_c)\partial_{x_i} V = 0, \qquad (3.18)$$

where all matrix elements are constant. Applying the inverse constant matrix $A^{0^{-1}}$ from the left yields the non-symmetric linear system

$$\partial_t V + \sum_{i=1}^{2} A^{0^{-1}}(U_c)\nabla_u f_i(u_c)A^0(U_c)\partial_{x_i} V = 0. \qquad (3.19)$$

Equation (3.18) is our starting point. It can be rewritten as

$$LV := \sum_{i=0}^{2} A_i(U_c)\partial_{x_i} V = 0, \qquad (3.20)$$

where the A_i are evaluated at the constant state U_c, i.e.

$$
\begin{aligned}
A_0(U_c) &:= A^0(U_c), \\
A_i(U_c) &:= \nabla_u f_i(u_c)A^0(U_c), \quad i = 1,2.
\end{aligned}
$$

Comparing system (3.20) with the generic system (3.13) we see that our symmetrised, linearised Euler equations in a space-time prism is very simple. In comparison with (3.13) we have $C \equiv 0$ and matrices A_j which do not depend on space and time. However, we present the *a posteriori* error analaysis for the more general case (3.13).

Before we can proceed some remarks concerning boundary conditions and the kind of error which can be controlled by the residual are in order.

3.12.3   Cell error and transported error

We consider a space-time prism $P_{in} := (n\Delta t,(n+1)\Delta t) \times T_i$ for a triangle $T_i \in \mathcal{T}^h$ and introduce the matrix

$$B(P_{in}) := \sum_{j=0}^{2} A_j n_j,$$

where $n = (n_0,n_1,n_2)^T$ denotes the outer unit normal vector to ∂P_{in}. We shall suppose that $B(P_{in})$ is non-singular on ∂P_{in}, i.e. ∂P_{in} is a non-characteristic hypersurface for the operator L. Now B is splitted into a negative semi-definite part B^- and a positive semi-definite part $B^+ = B - B^-$. We call B^-u the inflow part of the vector field u and B^+u its outflow part. Given a sufficiently smooth function g on ∂P_{in},

$$B^-u = B^-g \quad \text{on } \partial P_{in}$$

defines an admissible boundary problem for the symmetric hyperbolic problem introduced above in the following sense. It was shown in [43] and [50] that symmetric hyperbolic systems have unique strong solutions subject to this boundary condition.

We proceed in assuming that we are given a numerical solution u^h on the space-time prism P_{in}, and that we have calculated, converting to entropy variables, $U^h = U(u^h)$. Following [53] we consider the following boundary value problem on P_{in}:

$$\begin{aligned} L\tilde{U}^h &= 0 \quad \text{on } P_{in} \\ B^-\tilde{U}^h\big|_{\partial P_{in}} &= B^-U^h\big|_{\partial P_{in}}. \end{aligned}$$

We interpret the function $\tilde{U}^h$ as follows. Suppose that we have constructed an approximation U^h to the analytical solution U of (3.20). Then the boundary data $B^-U^h|_{\partial P_{in}}$ is distorted by numerical errors upwind of cell P_{in}. Thus, $\tilde{U}^h$ is the exact solution of (3.20) under these distorted boundary values. Consequently,

$$e_{P_{in}}^{cell} := U^h - \tilde{U}^h$$

is the error in the numerical solution which is produced on P_{in}, while

$$e_{P_{in}}^{trans} := \tilde{U}^h - U$$

is the error which is created upwind of the cell P_{in} and is advected into the cell by the numerical method. We call $e_{P_{in}}^{cell}$ the *cell error* and $e_{P_{in}}^{trans}$ the *transported error*. Clearly,

$$e_{P_{in}} = e_{P_{in}}^{cell} + e_{P_{in}}^{trans}.$$

It is important to note that the residual has no direct control over the transported error. Indeed, a simple calculation shows that

$$Le_{P_{in}}^{trans} = L\tilde{U}^h - LU = 0,$$

subject to the boundary condition

$$B^-e_{P_{in}}^{trans} = B^-(U^h - U)\big|_{\partial P_{in}} = B^-e\big|_{\partial P_{in}},$$

while the cell error is governed directly by the residual via

$$r^h = Le_{P_{in}} = Le_{P_{in}}^{cell} \qquad \text{on } P_{in},$$

subject to the boundary condition

$$B^- e_{P_{in}}^{cell} = 0 \qquad \text{on } \partial P_{in}.$$

Note that the remarks of this section apply equally well to the non-symmetric system (3.19), where the corresponding boundary matrix is given by

$$\tilde{B}(P_{in}) = \sum_{j=0}^{2} n_j \tilde{A}_j. \tag{3.21}$$

3.12.4 A weak *a posteriori* error estimate

We return to our generic symmetric posistiv system (3.13). Let

$$B(x) = \sum_{j=0}^{2} n_j A_j(x),$$

and suppose that the matrix $B(x)$ is non-singular, almost everywhere on ∂P_{in}. Decomposing $B(x)$ as $B(x) = B^+(x) + B^-(x)$, where $B^+(x)$ is positive semi-definite on ∂P_{in}, $B^-(x)$ is negative semi-definite on ∂P_{in}, we consider the weighted graph-norm $\|\cdot\|_{D(L,P_{in})}$ on

$$D_-(L,P_{in}) := \{\phi \in L^2(P_{in}) \,|\, L\phi \in L^2(P_{in}), \quad B^-\phi = 0 \text{ on } \partial P_{in}\},$$

defined by

$$\|\phi\|_{D(L,P_{in})} = (\|w_{in}\phi\|_{L^2(P_{in})}^2 + \|w_{in}L\phi\|_{L^2(P_{in})}^2)^{1/2},$$

and the associated dual graph-norm

$$\|v\|_{D'(L,P_{in})} := \sup_{\phi \in D_-(L,P_{in})} \frac{|(v,\phi)_{P_{in}}|}{\|\phi\|_{D(L,P_{in})}},$$

where $(\cdot,\cdot)_{P_{in}}$ denotes the usual L^2 inner product on P_{in}. Similarly, by introducing the formal adjoint

$$L^*\phi := -\sum_{j=0}^{2} \partial_{x_j}(A_j\phi) + C^*\phi,$$

with

$$\phi \in D_+(L^*,P_{in}) := \{\phi \in L^2(P_{in}) \,|\, L^*\phi \in L^2(P_{in}), \quad B^+\phi = 0 \text{ on } \partial P_{in}\},$$

we may equip $D_+(L^*,P_{in})$ with the graph-norm

$$\|\phi\|_{D(L^*,P_{in})} = (\|w_{in}\phi\|^2_{L^2(P_{in})} + \|w_{in}L^*\phi\|^2_{L^2(P_{in})})^{1/2}.$$

The associated dual graph-norm is defined by

$$\|v\|_{D'(L^*,P_{in})} := \sup_{\phi \in D_+(\mathcal{L}^*,P_{in})} \frac{|(v,\phi)_{P_{in}}|}{\|\phi\|_{D(L^*,P_{in})}}.$$

The existence of the traces is not obvious and has to be proven. This was done in [53].

We have the following local *a posteriori* error bound for the L^2-norm of the cell error $e^{cell}_{P_{in}}$ on a space-time prism P_{in} in terms of the dual graph-norm of the local residual r^h.

Theorem 3.12.1 *Under hypotheses* (1) *and* (2) *stated above,*

$$(\min_{P_{in}} w_{in}) \|r^h\|_{D'(L^*,P_{in})} \le \|e_c\|_{L^2(P_{in})} \le \left(1 + \frac{1}{c^2_{in}}\right)^{1/2} (\max_{P_{in}} w_{in}) \|r^h\|_{D'(L^*,P_{in})}.$$

Proof: In order to simplify the notation, we shall write e_c instead of $e^{cell}_{P_{in}}$ throughout this proof. To begin, we show that the graph norm $\|\cdot\|_{D(L^*,P_{in})}$ is equivalent to the norm $\|w_{in}L^*\cdot\|_{L^2(P_{in})}$ on $D_+(L^*,P_{in})$. Then, in the second part of the proof, we shall use this result to deduce the stated two-sided bound on the cell error in terms of the dual graph-norm of the local residual.

A straightforward calculation based on integration by parts shows that, for any $\phi \in D_+(L^*,P_{in})$,

$$\begin{aligned}
\left(\partial_{x_j}(A_j\phi),w^2_{in}\phi\right)_{P_{in}} &= \frac{1}{2}\left(\hat{\nu}_j A_j\phi,w^2_{in}\phi\right)_{\partial P_{in}} \qquad (3.22)\\
&\quad + \frac{1}{2}\left(\left(\partial_{x_j}A_j - A_j\partial_{x_j}(\ln w^2_{in})\right)\phi,w^2_{in}\phi\right)_{P_{in}},
\end{aligned}$$

for $j = 0,1,2$, where $\hat{\nu} = (\hat{\nu}_0,\hat{\nu}_1,\hat{\nu}_2)$ denotes the unit outward normal to ∂P_{in}. To prove this equality we start from its right-hand side. It follows from integration by parts that

$$\begin{aligned}
&\frac{1}{2}\left((\partial_{x_j}A_j)\phi,w^2_{in}\phi\right)_{P_{in}} - \frac{1}{2}\left((\partial_{x_j}w^2_{in})A_j\phi,\phi\right)_{P_{in}}\\
&= \frac{1}{2}\left(n_j A_j\phi,w^2_{in}\phi\right)_{\partial P_{in}} - \left(A_j\phi,(\partial_{x_j}w^2_{in})\phi + w^2_{in}(\partial_{x_j}\phi)\right)_{P_{in}}.
\end{aligned}$$

Thus, the right-hand side of (3.22) becomes

$$\left(n_j A_j\phi,w^2_{in}\phi\right)_{\partial P_{in}} - \left(A_j\phi,(\partial_{x_j}w^2_{in})\phi + w^2_{in}(\partial_{x_j}\phi)\right)_{P_{in}}.$$

On the other hand, integration by parts on the left-hand side of (3.22) yields

$$\begin{aligned}
\left(\partial_{x_j}(A_j\phi),w^2_{in}\phi\right)_{P_{in}} &= \left(n_j A_j\phi,w^2_{in}\phi\right)_{\partial P_{in}} - \left(A_j\phi,\partial_{x_j}(w^2_{in}\phi)\right)_{P_{in}}\\
&= \left(n_j A_j\phi,w^2_{in}\phi\right)_{\partial P_{in}}\\
&\quad - \left(A_j\phi,(\partial_{x_j}w^2_{in})\phi + w^2_{in}(\partial_{x_j}\phi)\right)_{P_{in}}
\end{aligned}$$

which proves (3.22). Thence, with

$$B = \sum_{j=0}^{2} \hat{\nu}_j A_j,$$

and $B^+ + B^- = B$, we have that

$$\left(L^*\phi, w_{in}^2\phi\right)_{P_{in}} = \left(K^*\phi, w_{in}^2\phi\right)_{P_{in}} - \frac{1}{2}\sum_{j=0}^{2}\left(n_j A_j\phi, w_{in}^2\phi\right)_{\partial P_{in}},$$

yielding

$$(L^*\phi, w_{in}^2\phi)_{P_{in}} = (K^*\phi, w_{in}^2\phi)_{P_{in}} + \frac{1}{2}(-B^-\phi, w_{in}^2\phi)_{\partial P_{in}},$$

where we have made use of the fact that $B^+\phi = 0$ on ∂P_{in}. Now

$$
\begin{aligned}
(L^*\phi, w_{in}^2\phi)_{P_{in}} &= \frac{1}{2}\left[(L^*\phi, w_{in}^2\phi)_{P_{in}} + (w_{in}^2\phi, L^*\phi)_{P_{in}}\right] \\
&= \frac{1}{2}\left((K + K^*)\phi, w_{in}^2\phi\right)_{P_{in}} + \frac{1}{2}(-B^-\phi, w_{in}^2\phi)_{\partial P_{in}}.
\end{aligned}
$$

Recalling that the matrix $-B^-$ is positive semi-definite and exploiting hypothesis (1), we deduce that

$$(L^*\phi, w_{in}^2\phi)_{P_{in}} \geq c_{in}\|w_{in}\phi\|_{L^2(P_{in})}^2.$$

Applying the Cauchy-Schwarz inequality on the left-hand side yields

$$\|w_{in}L^*\phi\|_{L^2(P_{in})} \geq c_{in}\|w_{in}\phi\|_{L^2(P_{in})}.$$

Consequently,

$$\|\phi\|_{D(L^*, P_{in})} \leq \left(1 + \frac{1}{c_{in}^2}\right)^{1/2}\|w_{in}L^*\phi\|_{L^2(P_{in})}. \tag{3.23}$$

Since, by the definition of the graph norm,

$$\|w_{in}L^*\phi\|_{L^2(P_{in})} \leq \|\phi\|_{D(L^*, P_{in})}, \tag{3.24}$$

recalling (3.23) we obtain the two-sided bound

$$\|w_{in}L^*\phi\|_{L^2(P_{in})} \leq \|\phi\|_{D(L^*, P_{in})} \leq \left(1 + \frac{1}{c_{in}^2}\right)^{1/2}\|w_{in}L^*\phi\|_{L^2(P_{in})}, \tag{3.25}$$

for any $\phi \in D_+(L^*, P_{in})$. We shall exploit this pair of inequalities to derive a bound on the dual graph-norm of the residual r^h. Since

$$
\begin{aligned}
L e_c &= r^h &&\text{on } P_{in} \\
B^- e_c &= 0 &&\text{on } \partial P_{in},
\end{aligned}
$$

it follows that

$$
\begin{aligned}
\|r^h\|_{D'(L^*,P_{in})} &= \sup_{\phi \in D_+(L^*,P_{in})} \frac{|(Le_c,\phi)_{P_{in}}|}{\|\phi\|_{D(L^*,P_{in})}} \\
&= \sup_{\phi \in D_+(L^*,P_{in})} \frac{|(e_c,L^*\phi)_{P_{in}}|}{\|\phi\|_{D(L^*,P_{in})}} \\
&= \sup_{\phi \in D_+(L^*,P_{in})} \frac{|(w_{in}^{-1}e_c,w_{in}L^*\phi)_{P_{in}}|}{\|\phi\|_{D(L^*,P_{in})}} \\
&\leq \sup_{\phi \in D_+(L^*,P_{in})} \frac{\|w_{in}^{-1}e_c\|_{L^2(P_{in})}\|w_{in}L^*\phi\|_{L^2(P_{in})}}{\|\phi\|_{D(L^*,P_{in})}}.
\end{aligned}
$$

Thus, by virtue of (3.24), we obtain

$$
\|r^h\|_{D'(L^*,P_{in})} \leq \|w_{in}^{-1}e_c\|_{L^2(P_{in})} \leq (\min_{P_{in}} w_{in})^{-1} \|e_c\|_{L^2(P_{in})},
$$

and hence

$$
(\min_{P_{in}} w_{in}) \|r^h\|_{D'(L^*,P_{in})} \leq \|e_c\|_{L^2(P_{in})}, \tag{3.26}
$$

which is the desired lower bound on the cell error. In order to prove the upper bound on the L^2-norm of the cell error, we consider the auxiliary problem

$$
\begin{aligned}
L^*\psi &= e_c &&\text{on } P_{in} \\
B^+\psi &= 0 &&\text{on } \partial P_{in};
\end{aligned}
$$

this has a unique solution $\psi \in D_+(L^*,P_{in})$ satisfying (3.25) (see [50]). Thus,

$$
\begin{aligned}
\|r^h\|_{D'(L^*,P_{in})} &= \sup_{\phi \in D_+(L^*,P_{in})} \frac{|(Le_c,\phi)_{P_{in}}|}{\|\phi\|_{D(L^*,P_{in})}} \\
&= \sup_{\phi \in D_+(L^*,P_{in})} \frac{|(e_c,L^*\phi)_{P_{in}}|}{\|\phi\|_{D(L^*,P_{in})}} \\
&\geq \frac{|(e_c,L^*\psi)_{P_{in}}|}{\|\psi\|_{D(L^*,P_{in})}} \\
&= \frac{\|e_c\|_{L^2(P_{in})}\|L^*\psi\|_{L^2(P_{in})}}{\|\psi\|_{D(L^*,P_{in})}} \\
&= \frac{\|e_c\|_{L^2(P_{in})}\|w_{in}^{-1}w_{in}L^*\psi\|_{L^2(P_{in})}}{\|\psi\|_{D(L^*,P_{in})}} \\
&\geq (\max_{P_{in}} w_{in})^{-1} \frac{\|e_c\|_{L^2(P_{in})}\|w_{in}L^*\psi\|_{L^2(P_{in})}}{\|\psi\|_{D(L^*,P_{in})}}.
\end{aligned}
$$

Recalling (3.23), it follows that

$$
\|r^h\|_{D'(L^*,P_{in})} \geq \left(1 + \frac{1}{c_{in}^2}\right)^{-1/2} (\max_{P_{in}} w_{in})^{-1}\|e_c\|_{L^2(P_{in})},
$$

which yields the required upper bound on the L^2-norm of the cell error in terms of the dual graph norm of the local residual r^h. $\qquad\square$

3.12.5 A discrete dual graph-norm indicator

There are in principle two different strategies for the implementation of the dual graph norm indicator: both of them are based on approximating $\|r^h\|_{D'(L^*,P_{in})}$ using a subdivision of P_{in}. On a particular subdivision we exploit a finite element basis $\{\Phi\}$ that satisfies the local boundary condition $B^+\Phi = 0$ on ∂P_{in}. We start by describing the imposition of boundary conditions. In order to derive an applicable error indicator we start with the non-symmetric system (3.19) which is nothing but the symmetric system (3.18) in a form more convenient for computational purposes. Consider a space-time prism P_{in} as in figure 3.79. Recall that the boundary condition matrix for system (3.19) is given by

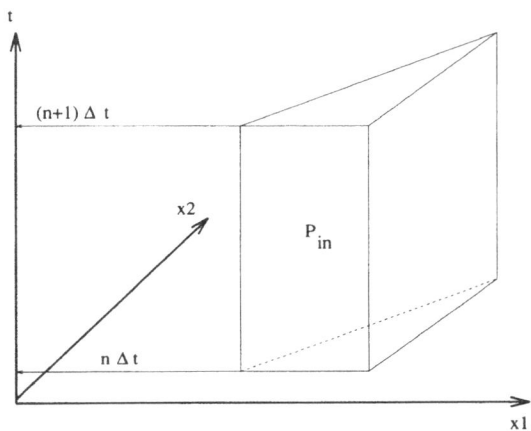

Figure 3.79 Space-time prism P_{in}

$$\tilde{B}(P_{in}) = \sum_{j=0}^{2} n_j \tilde{A}_j = \sum_{j=0}^{2} n_j A^{0^{-1}}(U_{ci}) \nabla_u f_j(u_{ci}) A^0(U_{ci})$$

$$= n_0 I + \sum_{j=1}^{2} n_j A^{0^{-1}}(U_{ci}) \nabla_u f_j(u_{ci}) A^0(U_{ci}),$$

where U_{ci}, u_{ci} denote constant mean states of entropy and conservative variables, respectively, within the prism P_{in}.

In order to distinguish inflow from outflow boundaries the matrix $\tilde{B}(P_{in})$ is split into a positive semi-definite part $\tilde{B}^+(P_{in})$, the outflow part, and a negative semi-definite part $\tilde{B}^-(P_{in})$, the inflow part, so that

$$\tilde{B}(P_{in}) = \tilde{B}^+(P_{in}) + \tilde{B}^-(P_{in}).$$

Since the matrices $\tilde{A}_j$ are similar to $\nabla_u f_j$, for $j = 1,2$, and $\tilde{A}_0$ is similar to I, the set of eigenvalues of $\tilde{B}(P_{in})$ is given by

$$ev(\tilde{B}(P_{in})) = n_0 + ev\left(\sum_{j=1}^{2} n_j \nabla_u f_j(u_{ci})\right).$$

Now it is well known (see [42]) that there exists an invertible matrix $P(u_{ci},n_1,n_2) \in \mathbb{R}^{4\times4}$ such that

$$P(u_{ci},n_1,n_2)^{-1} \left(\sum_{j=1}^{2} n_j \nabla_u f_j(u_{ci}) \right) P(u_{ci},n_1,n_2) = \Lambda(u_{ci},n_1,n_2),$$

where Λ is the diagonal matrix

$$\Lambda(u_{ci},n_1,n_2) \;=\; \mathrm{diag}\left\{ \sum_{j=1}^{2} n_j v_{ci,j}, \sum_{j=1}^{2} n_j v_{ci,j}, \sum_{j=1}^{2} n_j v_{ci,j} + a_{ci}|(n_1,n_2)|, \right.$$
$$\left. \sum_{j=1}^{2} n_j v_{ci,j} - a_{ci}|(n_1,n_2)| \right\},$$

$a_{ci} := \sqrt{\kappa \frac{p_{ci}}{\rho_{ci}}}$ denoting the mean constant speed of sound in P_{in}. We split Λ into a matrix Λ^+ containing the positive eigenvalues, and Λ^- containing the negative eigenvalues, i.e.

$$\Lambda(u_{ci},n_1,n_2) = \Lambda^+(u_{ci},n_1,n_2) + \Lambda^-(u_{ci},n_1,n_2).$$

Thus, $\sum_{j=1}^{2} n_j \nabla_u f_j(u_{ci})$ can be represented as a sum of a positive semi-definit and a negative semi-definite part, i.e.

$$\sum_{j=1}^{2} n_j \nabla_u f_j(u_{ci}) = P\lambda^+ P^{-1}(u_{ci},n_1,n_2) + P\lambda^- P^{-1}(u_{ci},n_1,n_2)$$

and we end up with a representation for the boundary matrix $B(P_{in})$ of the form

$$\tilde{B}(P_{in}) = n_0 I \;\; + \;\; A^{0^{-1}}(U_{ci}) \left(P\Lambda^+ P^{-1}(u_{ci},n_1,n_2) \right) A^0(U_{ci})$$
$$+ \;\; A^{0^{-1}}(U_{ci}) \left(P\Lambda^- P^{-1}(u_{ci},n_1,n_2) \right) A^0(U_{ci}).$$

Note that on the bottom face $T_i \cap \{n\Delta t\}$ of the space-time prism P_{in} there holds $n = (-1,0,0)$ and thus

$$\tilde{B}(P_{in}) = -I = \tilde{B}^-(P_{in}).$$

Therefore, the bottom face is an inflow boundary of P_{in}. Analogously, on the top face $T_i \cap \{(n+1)\Delta t\}$ we have $n = (1,0,0)$ and thus

$$\tilde{B}(P_{in}) = I = \tilde{B}^+(P_{in}),$$

i.e. this is an outflow face. On the three side faces of P_{in}, $n_0 = 0$ is valid and thus $\tilde{B}(P_{in})$ is split according to the signs of eigenvalues in Λ. Note that this corresponds to the flux vector splitting of Steger and Warming, see [58]. Thus, if $\phi \in D_+(L^*,P_{in})$ is sought then the components of ϕ have to be chosen in order to cancel the expression

$$\tilde{B}^+(P_{in})\phi = 0$$

on the face under consideration. This is achieved by splitting Λ according to $\Lambda = \Lambda^+ + \Lambda^-$ and looking for ϕ which satisfy the condition

$$\Lambda^+\phi = 0.$$

3.12.6 The error indicator

As was already noted, two different strategies exist in principle for setting up the graph norm error indicator. In both approaches the space-time prisms P_{in} have to be subdivided and a finite set of test functions ϕ has to be defined on this subdivision. In the first approach one uses test functions $\phi \in D_+(L^*, P_{in})$ in linear-linear tensor product form

$$\phi(x,t) = \alpha_{000} + \alpha_{100}t + \alpha_{010}x_1 + \alpha_{001}x_2 + \alpha_{110}x_1t + \alpha_{101}x_2t,$$

compare [41]. On T_i the Lagrange interpolants $u_{T_i}^h$ of the cell averages are computed at times $n\Delta t$ and $(n+1)\Delta t$ and u_{ci} is defined to be the average of the six values of $u_{T_i}^h$ at the nodes of T_i at the two time levels. Using quadrature rules the dual graph norm of the residual may be computed within P_{in} on the partition of the space-time prisms.

For the sake of simplicity and to save computing time we shall exploit an alternative to this approach which seems to be better suited to explicit time stepping schemes. We aim to compute the dual graph norm indicator from data which is available at time $n\Delta t$. Note that after one flux balance we know not only the values $\bar{u}_l(n\Delta t)$ but also the values of the temporal change

$$\frac{\bar{u}_l((n+1)\Delta t) - \bar{u}_l(n\Delta t)}{\Delta t} = Q_l,$$

where Q_l denotes the spatial finite volume discretisation. On triangle T_i a subdivision of the form shown in figure 3.80 is established and we assume our test functions to be constant in time, piecewise linear in space, i.e. of the form

$$\phi(x,t) = \phi(x) = \alpha_{00} + \alpha_{10}x_1 + \alpha_{01}x_2.$$

The subdivision of T_i gives rise to 15 different test functions $\phi_k, k = 1, \ldots, 15$, which we

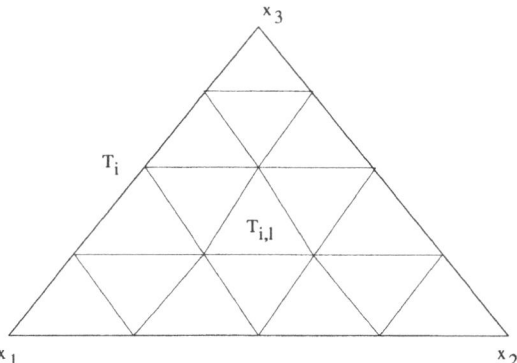

Figure 3.80 Local subdivision of triangle T_i

define to be the linear hat functions characterised by $\phi_k(x_j) = \delta_j^k$. To take into account the boundary conditions we proceed as follows. If element λ_{ll}^+ of Λ^+ is non-zero at one of the boundary points on ∂T_i the corresponding component of ϕ_k, i.e. $\phi_{k,l}$, is set to zero so that $\Lambda^+ \phi_k = 0$ holds for all ϕ_k. Obviously, $\phi_k \in D_+(L^*, P_{in})$. Thus, the dual graph norm is approximated by

$$\|r^h\|_{D'(L^*,P_{in})} \approx \max_{k=1,\dots,15} \frac{|(r^h,\phi_k)_{P_{in}}|}{\|\phi_k\|_{D(L^*,P_{in})}},$$

with ϕ_k as previously defined. To compute the numerator and the denominator in the approximated dual graph norm we choose the mean constant state in P_{in} to be

$$u_{ci} := \frac{1}{3}(\bar{u}_1(n\Delta t) + \bar{u}_2(n\Delta t) + \bar{u}_3(n\Delta t)), \tag{3.27}$$

where the nodes of T_i are labelled 1,2,3 for simplicity. Note that this gives an $\mathcal{O}(\Delta t)$-approximation to a mean constant state which also includes the values at time level $(n+1)\Delta t$, provided we assume smoothness of the solution. Thus,

$$u^h(x,n\Delta t)|_{T_i} - u_{ic} \tag{3.28}$$

is identified as linear perturbation of the mean constant state at time $[n\Delta t,(n+1)\Delta t]$. From (3.27) the mean constant state of entropy variabels U_{ci} is computed using (3.17) and from (3.28) the linear perturbation

$$V(x,t) = U(x,t)|_{T_i} - U_{ci}, \quad t \in [n\Delta t,(n+1)\Delta t],$$

follows. According to (3.20) the residual is now computed from

$$r^h = LV = \partial_t V + \sum_{j=1}^{2} A^{0^{-1}}(U_{ci})\nabla_u f_i(u_{ci})A^0(U_{ci})\partial_{x_i} V.$$

Taking the L^2-inner product with our test functions $\phi_k \in D_+(L^*,P_{in})$ leads to

$$(r^h,\phi_k)_{P_{in}} = \int_{T_i}\int_{n\Delta t}^{(n+1)\Delta t} \partial_t V \cdot \phi_k \, dt \, dx$$
$$+ \sum_{j=1}^{2} A^{0^{-1}}(U_{ci})\nabla_u f_j(u_{ci})A^0(U_{ci}) \int_{T_i}\int_{n\Delta t}^{(n+1)\Delta t} \partial_{x_j} V \cdot \phi_k \, dt \, dx.$$

In order to stay at the n-th time level in the spatial part we use a simple one-point quadrature rule where the quadrature point is $n\Delta t$. This gives

$$(r^h,\phi_k)_{P_{in}} = \int_{T_i} \phi_k(x) \cdot (V(x,(n+1)\Delta t) - V(x,n\Delta t)) \, dx$$
$$+ \sum_{j=1}^{2} A^{0^{-1}}(U_{ci})\nabla_u f_j(u_{ci})A^0(U_{ci})\Delta t \int_{T_i} \partial_{x_j} V(x,n\Delta t) \cdot \phi_k(x) \, dx$$
$$+ \mathcal{O}(\Delta t^2).$$

For each of the nodes $x_l, l = 1,2,3$, of the triangle T_i we know the value Q_l of

$$\frac{\bar{u}_l((n+1)\Delta t) - \bar{u}_l(n\Delta t)}{\Delta t} = Q_l.$$

Since Q_l can be expressed as the difference of conservative variables we can switch to entropy data by means of transformation (3.17) yielding

$$\overline{U}_l := U\left(\overline{u}_l((n+1)\Delta t)\right) - U\left(\overline{u}_l(n\Delta t)\right)$$
$$= U\left(\overline{u}_l(n\Delta t) + \Delta t Q_l\right) - U\left(\overline{u}_l(n\Delta t)\right)$$

The linear interpolant of $\overline{U}_l$ on T_i given by

$$U_{T_i}^h(x) := \sum_{l=1}^{3} \overline{U}_l \varphi_l(x),$$

$\varphi_l(x_m) = \delta_j^m$, can be calculated and this yields the linear perturbation

$$V_{T_i}^h(x) := U_{T_i}^h(x) - U_{ci} = \sum_{l=1}^{3} \overline{U}_l \varphi_l(x) - U_{ci}$$

in entropy variables. Thus, the time difference $V(x,(n+1)\Delta t) - V(x,n\Delta t)$ can be replaced by a transformed spatial flux balance resulting in

$$(r^h, \phi_k)_{P_{in}} \doteq \int_{T_i} \phi_k(x) \cdot V_{T_i}^h(x)\, dx$$
$$+ \sum_{j=1}^{2} A^{0^{-1}}(U_{ci}) \nabla_u f_j(u_{ci}) A^0(U_{ci}) \Delta t \int_{T_i} \partial_{x_j} V(x,n\Delta t) \cdot \phi_k(x)\, dx$$

where we introduce the symbol $\doteq$ meaning equality up to terms of order $\mathcal{O}(\Delta t^2, h|T_i|)$. Now $\partial_{x_j} V$ is a constant on T_i and can be removed from the integral. Since $T_i = \cup_{l=1}^{16} T_{i,l}$ and $|T_{i,l}| = |T_i|/16$ we arrive at

$$(r^h, \phi_k)_{P_{in}} \doteq \sum_{l=1}^{16} \int_{T_{i,l}} \phi_k(x) \cdot V_{T_i}^h(x)\, dx$$
$$+ \sum_{j=1}^{2} A^{0^{-1}}(U_{ci}) \nabla_u f_j(u_{ci}) A^0(U_{ci}) \Delta t\, \partial_{x_j} V\big|_{T_i} \sum_{l=1}^{16} \int_{T_{i,l}} \phi_k(x)\, dx.$$

If we apply the one-point quadrature rule in each subtriangle $T_{i,l}$ we finally end up with

$$(r^h, \phi_k)_{P_{in}} \doteq \frac{|T_i|}{16} \sum_{l=1}^{16} \phi_k(c_{i,l}) \cdot V_{T_i}^h(c_{i,l})$$
$$+ \sum_{j=1}^{2} A^{0^{-1}}(U_{ci}) \nabla_u f_j(u_{ci}) A^0(U_{ci}) \Delta t\, \partial_{x_j} V\big|_{T_i} \frac{|T_i|}{16} \sum_{l=1}^{16} \phi_k(c_{i,l}).$$

The denominator in the dual graph norm indicator is given by

$$\|\phi_k\|_{D(L^*, P_{in})} = \sqrt{\|w_{in}\phi_k\|_{L^2(P_{in})}^2 + \|w_{in}L^*\phi_k\|_{L^2(P_{in})}^2}$$

where $w_{in}(x) = e^{\xi_{in} \cdot (x - x_{in}^c)}$ is the local scaling (weight) function. Note that using this weight function with the time-like direction ξ_{in} ensures that hypotheses (1) in subsection 3.12.4 holds. We have

$$\|w_{in}\phi_k\|^2_{L^2(P_{in})} = \int_{T_i} \int_{n\Delta t}^{(n+1)\Delta t} (w_{in}\phi_k)^2 \, dt \, dx = \int_{T_i} \phi_k^2 \int_{n\Delta t}^{(n+1)\Delta t} w_{in}^2 \, dt \, dx$$

since our test functions were assumed to be constant in time. Using a one-point quadrature in the space-time barycenter x_{in}^c yields

$$\|w_{in}\phi_k\|^2_{L^2(P_{in})} \doteq \Delta t \int_{T_i} \phi_k^2 \, dx \doteq \Delta t \sum_{l=1}^{16} \int_{T_{i,l}} \phi_k^2 \, dx \doteq \frac{|T_i|}{16} \sum_{l=1}^{16} (\phi_k(c_{i,l}))^2 \, .$$

The formal adjoint of the linearised operator

$$L = \partial_t + \sum_{j=1}^{2} A^{0^{-1}}(U_{ci})\nabla_u f_j(u_{ci})A^0(U_{ci})\partial_{x_j}$$

is simply given by

$$L^* = -L.$$

Thus,

$$\|w_{in}L^*\phi_k\|^2_{L^2(P_{in})} = \int_{T_i} \int_{n\Delta t}^{(n+1)\Delta t} (w_{in}L\phi_k)^2 \, dt \, dx.$$

Since ϕ_k are assumed not to change with time within P_{in} we arrive at

$$\begin{aligned}
\|w_{in}L^*\phi_k\|^2_{L^2(P_{in})} &= \int_{T_i} \left(\sum_{j=1}^{2} A^{0^{-1}}(U_{ci})\nabla_u f_j(u_{ci})A^0(U_{ci})\partial_{x_j}\phi_k \right)^2 \times \\
&\quad \int_{n\Delta t}^{(n+1)\Delta t} w_{in}^2 \, dt \, dx \\
&\doteq \Delta t \int_{T_i} \left(\sum_{j=1}^{2} A^{0^{-1}}(U_{ci})\nabla_u f_j(u_{ci})A^0(U_{ci})\partial_{x_j}\phi_k \right)^2 dx
\end{aligned}$$

where the time integral was again approximated by a one-point rule with quadrature point x_{in}^c. Thus, we end up with

$$\|w_{in}L^*\phi_k\|^2_{L^2(P_{in})} \doteq \Delta t \frac{|T_i|}{16} \sum_{l=1}^{16} \left(\sum_{j=1}^{2} A^{0^{-1}}(U_{ci})\nabla_u f_j(u_{ci})A^0(U_{ci})\partial_{x_j}\phi_k(c_{i,l}) \right)^2 .$$

Gathering our partial approximations gives

$$\|r^h\|_{D'(L^*,P_{in})} \approx \|r^h\|_{\Delta'(L^*,P_{in})} :=$$

$$\frac{\frac{|T_i|}{16} \left| \sum_{l=1}^{16} \phi_k(c_{i,l}) \cdot \left\{ V_{T_i}^h(c_{i,l}) + \Delta t \sum_{j=1}^{2} A^{0^{-1}}(U_{ci})\nabla_u f_j(u_{ci})A^0(U_{ci}) \, \partial_{x_j} V|_{T_i} \right\} \right|}{\sqrt{\frac{|T_i|}{16} \sum_{l=1}^{16} \left\{ (\phi_k(c_{i,l}))^2 + \Delta t \left(\sum_{j=1}^{2} A^{0^{-1}}(U_{ci})\nabla_u f_j(u_{ci})A^0(U_{ci})\partial_{x_j}\phi_k(c_{i,l}) \right)^2 \right\}}}$$

as an approximation to the dual graph norm error indicator.

The adaptive procedure using this refinement indicator is as follows. Given two tolerances $\mathrm{TOL}_{\mathrm{refine}}$ and $\mathrm{TOL}_{\mathrm{coarse}}$ the adaptive algorithm sweeps through the grid at certain times and refines all those triangles, for which

$$\|r^h\|_{\Delta'(L^*,P_{in})} < \mathrm{TOL}_{\mathrm{refine}}$$

is valid, while triangles are deleted from the mesh for which

$$\|r^h\|_{\Delta'(L^*,P_{in})} > \mathrm{TOL}_{\mathrm{coarse}},$$

holds.

3.12.7 Numerical experiments

The discrete approximation to the dual graph-norm was tested on several flow fields, see [57]. In the case of the transonic NACA0012 flow considered above the adapted grid after three adaption cycles can be seen in figure 3.81. The flow features are nicely captured

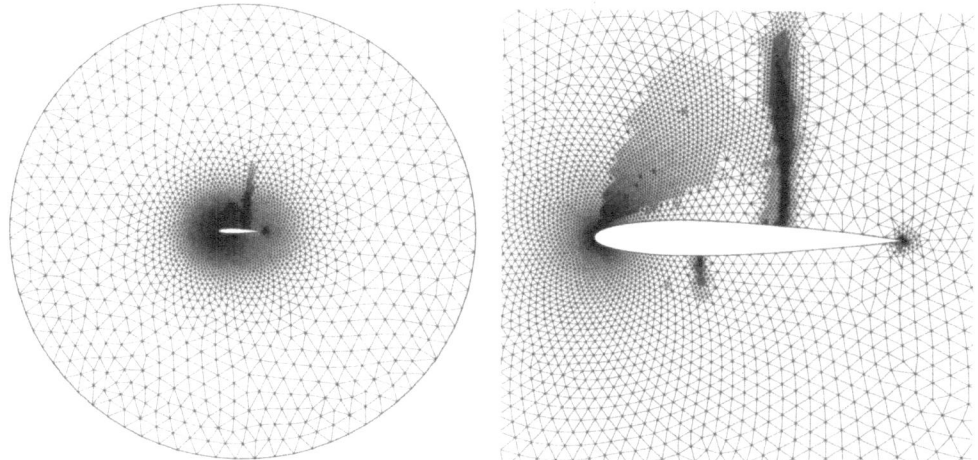

Figure 3.81 Grid after three adaption cycles for transonic flow about NACA0012 airfoil

and the indicator has started to detect the supersonic region on the upper side. Another nice feature in comparison with the H^{-1}-indicator described above is the absence of any noise spoiling the grid far away from the obstacle.

Considering $\mathrm{Ma}_\infty = 3$ flow about a cylinder results in the adapted grid and density distribution as shown in figure 3.82. The bow shock is captured as is the influence of poor outflow boundary conditions (simple extrapolation of 0th order was used). Note that the density isolines show a slight carbuncle phenomenon typical for the approximate Riemann solver of Roe. Note that the indicator adapted the vicinity of the stagnation point. Whether this behaviour depends on the carbuncle phenomenon was not studied until now but the sensitivity of error indicators concerning numerical anomalies seems to comprise an interesting area of research in CFD.

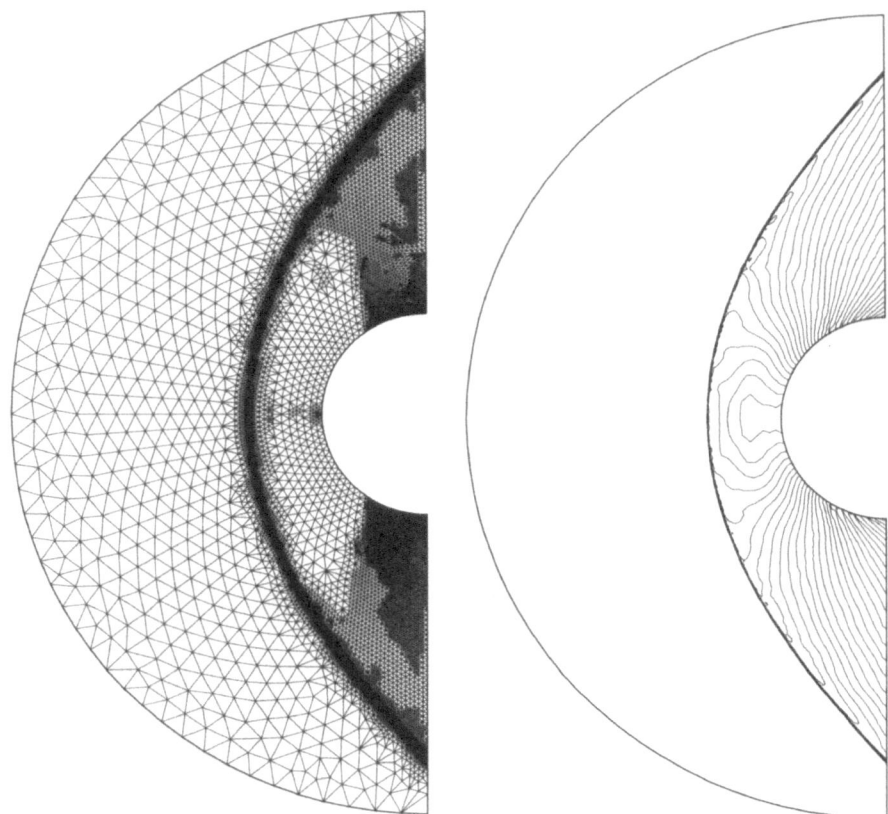

Figure 3.82 Adapted grid and density distribution

As an unsteady test case we consider again the $Ma_\infty = 3$ channel with forward facing step. Figure 3.83 shows the grid at time $t = 0.119$, where the bow shock has just detached from the step. A detail of the grid and the corresponding density distribution can be seen

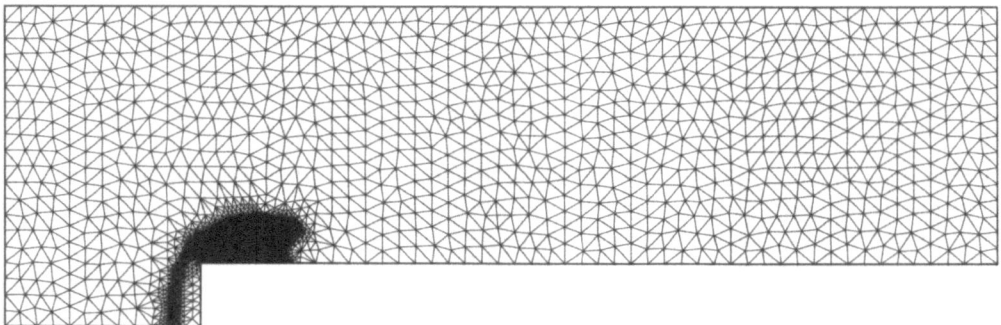

Figure 3.83 Adapted grid and detail of the grid at $t = 0.119$

in figure 3.84. At time $t = 0.67$ grid and density distribution are shown in figure 3.85. At time $t = 1.44$ the bow shock is already attached to the upper wall of the channel. As can be seen from figure 3.86 the refinement of the grid follows the development of the

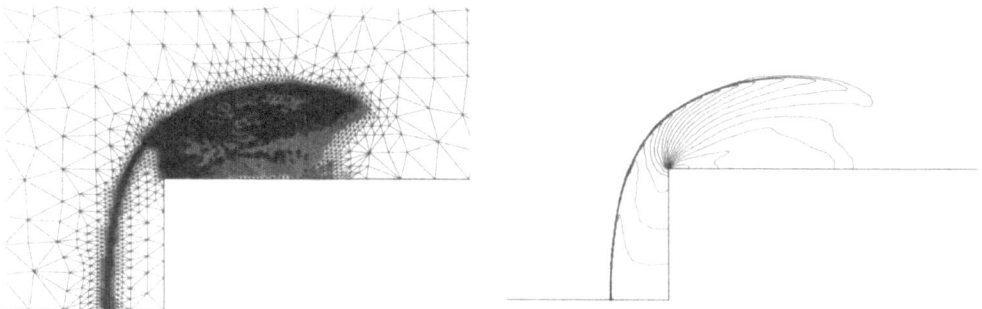

Figure 3.84 Detail of the adapted grid and density distribution at $t = 0.119$

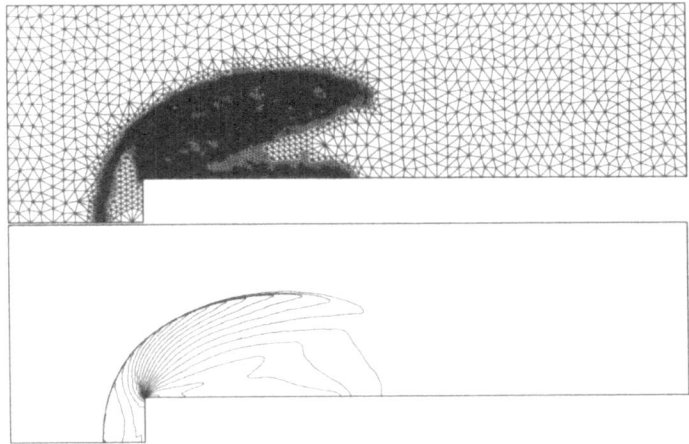

Figure 3.85 Adapted grid and density distribution at $t = 0.67$

flow phenomena accurately. A reflected shock is now moving downstream. Figure 3.87 shows the grid and solution at time $t = 2.48$ where the first reflected shock just touches the lower wall. Approximately at time $t = 4$ the overall shock structure is established in the channel. From that time on the dynamics of the flow field change drastically and start to influence the refinement indicator. Before $t = 4$ a drastic temporal change of the flow field can be observed in the sense that flow phenomena, like shocks, move with high relative velocity. After $t = 4$ temporal changes are negligible and all structures of the flow field move very slowly and do not change very much between $t = 4$ and $t = 8$. As a consequence, the detected error exceeds the user-given tolerances TOL_{refine} and TOL_{coarse}. To take this behaviour of the flow field into account we had to change the refinement tolerance bounds within the computation. The adaptive tuning of the tolerance bounds makes sense since our refinement indicator is a space-time indicator. Thus, temporal as well as spatial errors contribute to the overall error.

Figure 3.88 shows the grid and the density distribution at time $t = 4.77$. As can be seen the structural details of the solution are already fully developed. Note that the second reflected shock is already poorly treated by the adaption process. Changing the refinement limits we get the final solution and grid at time $t = 8$ shown in figure 3.89.

Figure 3.90 shows the Mach number distribution on the final grid. The results compare

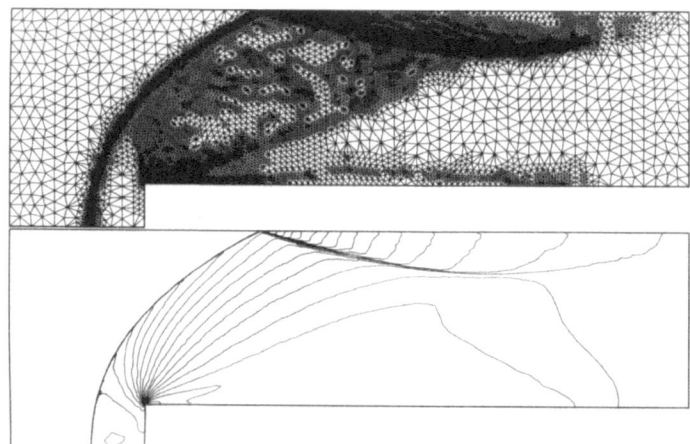

Figure 3.86 Adapted grid and density distribution at $t = 1.44$

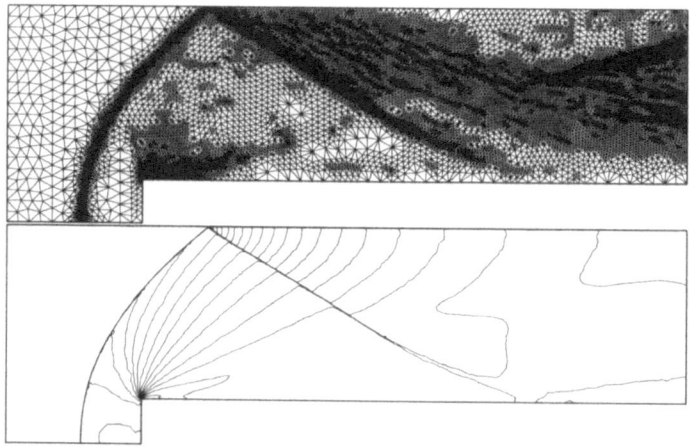

Figure 3.87 Adapted grid and density distribution at $t = 2.48$

well with results obtained with other residual based error indicators but in contrast with them, the dual graph norm indicator is a true rather than *ad hoc* error indicator.

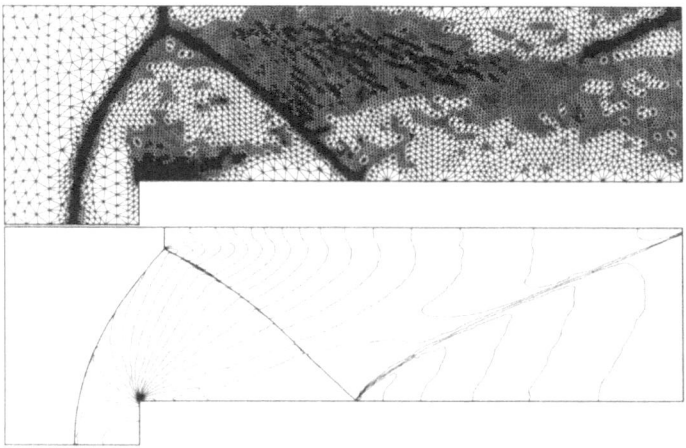

Figure 3.88 Adapted grid and density distribution at $t = 4.77$

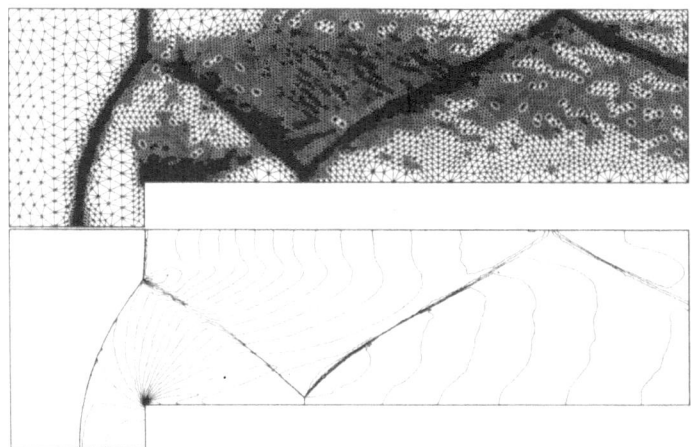

Figure 3.89 Adapted grid and density distribution at $t = 8$

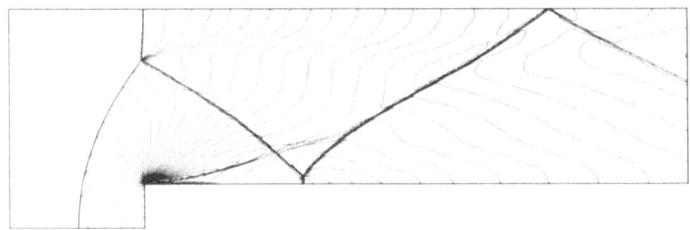

Figure 3.90 Mach number distribution at $t = 8$

3.12.8   Problems with the dual graph-norm

Although the results presented look quite promising there are some problems associated with the dual graph-norm. First of all, it is very expensive to compute in comparison with the weighted L^2-indicator. The second problem is concerned with the sensitivity of the indicator. It turns out that a small change in tolerance can influence the adapted grid enormously, which makes this indicator hard to use for practical purposes. The main problem lies in the inability of our discrete dual graph-norm indicator to detect contact discontinuities. As can be seen already from figures 3.88 and 3.89 the greater vicinity of the contact seems to be refined but there is no trace of the contact discontinuity at all.

In order to determine the source of this trouble we made numerical experiments with different tolerances. As can be seen from the sequence of different grids and corresponding density, all shown at final time $t = 8$, clearly indicates that the absence of the contact discontinuity is not a problem of choosing the 'right' tolerance.

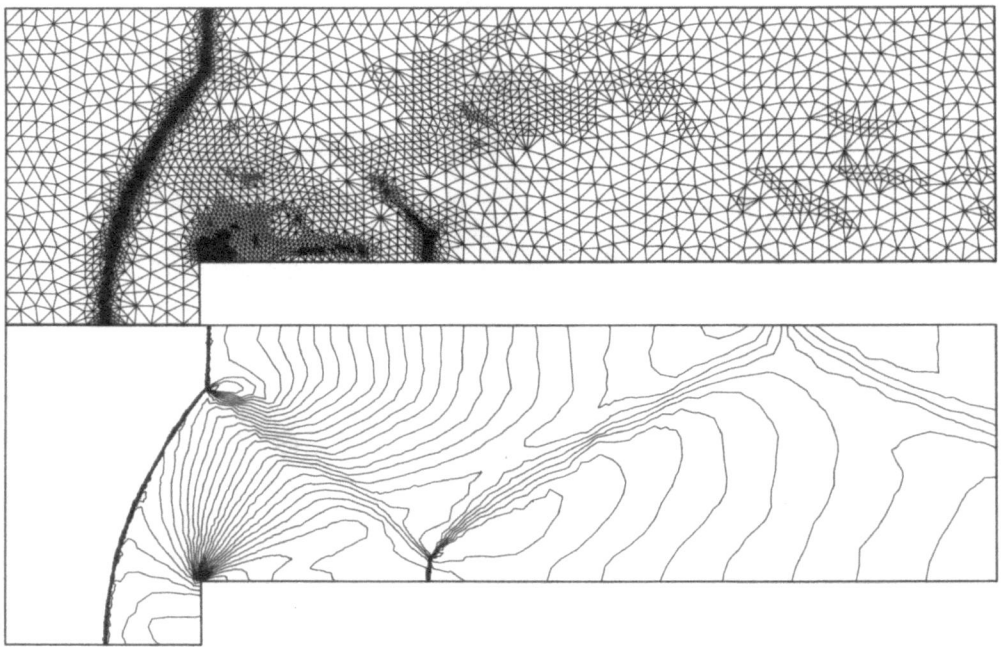

Figure 3.91 TOL$_{\text{refine}} = 10^{-2}$, TOL$_{\text{coarse}} = 10^{-3}$, # triangles $= 9530$, # points $= 4877$

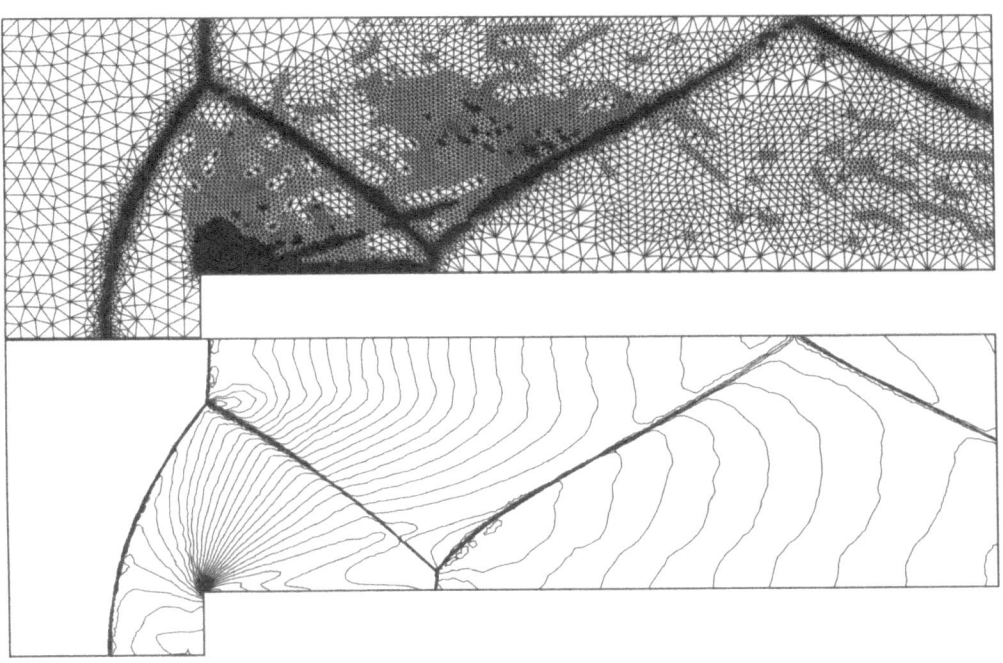

Figure 3.92 $\text{TOL}_{\text{refine}} = 0.5 \cdot 10^{-2}$, $\text{TOL}_{\text{coarse}} = 0.5 \cdot 10^{-3}$, # triangles = 29848, # points = 15095

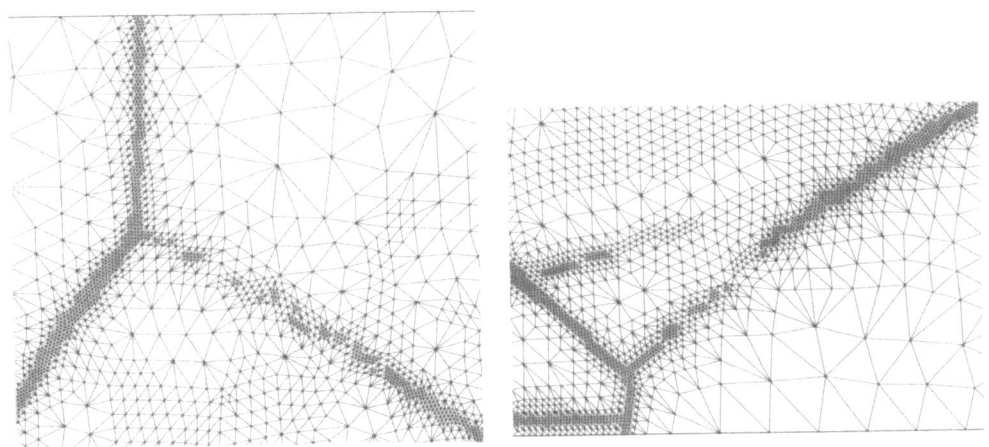

Figure 3.93 $\text{TOL}_{\text{refine}} = 0.5 \cdot 10^{-2}$, $\text{TOL}_{\text{coarse}} = 0.5 \cdot 10^{-3}$, # triangles = 29848, # points = 15095. Details at Mach stem and lower reflection point

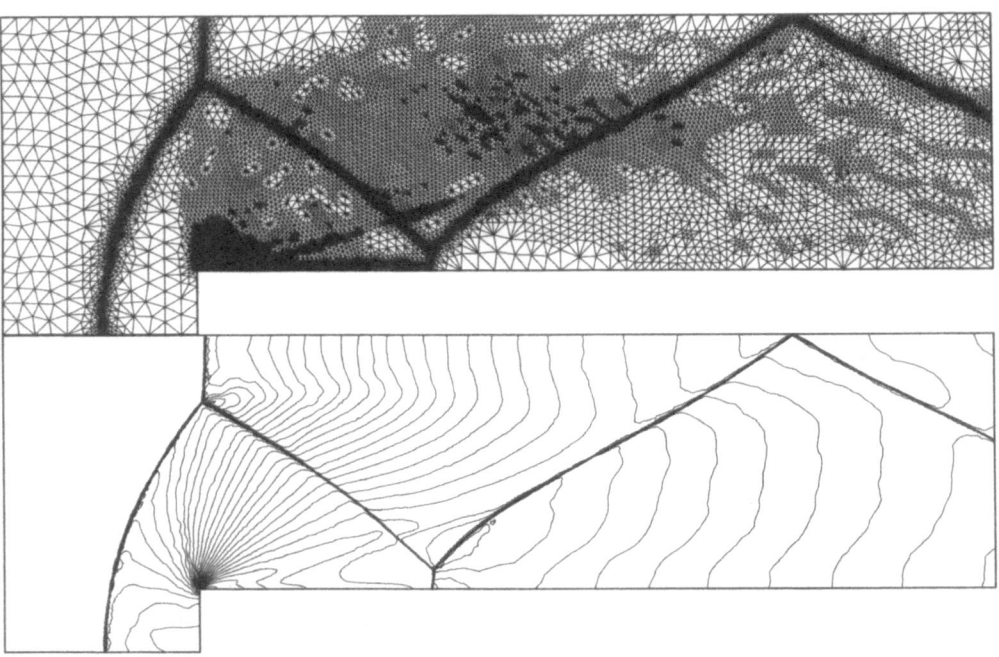

Figure 3.94 TOL $_{\text{refine}}$ = $0.3 \cdot 10^{-2}$, TOL $_{\text{coarse}}$ = $0.3 \cdot 10^{-3}$, # triangles = 37510, # points = 18947

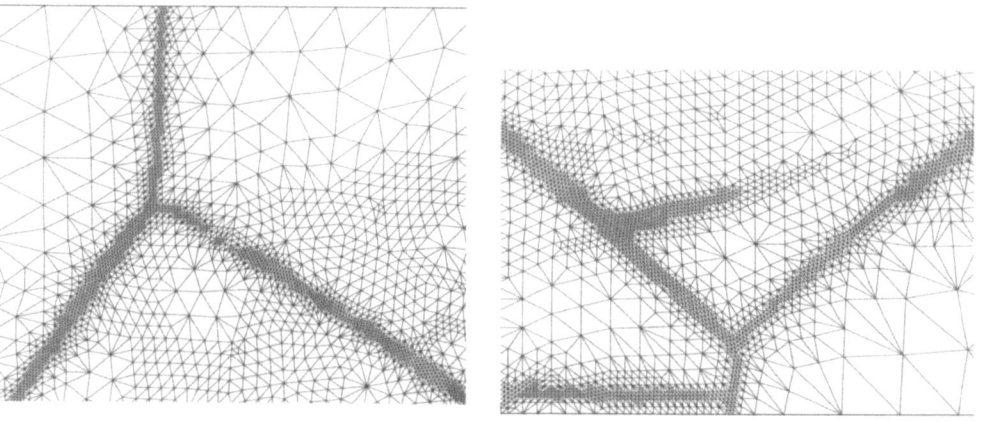

Figure 3.95 TOL $_{\text{refine}}$ = $0.3 \cdot 10^{-2}$, TOL $_{\text{coarse}}$ = $0.3 \cdot 10^{-3}$, # triangles = 37510, # points = 18947. Details at Mach stem and lower reflection point

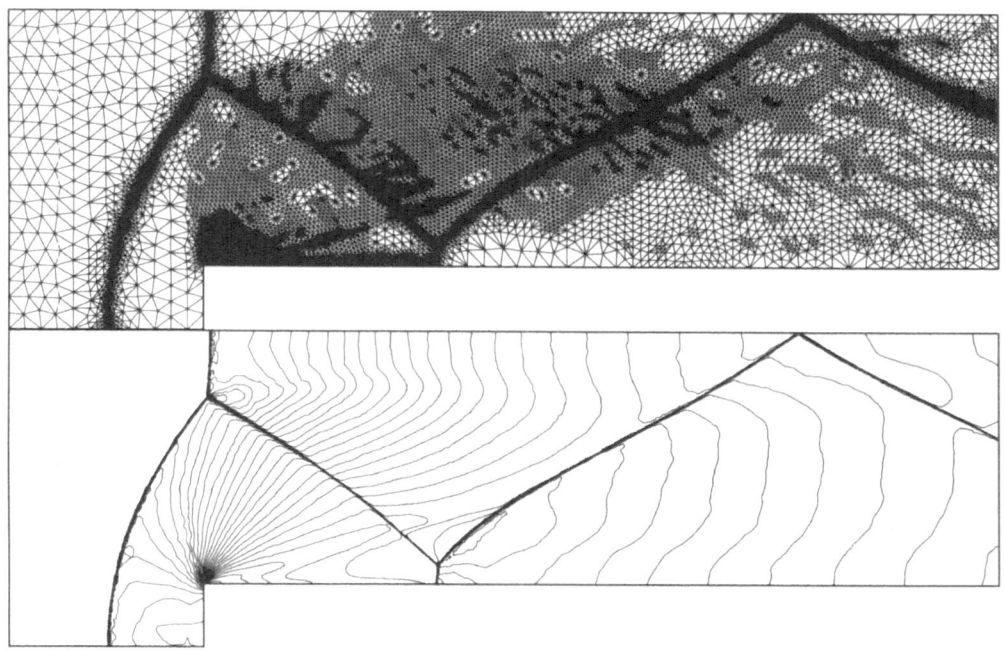

Figure 3.96 TOL $_{\text{refine}}$ = $0.2 \cdot 10^{-2}$, TOL $_{\text{coarse}}$ = $0.2 \cdot 10^{-3}$, # triangles = 42624, # points = 21513

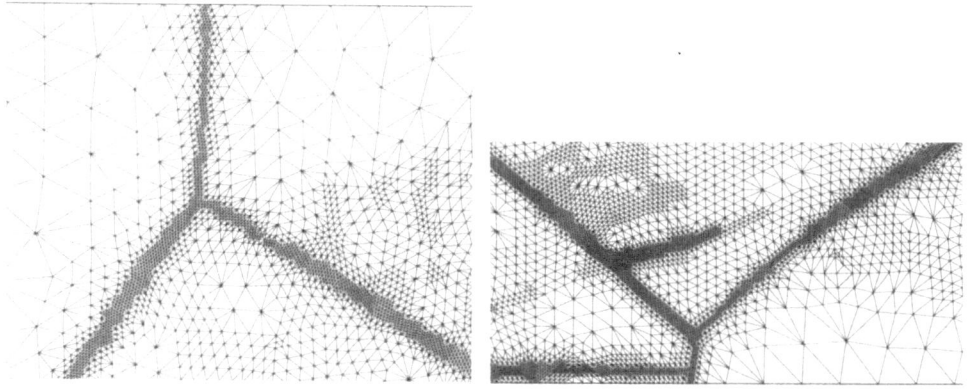

Figure 3.97 TOL $_{\text{refine}}$ = $0.2 \cdot 10^{-2}$, TOL $_{\text{coarse}}$ = $0.2 \cdot 10^{-3}$, # triangles = 42624, # points = 21513. Details at Mach stem and lower reflection point

At this point we can only conjecture that the loss of the contact discontinuity is related to a certain (not yet exploited) orthogonality relation between the residual and the test functions. Note that in the dual graph-norm an inner product $(r^h, \phi)_{P_{in}}$ has to be evaluated on the space-time prism P_{in}. If r^h is orthogonal to ϕ with respect to the L^2-inner product the dual graph-norm of the residual vanishes. This could be one possible explanation for the inability of the discrete dual graph-norm indicator to detect contact discontinuities.

3.13   Closing of the circle: L^2 meets dual graph norm

As it stands the theory of *a posteriori* error control in finite volume approximations of Friedrichs systems is quite satisfactory, but the treatment of the Euler equations suffers from certain problems which were already described above. It was Süli in [56] who proved that the dual graph-norm of the residual is essentially equivalent to the weighted L^2-norm for Friedrichs systems. Thus, via symmetrisation and linearisation we can close the circle and can finally state that the indicator we started with is indeed a reliable *a posteriori* error indicator.

The following results are published as Theorem 2.5 and Theorem 2.6 in [56] by Endre Süli. They were derived by Süli and are taken without change (modulo notational details) from [56] in which the corresponding proofs can be found.

Instead of hypothesis (1) Süli used the hyperbolicity in the sense of Lax as explained in 3.14. Then one can prove that the weighted L^2-norm is an upper bound of the dual graph-norm. This gives the following

Theorem 3.13.1 *For the cell error there holds the following* a posteriori *upper bound:*

$$\|e_c\|_{L^2(P_{in})} \leq Ch\|r^h\|_{L^2(P_{in})},$$

where h denotes the diameter of P_{in}.

Note that this results holds for Friedrichs systems again, i.e. if one would apply it to the Euler equations via symmetrisation and local linearisation one had to compute the L^2-norm of the residual in entropy variables.

Fortunately, there is also an efficiency estimate concerning the weighted L^2-norm. As a technical difficulty (but nothing but a technical difficulty) one has to consider a micropartition of P_{in} with corresponding diameter h_0 and an orthogonal projector P_{h_0} which maps L^2-functions onto the finite element space of the micropartition. Then one can prove

Theorem 3.13.2 *For the cell error there holds the following* a posteriori *lower bound:*

$$Ch_0\|P_{h_0}r^h\|_{L^2(P_{in})} \leq \|e_c\|_{L^2(P_{in})}.$$

Thus, the weighted L^2-norm is a true *a posteriori* error indicator for finite volume approximations of Friedrichs system.

Bibliography

[1] P. Alexandroff, H. Hopf - Topologie I. *Springer Verlag, Grundlehren der math. Wissenschaften Band 45 (1974)*

[2] T.J. Barth, D.C. Jespersen - The design and application of upwind schemes on unstructured meshes. *AIAA paper 89-0366, (1989)*

[3] P.G. Ciarlet - The Finite Element Method for Elliptic Problems. *North-Holland, 2nd edt. (1987)*

[4] O. Friedrich - A New Method for Generating Inner Points of Triangulations in Two Dimensions. *Comp. Meth. Appl. Mech. Eng.* **104**, *77-86, (1993)*

[5] M. Geiben, D. Kröner, M. Rokyta - A Lax-Wendroff Type Theorem for Cell-Centered, Finite Volume Schemes in 2-D. *Report No.278, Sonderforschungsbereich 256, Rheinische Friedrich-Wilhelms-Universität, Bonn, (1993)*

[6] B. Heinrich - Finite Difference Methods on Irregular Networks. *Birkhäuser Verlag, ISNM Vol.82, (1987)*

[7] C. Hirsch - Numerical Computation of Internal and External Flows, Volume 1: Fundamentals of Numerical Discretization. *John Wiley & Sons, (1988)*

[8] C. Hirsch - Numerical Computation of Internal and External Flows, Volume 2: Computational Methods for Inviscid and Viscous Flows. *John Wiley & Sons, (1990)*

[9] A. Meister - Ein Beitrag zum DLR-τ-Code: Ein explizites und implizites Finite-Volumen-Verfahren zur Berechnung instationärer Strömungen auf unstrukturierten Gittern. *DLR Interner Bericht IB 223-94 A 36, Göttingen, (1994)*

[10] A. Meister - Zur zeitgenauen numerischen Simulation reibungsbehafteter, kompressibler, turbulenter Strömungsfelder mit einer impliziten Finite-Volumen-Methode vom Box-Typ. *Dissertation TH Darmstadt, 1996*

[11] Y. Saad, M. H. Schulz - GMRES: A generalized minimal residual algorithm for solving nonsymmetric linear systems. *SIAM J. Sci. Stat. Comp.* **7**, *856-869, (1986)*

[12] C.-W. Shu, S. Osher - Efficient Implementation of Essentially Non-Oscillatory Shock-Capturing Schemes. *J. Comp. Phys.* **77**, *439-471, (1988)*

[13] Th. Sonar - On the design of an upwind scheme for compressible flow on general triangulations. *Numerical Algorithms* **4**, *135-149, (1993)*

[14] J. Stoer, R. Bulirsch - Introduction to Numerical Analysis. *Springer Verlag, (1980)*

[15] P. Vankeirsbilck - Algorithmic Developments for the Solution of Hyperbolic Conservation Laws on Adaptive Unstructured Grids. *Doktorarbeit, Katholieke Universiteit Leuven, Faculteit Toegepaste Wetenschappen, Afdeling Numerieke Analyse en Toegepaste Wiskunde, Celestijnenlaan 200A, 3001 Leuven (Heverlee), (1993)*

[16] R. Abgrall - Design of an Essentially Nonoscillatory Reconstruction Procedure on Finite-Element-Type Meshes. *Preprint (1994)*

[17] R. Abgrall - An Essentially Non-Oscillatory Reconstruction Procedure on Finite-Element Type Meshes: Application to Compressible Flows. *Comput. Methods Appl. Mech. Engrg.* **116**, *95-101, (1994)*

[18] R. Abgrall - On Essentially Non-Oscillatory Schemes on Unstructured Meshes: Analysis and Implementation. *J. Comp. Phys.* **114**, *45-58, (1994)*

[19] R. Abgrall, F.C. Lafon - ENO Schemes on Unstructured Meshes. *Lecture Series 1993-04, Computational Fluid Dynamics, von Karman Institute for Fluid Dynamics, (1993)*

[20] R. Abgrall, Th. Sonar - On the Use of Mühlbach Expansions in the Recovery Step of ENO Methods. *Numerische Mathematik* **76**, *1-25, (1997)*

[21] L.J. Durlofsky, B. Engquist, S. Osher - Triangle Based Adaptive Stencils for the Solution of Hyperbolic Conservation Laws. *J. Comp. Phys.* **98**, *64-73, (1992)*

[22] M. Golomb, H.F. Weinberger - Optimal Approximation and Error Bounds. in: *On Numerical Approximation, Edt.: R.E. Langer, The University of Wisconsin Press, Madison, (1959)*

[23] A. Harten - On High-Order Accurate Interpolation for Non-Oscillatory Shock Capturing Schemes. in: *IMA Vol.2, Oscillation Theory, Computation and Methods of Compensated Compactness, Springer Verlag, (1987)*

[24] A. Harten, S.R. Chakravarthy - Multi-Dimensional ENO Schemes for General Geometries. *ICASE Report No. 91-76, (1991)*

[25] A. Harten, S. Osher - Uniformly High Order Accurate Nonoscillatory Schemes I. *SIAM J. Num. Anal.* **24**, *279-309, (1987)*

[26] A. Harten, S. Osher, B. Engquist, S.R. Chakravarthy - Some Results on Uniformly High-Order Accurate Essentially Nonoscillatory Schemes. *Appl. Num. Math.* **2**, *347-377, (1986)*

[27] A. Harten, B. Engquist, S. Osher, S.R. Chakravarthy - Uniformly High Order Accurate Essentially Non-Oscillatory Schemes III. *J. Comp. Phys.* **71**, *231-303, (1987)*

[28] A. Iske, T. Sonar - On the Structure of Function Spaces in Optimal Recovery of Point Data for ENO-Schemes by Radial Basis Functions. *Numerische Mathematik* **74**, *177-201, (1996)*

[29] C.A. Micchelli, T.J. Rivlin - A Survey of Optimal Recovery. in: *Optimal Estimation in Approximation Theory, Eds.: C.A. Micchelli, T.J. Rivlin, Plenum Press, 1-54, (1977)*

[30] C.A. Micchelli, T.J. Rivlin - Lectures on Optimal Recovery. in: *Numerical Analysis, Lancaster 1984, Ed.: P.R. Turner, Springer Verlag, Lecture Notes in Mathematics 1129, 12-93, (1984)*

[31] C.A. Micchelli, T.J. Rivlin - Optimal Recovery of Best Approximations. *Resultate der Mathematik* **3**, *25-32, (1978)*

[32] R. Schaback, H. Wendland - Special Cases of Compactly Supported Radial Basis Functions. *Manuskript, Institut für Numerische und Angewandte Mathematik, Universität Göttingen, (1994)*

[33] T. Sonar - Optimal Recovery Using Thin Plate Splines in Finite Volume Methods for the Numerical Solution of Hyperbolic Conservation Laws. *IMA J. Num. Anal.* **16**, *549-581, (1996)*

[34] Th. Sonar - On the Construction of Essentially Non-Oscillatory Finite Volume Approximations to Hyperbolic Conservation Laws on General Trinagulations: Polynomial Recovery, Accuracy, and Stencil Selection. *Comp. Meth. Appl. Mech. Eng.* **140**, *157-181, (1997)*

[35] Th. Sonar - Multivariate Rekonstruktionsverfahren zur numerischen Lösung hyperbolischer Erhaltungsgleichungen. *Habilitationsschrift, Technische Hochschule Darmstadt, (1995)*

[36] J.L. Synge - The Hypercircle in Mathematical Physics. *Cambridge University Press, (1957)*

[37] H. Wendland - Ein Beitrag zur Interpolation mit radialen Basisfunktionen. *Diplomarbeit, Institut für Numerische und Angewandte Mathematik, Universität Göttingen, (1994)*

[38] Z.-M. Wu - Multivariate Compactly Supported Positive Definite Radial Basis Functions. *Manuskript, Inst. Num. Ang. Math., Universität Göttingen, (1994)*

[39] E. Bänsch. *An adaptive finite-element-strategy for the 3-D time-dependent Navier-Stokes-equations. J. Comput. Appl. Math.* **36**, *3-28, (1991).*

[40] F. Bornemann, B. Erdmann, R. Kornhuber. *Adaptive Multilevel-Methods in Three Space Dimensions. Int. J. Num. Meth. in Eng.* **36**, *3187–3203, (1993).*

[41] P.G. Ciarlet - The Finite Element Method for Elliptic Problems. *North-Holland, 2nd edt. (1987)*

[42] M. Feistauer - Mathematical Methods in Fluid Dynamics. *Longman Scientific & Technical, Harlow, (1993)*

[43] K.O. Friedrichs - Symmetric Positive Linear Differential Operators. *Comm. Pure Appl. Math.* **11**, *333-418 (1958)*

[44] O. Friedrich - Weighted Essentially Non-Oscillatory Schemes for the Interpolation of Mean Values on Unstructured Grids. (J. Comp. Phys. **144**, 194-212, 1998)

[45] O. Friedrich - Gewichtete wesentlich nicht-oszilierende Verfahren auf unstrukturi-.erten Gittern. (PhD thesis, Department of Mathematics, University of Hamburg, 1999)

[46] P. Hansbo, C. Johnson - Adaptive Streamline Diffusion Methods for Compressible Flow Using Conservation Variables. *Comp. Meth. Appl. Mech. and Eng.* **87**, *267-280, (1991)*

[47] D. Hempel. *Local Mesh Adaption in Two Space Dimensions. IMPACT Comput. Sci. Eng.* **5**, *309-317, (1993)*.

[48] D. Hempel. *Isotropic Refinement and Recoarsening in 2 Dimensions. DLR IB 223-95 A 35, (1995)*.

[49] T.J.R. Hughes, L.P. Franca, M. Mallet - A New Finite Element Formulation for Computational Fluid Dynamics: I. Symmetric Forms of the Compressible Euler and Navier-Stokes Equations and the Second Law of Thermodynamics. *Comp. Meth. Appl. Mech. Eng.* **54**, *223-234, (1986)*

[50] P.D. Lax, R.S. Phillips - Local Boundary Conditions for Dissipative Symmetric Linear Differential Operators. *Comm. Pure Appl. Math.* **13**, *427-455 (1960)*

[51] X.-D. Liu, S. Osher, T. Chan - Weighted Essentially Non-oscillatory Schemes. (J. Comput. Phys., **115**, 200–212, 1994)

[52] J.A. Mackenzie, E. Süli, G. Warnecke - A Posteriori Analysis for Petrov-Galerkin Approximations of Friedrichs Systems. *Oxford University Computing Laboratory, Report No.95/01, (1995)*

[53] J.A. Mackenzie, E. Süli, G. Warnecke - A Posteriori Error Estimates for the Cell-Vertex Finite Volume Method. *in: Adaptive Methods: Algorithms, Theory and Applications, Eds.: W. Hackbusch, G. Wittum, Vieweg, Braunschweig, 221-235, (1994)*

[54] T. Sonar - Strong and Weak Norm Refinement Indicators Based on the Finite Element Residual for Compressible Flow Computation. *Impact of Comp. in Science and Eng.* **5**, *111-127, (1993)*

[55] T. Sonar, V. Hannemann, D. Hempel - Dynamic Adaptivity and Residual Control in Unsteady Compressible Flow Computation. *Math. and Comp. Modelling* **20**, *201-213, (1994)*

[56] E. Süli - A posteriori error analysis and global error control for adaptive finite element approximations of hyperbolioc problems. *Oxford University Computing Laboratory Technical Report 95/14, 1995. Also in: (D.F. Griffiths, ed.) Proceedings of the 16th International COnference on Numerical Analysis, Dundee, June 1995, Longmans*

[57] T. Sonar, E. Süli - A Dual Graph-Norm Refinenemt Indicator for Finite Volume Approximations of the Euler Equations. *Numerische Mathematik* **78**, *619-658, (1998)*

[58] J.L. Steger, R.F. Warming - Flux Vector Splitting of the Inviscid Gas Dynamic Equations with Applications to Finite Difference Methods. *J. Comp. Phys.* **40**, *263-293, (1980)*

[59] P. Woodward, P. Colella - The Numerical Simulation of Two-Dimensional Fluid Flow with Strong Shocks. *J. Comp. Phys.* **54**, *115-173, (1984)*

4 Pressure-Correction Methods for all Flow Speeds

4.1 Introduction

Most of computational methods for compressible flows use time-marching techniques and compute density from the mass-conservation equation; the pressure is then computed from an equation of state. A vast literature is available on the subject, including several books [1, 5, 6]. The extension of these methods to low Mach-number flows is possible, but they are then usually not as efficient as the methods developed specifically for incompressible flows.

Most of the methods designed for incompressible flows compute pressure or pressure-correction from an equation derived by combining the mass- and momentum-conservation equations. They are usually implicit and are tuned for efficient computation of steady flows. It has been recognized long ago that these methods can be extended to flows at all Mach numbers [8, 2], and although the most popular commercial CFD codes use this approach, this fact is still not widely appreciated.

This chapter is devoted to pressure-correction methods for all flow speeds. We start by presenting the Navier-Stokes equations and the derivation of pressure and pressure-correction equation. The discretization is first left aside and only the principles of the various methods are outlined using symbolic notation. Later, a particular finite-volume method for all flow speeds is described. Some application examples are also presented to demonstrate the main features of the method.

4.2 Conservation Equations

The conservation equations for mass, momentum and energy in differential form read:

$$\frac{\partial \rho}{\partial t} + \boldsymbol{\nabla} \cdot (\rho \mathbf{v}) = 0 \,, \tag{4.1}$$

$$\frac{\partial (\rho \mathbf{v})}{\partial t} + \boldsymbol{\nabla} \cdot (\rho \mathbf{v} \mathbf{v}) = -\boldsymbol{\nabla} p + \boldsymbol{\nabla} \cdot \boldsymbol{S} + \rho \mathbf{b} \,, \tag{4.2}$$

$$\frac{\partial (\rho h)}{\partial t} + \boldsymbol{\nabla} \cdot (\rho \mathbf{v} h) = -\boldsymbol{\nabla} \cdot \mathbf{q} + \mathbf{v} \cdot \boldsymbol{\nabla} p + \boldsymbol{S} : \boldsymbol{\nabla} \mathbf{v} + \frac{\partial p}{\partial t} \,. \tag{4.3}$$

The integral form of these equations, written for a control volume V bounded by the surface S with the unit outward normal vector $\mathbf{n}$, read:

$$\frac{\partial}{\partial t} \int_V \rho \, dV + \int_S \rho \mathbf{v} \cdot \mathbf{n} \, dS = 0 \,, \tag{4.4}$$

$$\frac{\partial}{\partial t} \int_V \rho \mathbf{v} \, dV + \int_S \rho \mathbf{v} \mathbf{v} \cdot \mathbf{n} \, dS = -\int_S p \mathbf{n} \, dS + \int_S \mathbf{S} \cdot \mathbf{n} \, dS + \int_V \rho \mathbf{b} \, dV \,, \tag{4.5}$$

$$\frac{\partial}{\partial t} \int_V \rho h \, dV + \int_S \rho h \mathbf{v} \cdot \mathbf{n} \, dS = -\int_S \mathbf{q} \cdot \mathbf{n} \, dS + \int_V \left[\mathbf{v} \cdot \nabla p + \mathbf{S} : \nabla \mathbf{v} \right] dV + \frac{\partial}{\partial t} \int_V p \, dV \,. \tag{4.6}$$

In these equations, ρ is fluid density, $\mathbf{v}$ is the velocity vector, p is the static pressure, $\mathbf{S}$ is the viscous part of the stress tensor, $\mathbf{b}$ is the body force, h is the enthalpy and $\mathbf{q}$ is the heat flux. In order to close the system of equations (4.1) to (4.3), or (4.4) to (4.6), the Stokes' law (for a Newtonian fluid):

$$\mathbf{S} = \mu \left[\nabla \mathbf{v} + (\nabla \mathbf{v})^{\mathrm{T}} \right] - \frac{2}{3} \mu \nabla \cdot \mathbf{v} \mathbf{I} \,, \tag{4.7}$$

and the Fourier's law:

$$\mathbf{q} = -k \, \nabla T \tag{4.8}$$

are required. In addition, we need an equation of state, e.g. for an ideal gas:

$$h = c_p T \,, \quad \rho = \frac{p}{RT} \,. \tag{4.9}$$

In the Eqs. (4.7) to (4.9) μ is the viscosity, k is the thermal conductivity, T stands for the absolute temperature, c_p is the specific heat at constant pressure, and R is the gas constant.

The most important issue in computing incompressible flows is the calculation of pressure. In compressible flows, the continuity equation can be used to determine the *density* and the pressure is then calculated from an equation of state. In incompressible flows, the continuity equation does not have a dominant variable – it is a kinematic constraint on the velocity field rather than a dynamic equation. The question is therefore: How to compute the velocity and pressure fields in an incompressible flow, so that both mass and momentum conservation equations are satisfied? There are many ways of achieving this goal; while they are known in literature under various names and often treated as new or different methods, they actually all have the same roots, as we shall show below. We shall deal with this issue first and then show how the method which is well suited for computing incompressible flows can be extended to all flow speeds.

The mass conservation (or continuity) equation requires that, in the limit of an incompressible fluid, the velocity filed must be divergence-free. Since in a sequential (segregated) solution algorithm (as the one considered here) the velocity is computed from momentum equations, this suggests that we take the divergence of the momentum equation (4.2, which leads to a Poisson-type equation for pressure:

$$\boldsymbol{\nabla}\cdot(\boldsymbol{\nabla} p) = -\boldsymbol{\nabla}\cdot\left[\boldsymbol{\nabla}\cdot(\rho\mathbf{v}\mathbf{v} - \boldsymbol{S}) - \rho\mathbf{b} + \frac{\partial(\rho\mathbf{v})}{\partial t}\right]\,. \tag{4.10}$$

For the case of constant density and viscosity, this equation gets the following simplified form:

$$\boldsymbol{\nabla}\cdot(\boldsymbol{\nabla} p) = \boldsymbol{\nabla}\cdot\left[\boldsymbol{\nabla}\cdot(\rho\mathbf{v}\mathbf{v})\right]\,. \tag{4.11}$$

This equation allows us to compute the pressure if the velocity field (which satisfies the continuity equation) is known; on the other hand, if the three velocity components and the pressure are considered a vector of unknowns, than the three momentum component equations and the pressure equation form a closed system of four equations with four unknowns.

The above pressure equation is sometimes used to compute the pressure when the velocities are computed from the streamfunction-vorticity formulation of the Navier-Stokes equations. It could – although this is seldom done – be also used together with the momentum equations to compute incompressible or compressible flows using primitive variables.

There are several reasons why the pressure-correction formulation to be presented below is preferred; one is that the boundary conditions for the pressure equation are more difficult to obtain and the other is that numerical difficulties are often encountered.

One important aspect of the pressure equation is worth mentioning. We have written the Laplacian of pressure on the left-hand side of Eqs. (4.10) and (4.11) as a combination of divergence and gradient operators, which follows the way of deriving this equation. In the course of numerical solution, it is essential that the individual terms are treated in exactly the same way as in the momentum and continuity equations, from which they stem. In addition, the outer divergence operator, which resembles the continuity equation, must be approximated in the way the continuity is checked. If one fails to observe these consistency requirements, the computed velocity field (if one manages to obtain the solution) will probably not be divergence-free.

When density and viscosity are not constant, the pressure equation becomes rather complicated; see Eq. (4.10). This is why the derivation of a pressure-correction equation using discretized momentum and continuity equations is a more popular approach of computing the pressure, since in this case the resulting equation is simpler and the boundary conditions are easier to specify.

There are many approaches that have been suggested so far in literature; it is beyond the scope of this Chapter to even try to describe many of them. We shall therefore give details of the method that is the most widely spread and is also preferred by the present authors. It is based on a sequential iterative solution of the linearized conservation equations for individual variables, which has proven to be both efficient and versatile, in particular with respect to introducing additional coupled equations to take into account turbulence, heat and mass transfer, moving boundaries and complicated boundary conditions. A more detailed description of some of the aspects of the method can be found in [4].

4.3   Pressure-Correction Equation for Incompressible Flows

From the three component momentum equations, we compute the three velocity components. These equations – see Eqs. (4.2) and (4.5) – are non-linear and coupled. Whether finite-volume (FV) or finite-difference (FD) method is used to discretize the momentum equations, and whether the grid is structured or unstructured, is not relevant for the analysis that follows, while a particular FV-method will be introduced later. We shall assume that an implicit method is used for time integration, which is the usual case when steady solutions are sought; modifications for the (simpler) explicit schemes are straightforward.

In the process of numerical solution, we have to linearize the equations; usually, they are also de-coupled, so that one can consider equations for each variable one at a time. The discretized momentum equation for the velocity component u_i can be written for one grid point as:

$$A_C^{u_i} u_{i,C}^{n+1} + \sum_k A_k^{u_i} u_{i,k}^{n+1} = Q_{u_i}^{n+1} - \left(\frac{\delta p^{n+1}}{\delta x_i} \right)_C .$$ (4.12)

Here C denotes the node for which the equation is written and k runs over neighbor nodes; how many nodes are involved depends on discretization in space, which is unimportant at this stage. The superscript n stands for a time-step counter and the pressure term has been extracted from the source term to show explicitly the dependence of velocity on pressure gradient. The coefficients A_k contain contributions from convective and diffusive terms; the coefficient of the central node may in addition contain contribution from unsteady and source terms. Q_{u_i} contains all other terms of the discretized equation not explicitly shown. Even when the FV-method is used and pressure forces on control-volume (CV) faces are computed, the net force in x_i direction can be expressed as the volume integral of the pressure derivative $\partial p/\partial x_i$, so the notation used is still appropriate to illustrate the derivation of pressure-correction methods.

For the solution domain as a whole, a system of equations results. Implicit time-integration methods and the non-linearity of the underlying differential equation require iterative solution of this equation system. A typical solution method implies two kinds of iterations:

- *outer iterations*, in which the coefficient and source matrices are updated to account for non-linearity and inter-equation coupling (they can be also viewed as steps in a pseudo-time), and

- *inner iterations*, performed on linear systems with fixed coefficients.

For the m th outer iteration on the new time level t_{n+1} , the linear equation solved is:

$$A_C^{u_i} u_{i,C}^{m*} + \sum_k A_k^{u_i} u_{i,k}^{m*} = Q_{u_i}^{m-1} - \left(\frac{\delta p^{m-1}}{\delta x_i} \right)_C ,$$ (4.13)

where the coefficients A and the source term Q are based on the solution from the previous outer iteration. The velocity components u_i^{m*} computed at this step do not in general satisfy the continuity equation. The latter is to be enforced by correcting the velocity field; our aim is to express the velocity corrections through gradients of pressure correction, which leads to a pressure-correction equation.

The dependence of velocity on pressure is obvious from the (discretized) momentum equations; once the equation (4.13) is solved, we can express the velocity at any computational node as:

$$u_{i,\mathrm{C}}^{m*} = \frac{Q_{u_i}^{m-1} - \sum_k A_k^{u_i} u_{i,k}^{m*}}{A_{\mathrm{C}}^{u_i}} - \frac{1}{A_{\mathrm{C}}^{u_i}} \left(\frac{\delta p^{m-1}}{\delta x_i} \right)_{\mathrm{C}} . \tag{4.14}$$

However, the velocity at any grid point depends not only on the pressure gradient, but also on the velocity at surrounding nodes, which makes the task of deriving the pressure-correction equation difficult.

When we insert the velocity field u_i^{m*} into discretized continuity equation, we find that an imbalance results (except at the converged state):

$$\left[\frac{\delta(\rho u_i^{m*})}{\delta x_i} \right]_{\mathrm{C}} = \Delta \dot{m}_{\mathrm{C}} . \tag{4.15}$$

We now look for the corrected velocity field, $u_i^m = u_i^{m*} + u_i'$, which is requested to satisfies the continuity equation:

$$\left[\frac{\delta(\rho u_i^m)}{\delta x_i} \right]_{\mathrm{C}} = \Delta \dot{m}_{\mathrm{C}} + \left[\frac{\delta(\rho u_i')}{\delta x_i} \right]_{\mathrm{C}} = 0 . \tag{4.16}$$

Several methods of achieving this goal are described below.

4.3.1 SIMPLE-Method

We can postulate that the required velocity correction can be obtained by correcting the pressure, $p^m = p^{m-1} + p'$, such that both the continuity and (linearized) momentum equations are satisfied; by analogy to Eq. (4.14), we can write:

$$u_{i,\mathrm{C}}^{m} = \frac{Q_{u_i}^{m-1} - \sum_k A_k^{u_i} u_{i,k}^{m*}}{A_{\mathrm{C}}^{u_i}} - \frac{1}{A_{\mathrm{C}}^{u_i}} \left(\frac{\delta p^{m}}{\delta x_i} \right)_{\mathrm{C}} . \tag{4.17}$$

Here, an approximation has been introduced: $u_{i,k}^{m*}$ remained uncorrected on the right-hand side, so that corrections to velocities at neighbor nodes are ignored. This approximation (which is the key feature of the SIMPLE-method [13]) allows a simple relation between velocity and pressure corrections, but it has consequences on the performance of the method, as will be shown below.

By subtracting Eq. (4.14) from (4.17) we find that the velocity correction at a grid point is a function of the gradient of pressure correction at the same point:

$$u_{i,\mathrm{C}}' = -\frac{1}{A_{\mathrm{C}}^{u_i}} \left(\frac{\delta p'}{\delta x_i} \right)_{\mathrm{C}} . \tag{4.18}$$

Now we can insert this relation into the continuity equation (4.16) which leads to:

$$\frac{\delta}{\delta x_i} \left[\frac{\rho}{A_{\mathrm{C}}^{u_i}} \left(\frac{\delta p'}{\delta x_i} \right) \right]_{\mathrm{C}} = \Delta \dot{m}_{\mathrm{C}} . \tag{4.19}$$

This is the Poisson-like pressure-correction equation of the SIMPLE method. The coefficients in this equation depend on the discretization method, but for incompressible flows the matrix A is symmetric and diagonally dominant, which makes the use of some efficient iterative solution methods possible.

Upon solving the pressure-correction equation, both velocities and pressure are corrected. Due to approximations introduced above, the velocity field will only satisfy the continuity equation (if the pressure-correction equation is solved to a tight tolerance); however, since outer iterations are needed anyway due to non-linearity and inter-equation coupling, the approximation from Eq. (4.17) can be also corrected through iterations.

In order to achieve a good convergence of outer iterations, the method requires a well-tuned under-relaxation strategy; only an α_p-fraction of p' is added to p^{m-1}, as will be discussed below.

4.3.2 SIMPLEC-Method

If one does not neglect the velocity corrections on the right-hand side of Eq. (4.17) and writes instead

$$u_{i,\mathrm{C}}^m = \frac{Q_{u_i}^{m-1} - \sum_k A_k^{u_i} u_{i,k}^m}{A_{\mathrm{C}}^{u_i}} - \frac{1}{A_{\mathrm{C}}^{u_i}} \left(\frac{\delta p^m}{\delta x_i}\right)_{\mathrm{C}} , \tag{4.20}$$

we obtain the following link between velocity and pressure corrections by subtracting Eq. (4.14) from Eq. (4.20):

$$u_{i,\mathrm{C}}' = -\frac{\sum_k A_k^{u_i} u_{i,k}'}{A_{\mathrm{C}}^{u_i}} - \frac{1}{A_{\mathrm{C}}^{u_i}} \left(\frac{\delta p'}{\delta x_i}\right)_{\mathrm{C}} . \tag{4.21}$$

The problem is that, when we insert this expression for velocity corrections into the continuity equation (4.16), it will contain both u_i' and p' on the right-hand side, so an approximation is needed. However, instead of neglecting the terms involving u_i' (as is done in the SIMPLE method), we can approximate them in such a way that Eq. (4.21) contains only $u_{i,\mathrm{C}}'$ and p'. This is indeed possible; in [16], it was assumed that u_i' at any point can be reasonably approximated by a weighted mean of its neighbors:

$$u_{i,\mathrm{C}}' \approx \frac{\sum_k A_k^{u_i} u_{i,k}'}{\sum_k A_k^{u_i}} . \tag{4.22}$$

With this approximation, the following substitution can be made in Eq. (4.21):

$$-\frac{\sum_k A_k^{u_i} u_{i,k}'}{A_{\mathrm{C}}^{u_i}} = -u_{i,\mathrm{C}}' \frac{\sum_k A_k^{u_i}}{A_{\mathrm{C}}^{u_i}} , \tag{4.23}$$

which leads to the following expression for u_i':

$$u_{i,\mathrm{C}}' = -\frac{1}{A_{\mathrm{C}}^{u_i} + \sum_k A_k^{u_i}} \left(\frac{\delta p'}{\delta x_i}\right)_{\mathrm{C}} . \tag{4.24}$$

When this expression is inserted into the continuity equation (4.16), we obtain the following pressure-correction equation:

$$\frac{\delta}{\delta x_i} \left[\frac{\rho}{A_C^{u_i} + \sum_k A_k^{u_i}} \left(\frac{\delta p'}{\delta x_i} \right) \right]_C = \Delta \dot{m}_C \ . \tag{4.25}$$

This method is known under the acronym SIMPLEC [16]. The only difference compared to the corresponding equation of the SIMPLE method lies in the denominator of the coefficient. Since $A_k^{u_i}$ are normally negative, $A_C^{u_i} + \sum_k A_k^{u_i}$ is smaller than $A_C^{u_i}$, so the coefficients of the p'-equation are larger than in the SIMPLE method. The computed corrections are substantially smaller than in the SIMPLE method – there is no need to use under-relaxation of pressure correction. However, one can derive an optimum under-relaxation factor for the SIMPLE method following the same arguments as in Eq. (4.22), and when this factor is used, the performance of the SIMPLE and SIMPLEC methods is almost identical.

4.3.3 PISO-Method

The effect of the neglected term in the SIMPLE method can also be taken into account through a predictor-corrector procedure. One such approach has been suggested by Issa [7]. In this method (known under the acronym PISO) the first step is identical to SIMPLE: velocity correction at one point is computed by neglecting the effects of corrections at neighbor points, see Eq. (4.18). We then seek a *corrected* velocity correction, which will take into account the effect of corrections at neighbor nodes, as required by Eq. (4.21):

$$u'_{i,C} + u''_{i,C} = -\frac{\sum_k A_k^{u_i} u'_{i,k}}{A_C^{u_i}} - \frac{1}{A_C^{u_i}} \left(\frac{\delta p'}{\delta x_i} + \frac{\delta p''}{\delta x_i} \right)_C \ . \tag{4.26}$$

By subtracting Eq. (4.18) from Eq. (4.26) we obtain an expression for the second velocity correction as a function of the first velocity correction and the second pressure correction:

$$u''_{i,C} = -\frac{\sum_k A_k^{u_i} u'_{i,k}}{A_C^{u_i}} - \frac{1}{A_C^{u_i}} \left(\frac{\delta p''}{\delta x_i} \right)_C \ . \tag{4.27}$$

Assuming that the velocity after the first correction satisfies the continuity equation, and by requiring that the continuity equation remains satisfied after the second correction, leads to:

$$\left[\frac{\delta(\rho u''_i)}{\delta x_i} \right]_C = 0 \ . \tag{4.28}$$

When we introduce the expression for u''_i, Eq. (4.27), into the continuity equation (4.28), the following equation for the second pressure correction results:

$$\frac{\delta}{\delta x_i} \left[\frac{\rho}{A_C^{u_i}} \left(\frac{\delta p''}{\delta x_i} \right) \right]_C = \left[\frac{\delta(\rho \tilde{u}'_i)}{\delta x_i} \right]_C \ , \tag{4.29}$$

where

$$\tilde{u}'_i = -\frac{\sum_k A_k^{u_i} u'_{i,k}}{A_C^{u_i}} \ . \tag{4.30}$$

Since u_i' are available after the first step, one can compute $\tilde{u}_i'$ easily (but one needs to save the coefficient matrix of the momentum equation). The coefficients in Eq. (4.29) are the same as in the equation for p'; this can be exploited by saving the inverse iteration matrix after the first step and using it in the second step again.

Further corrector steps can be constructed in the same way. In the case of linear momentum equations (creeping flows, Re $\rightarrow$ 0), one would not need to solve momentum equations at all, just assemble the coefficients and then repeat PISO correction steps until both momentum and continuity equations are satisfied to a desired tolerance. In practice, when Reynolds numbers are large, more than two corrections are rarely applied, since the coefficients of the discretized momentum equation need to be updated. No under-relaxation of pressure correction is required; the two corrections are simply added to the pressure from previous outer iteration. One notices, though, that the second correction is usually smaller than the first one and of the opposite sign, so that $p' + p'' = \alpha_p p'$ holds, but α_p is different from point to point.

Issa [7] shows that, if the time step is small enough, one only needs to solve the momentum equations once per time step and perform one predictor and one corrector step of the PISO algorithm. The corrections that would be obtained through further outer iterations are proportional to $(\Delta t)^2$ so they are small if the time step is small; the definition of "small enough" is problem dependent.

4.3.4 SIMPLER-Method

Another approach to computing the pressure in incompressible flows has been proposed by Patankar [12]; he calls it SIMPLER-algorithm. Like in PISO, the first step is identical to the SIMPLE method; however, p' is only used to correct the velocity – the pressure is not corrected. The velocity fields now satisfies the continuity equation, but the linearized momentum equation is out of balance after velocity has been corrected – the pressure needs to be adapted to the corrected velocity field. This can be accomplished in the following way.

Equation (4.20) shows the corrected velocity as a function of corrected velocities at neighbor nodes and corrected pressure gradient. Since the corrected velocity satisfies the continuity equation, we can take the divergence of the right-hand side and require that it be zero; this leads to the following pressure equation:

$$\frac{\delta}{\delta x_i}\left[\frac{\rho}{A_C^{u_i}}\left(\frac{\delta p^m}{\delta x_i}\right)\right]_C = \left[\frac{\delta(\rho \tilde{u}_i^m)}{\delta x_i}\right]_C \, , \tag{4.31}$$

where the definition of $\tilde{u}_i^m$ follows from Eq. (4.20):

$$\tilde{u}_i^m = \frac{Q_{u_i}^{m-1} - \sum_k A_k^{u_i} u_{i,k}^m}{A_C^{u_i}} \, . \tag{4.32}$$

In SIMPLER-method we thus solve in the second step an equation for pressure, not pressure correction; the coefficients of Eq. (4.31) are the same as those of the pressure-correction equation from the first step, just the source term is different. The iterative solution of Eq. (4.31) starts from the pressure p^{m-1}.

Other variants of the SIMPLE-method have been proposed; the variations are of minor significance and the basic principle remains the same.

4.3.5   Under-Relaxation Strategies

It has been noted that the pressure changes need to be under-relaxed in the SIMPLE-method, i.e. only part of the computed pressure correction is added to pressure from the previous outer iteration:

$$p^m = p^{m-1} + \alpha_p p' , \qquad (4.33)$$

with $0 \leq \alpha_p \leq 1$. SIMPLEC, SIMPLER and PISO do not require under-relaxation of pressure correction, unless further approximations are made (e.g. by neglecting the grid non-orthogonality in the pressure-correction equation, as is usually done). However, none of the above methods will converge unless the velocity changes are under-relaxed. This is best done implicitly, as suggested by Patankar [12], by modifying the momentum equation prior to solving it:

$$\frac{A_C^{u_i}}{\alpha_u} u_{i,C}^{m*} + \sum_k A_k^{u_i} u_{i,k}^{m*} = Q_{u_i}^{m-1} + (1 - \alpha_u)\frac{A_C^{u_i}}{\alpha_u} u_{i,C}^{m-1} . \qquad (4.34)$$

When outer iterations converge, the terms involving α_u cancel out and the original equation (4.13) is solved. In general, the closer α_u lies to unity, the faster convergence is obtained – up to some value after which the convergence rate deteriorates and eventually the iterations diverge. This is demonstrated in Fig. 4.1; the dependence on grid fineness is also shown.

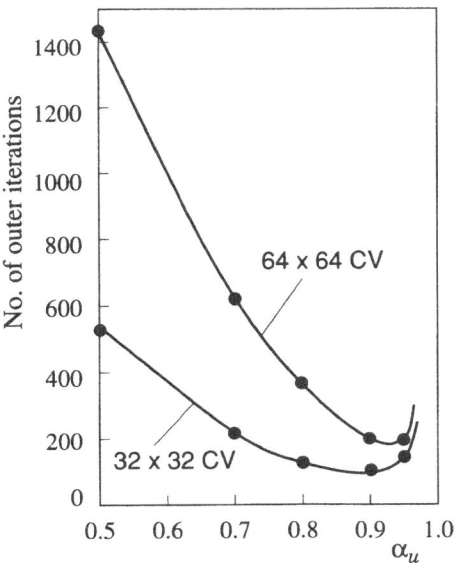
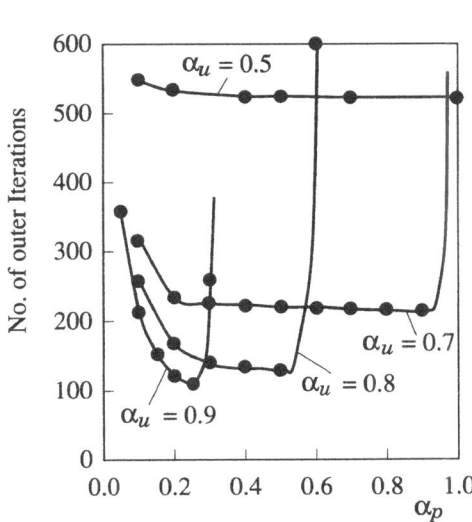

Figure 4.1 Numbers of outer iterations required to reduce the residual levels for all equations by three orders of magnitude using SIMPLE-method to compute a lid-driven cavity flow at Reynolds number Re = 1000: effects of under-relaxation factor α_u for two grids (left) and the influence of α_p for the grid with 32 × 32 CV (right)

The optimum combination of under-relaxation factors α_u and α_p for the SIMPLE-method can be obtained following the same arguments that led to SIMPLEC: the following relation is obtained (see [4] for more details on the derivation):

$$\alpha_p = 1 - \alpha_u \ . \tag{4.35}$$

It is interesting that, for a given value of α_u, one can vary α_p over some range without affecting the convergence of outer iterations; the range is wider for lower values of α_u, as shown in Fig. 4.1 for one example of incompressible flow. Typically, values of α_u between 0.7 and 0.9 are used, while α_p ranges between 0.2 and 1.0 (higher values are used for unsteady flows).

4.3.6 Fractional-Step Methods

Another class of popular methods for computing incompressible flows are the fractional-step methods; we present here one particular method, but again many variations are possible.

For the sake of simplicity we shall describe the method using a fully-implicit Euler scheme for time integration, which is suitable when marching towards the steady solution; when time-accurate simulations of unsteady flows are attempted, Crank-Nicolson or second order backward method are more appropriate.

In the first step, the velocity is advanced using pressure from the previous time step:

$$\frac{(\rho u_i)^* - (\rho u_i)^n}{\Delta t} = H(u_i^*) - \frac{\delta p^n}{\delta x_i} \ , \tag{4.36}$$

where $H(u_i^*)$ is an operator representing the discretized convective, diffusive, and source terms at the new time level. An asterisk is used instead of $(n+1)$ to indicate that the solution of the above equation needs further correction to represent the final solution at the time t_{n+1}. Since the equation is implicit and non-linear, iterative solution with outer and inner iterations is necessary, as described above. Outer iterations are often neglected when time-accurate computations with small time steps are performed, following the arguments discussed when presenting the PISO-method.

The final velocity at the new time level requires the gradient of the (as yet unknown) new pressure; it should thus satisfy the following form of the momentum equation:

$$\frac{(\rho u_i)^{n+1} - (\rho u_i)^n}{\Delta t} = H(u_i^*) - \frac{\delta p^{n+1}}{\delta x_i} \ . \tag{4.37}$$

Here the same approximation as in the SIMPLE-method is made, i.e. the operator H on the right-hand side refers to u_i^* instead of u_i^{n+1}. If we subtract Eq. (4.36) from (4.37), we obtain an equation for the velocity correction $u_i^{n+1} - u_i^*$ as a function of the gradient of pressure correction $p^{n+1} - p^n$:

$$\frac{(\rho u_i)^{n+1} - (\rho u_i)^*}{\Delta t} = -\frac{\delta p'}{\delta x_i} \ . \tag{4.38}$$

We now require that the new velocity satisfies the continuity equation,

$$\frac{\delta (\rho u_i)^{n+1}}{\delta x_i} = 0 \ , \tag{4.39}$$

which leads to a Poisson equation for pressure correction:

$$\frac{\delta}{\delta x_i}\left(\frac{\delta p'}{\delta x_i}\right) = \frac{1}{\Delta t}\frac{\delta(\rho u_i)^*}{\delta x_i} \ . \tag{4.40}$$

Upon solution of the pressure-correction equation, the new velocity field is obtained from Eq. (4.38). It satisfies the continuity equation and the momentum equation (4.37). The error due to keeping $H(u_i^*)$ instead of $H(u_i^{n+1})$ in Eq. (4.37) is of second order in time, as can be seen from Eq. (4.38), and thus consistent with other approximations:

$$u_i^{n+1} - u_i^* = -\frac{\Delta t}{\rho}\frac{\delta(p^{n+1} - p^n)}{\delta x_i} \approx \frac{(\Delta t)^2}{\rho}\frac{\delta}{\delta x_i}\left(\frac{\delta p}{\delta t}\right) \ . \tag{4.41}$$

Fractional step methods have become rather popular. There is a wide variety of them, due to a vast choice of approaches to time and space discretization; however, they are all based on the principles described above.

The major difference between the fractional-step method and pressure-correction methods of the SIMPLE-type is that in the former, the pressure-correction equation is solved only once per time step, while in the latter the pressure-correction equation is included in outer iterations on momentum equations. This is largely because fractional step methods are used mainly in unsteady flow simulations while the latter are used predominantly to compute steady flows. Also, in SIMPLE-type methods, the pressure-correction equation need not be solved accurately in each outer iteration (reduction of the residual level by one order of magnitude usually suffices), since the continuity equation is enforced only at the end of a time step. In fractional-step methods, the pressure-correction equation should be solved to a tight tolerance to ensure mass conservation.

The error referred to in Eq. (4.41) can be eliminated either by reducing the time step or by including the pressure-correction equation in outer iterations like in SIMPLE-type methods. However, if the splitting error is significant, the temporal discretization error is also large so the time step should be reduced if time-accurate solution is needed.

4.4 Pressure-Correction Equation for Compressible Flows

All of the above methods can be extended to compressible flows. The major difference is that, in compressible flows, both velocity *and* density variations affect the continuity equation; we shall show here how this can be taken into account in the SIMPLE-algorithm.

The linearized momentum equations are solved as before, using mass fluxes and pressure gradient from the previous outer iteration. The velocity field u_i^{m*} obtained by solving Eq. (4.13) needs to be corrected together with the density to satisfy the continuity equation (which now has an unsteady term due to density variation with time and which is – for the sake of simplicity – here discretized using the implicit Euler scheme):

$$\frac{\rho_C^m - \rho_C^n}{\Delta t} + \left[\frac{\delta(\rho^m u_i^m)}{\delta x_i}\right]_C = 0 \ . \tag{4.42}$$

If we introduce corrections $u_i' = u_i^m - u_i^{m*}$ and $\rho' = \rho^m - \rho^{m-1}$ and compute explicitly the imbalance before correction,

$$\frac{\rho_C^{m-1} - \rho_C^n}{\Delta t} + \left[\frac{\delta(\rho^{m-1} u_i^{m*})}{\delta x_i}\right]_C = \Delta \dot{m} , \tag{4.43}$$

the continuity equation can be expressed (by subtracting Eq. (4.43) from (4.42)) as:

$$\frac{\rho_C'}{\Delta t} + \left[\frac{\delta(\rho^{m-1} u_i' + \rho' u_i^{m*} + \rho' u_i')}{\delta x_i}\right]_C + \Delta \dot{m} = 0 . \tag{4.44}$$

The product of velocity and density corrections, $\rho' u_i'$, can be neglected because it tends to zero faster than other terms; as we approach converged solution, all corrections tend to zero so this approximation is justifiable (otherwise it can be taken into account iteratively as a kind of deferred correction).

The aim is now to express both velocity and density corrections through pressure correction, so that out of Eq. (4.44) a pressure-correction equation results.

If the approximation of the SIMPLE-method is adopted, the velocity correction can be expressed as a function of the gradient of pressure correction, see Eq. (4.18). For a link between density and pressure correction, an equation of state has to be used,

$$\rho' = \left(\frac{\partial \rho}{\partial p}\right)_T p' = C_\rho p' , \tag{4.45}$$

where, for an ideal gas, the coefficient C_ρ is:

$$C_\rho = \frac{1}{RT^{m-1}} . \tag{4.46}$$

Here we neglect the dependence of density correction on temperature changes. The justification lies in the fact that the final density will be computed from the equation of state after the energy equation is solved for temperature (or enthalpy, from which temperature can be computed). Also, since at converged state all corrections become zero, the value of C_ρ does not affect the final result – only the rate of convergence.

With the above approximations, the continuity equation (4.44) can be written as a pressure-correction equation as follows:

$$\frac{C_\rho p_C'}{\Delta t} + \left[\frac{\delta(C_\rho u_i^{m*} p')}{\delta x_i}\right]_C + \frac{\delta}{\delta x_i}\left[\frac{\rho^{m-1}}{A_C^{u_i}}\left(\frac{\delta p'}{\delta x_i}\right)\right]_C = -\Delta \dot{m} . \tag{4.47}$$

Important differences exist between this equation and the Poisson equation that was obtained for incompressible flows:

- In addition to the *diffusive part*, which stems from velocity correction (proportional to the gradient of p'), the compressible equation contains also a *convective part*, which stems from density correction (proportional to p' itself), as well as an unsteady term, thus resembling a wave equation.

- The ratio of convective to diffusive part is proportional to Ma^2 (where Ma is the local Mach number); in the limit $\text{Ma} \to 0$, the Poisson equation for incompressible flows is recovered, while for $\text{Ma} \gg 1$, the p'-equation resembles the density equation from classical methods for compressible flows.

The scheme is therefore self-adjusting and adapts itself automatically to the flow type. The method can thus be applied to flows containing both regions of highly compressible and virtually incompressible flow. Examples are valves with a small opening, wakes of bluff bodies etc.; some examples will be shown below. The method is reasonably efficient for both flow regimes; this too shall be demonstrated in example applications.

4.5 Solution Algorithm for all Flow Speeds

The solution algorithm based on a pressure-correction approach can be summarized as follows:

1. Start calculation of the fields at the new time t_{n+1} using the latest solution u_i^n, ρ^n, T^n, and p^n as starting estimates for the new solution;

2. Assemble and solve the linearized algebraic equation systems for the velocity components (momentum equations) to obtain u_i^{m*};

3. Assemble and solve the pressure-correction equation to obtain p';

4. Correct the velocities, density and pressure to obtain the velocity field u_i^m, which satisfies the continuity equation, and the new density and pressure fields, ρ^m and p^m;

   For the PISO algorithm, solve the second pressure-correction equation and correct velocities, density, and pressure again;

   For SIMPLER, solve the pressure equation for p^m after u_i^m is obtained above;

5. Assemble and solve the energy equation, using corrected velocity, density and pressure fields, to obtain new temperature T^m;

6. Compute new fluid properties (including density) that depend on temperature (and/or pressure);

7. Return to step 2 and repeat outer iterations until all corrections become negligibly small and the final values for u_i^{n+1}, ρ^{n+1}, T^{n+1}, and p^{n+1} are obtained;

8. Advance to the next time step.

In the following section we present a finite-volume method which incorporates SIMPLE-algorithm for all flow speeds and which has been used to compute flows in example applications.

4.6 FV-Method for Arbitrary Control Volumes

FV-methods use the conservation equations in integral form as a starting point, Eqs. (4.4) to (4.6). The method presented here is based on Ref. [3]. It is designed for arbitrary control volumes; Fig. 4.2 shows one polygonal CV in 2D. How is the solution domain subdivided into CVs and which shape they have is not relevant for the description of the method presented below.

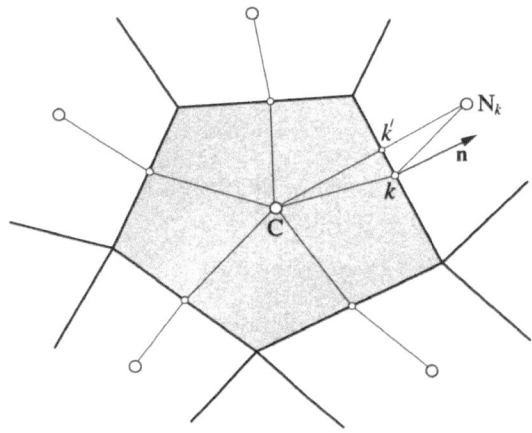

Figure 4.2 A two-dimensional control volume and notation used

The conservation equations are applied to all CVs; surface, volume, and time integrals are approximated using values of variables at computational nodes (which are assumed here to be located at the centroid of each CV), and at the centers of CV faces which coincide with solution domain boundaries. Many kinds of approximations for these terms exist.

The simplest integral approximation of the second order is the midpoint rule; it can be applied to arbitrary surface or volume elements and thus allows CVs of any shape. If CVs of polyhedral shape are allowed, integral approximations of higher order are not so simple and easy to develop; however, second order is in most practical cases sufficient, since largest errors usually come from models of phenomena that are not treated exactly (e.g. turbulence models).

When a cell-centered arrangement of variables is used (which is the case here), approximation of volume integrals is easy: since all data is stored at cell center, one only needs to multiply the integrand computed at this location with cell volume:

$$\int_V q \, dV \approx q_C \Delta V \ , \tag{4.48}$$

where ΔV is the volume of the CV. However, for surface integrals further approximations are necessary, since the value of the integrand is not known at cell-face center. For example, when computing convective fluxes,

$$\int_{S_k} \rho \phi \mathbf{v} \cdot \mathbf{n} \, dS \approx \phi_k \int_{S_k} \rho \phi \mathbf{v} \cdot \mathbf{n} \, dS = \phi_k \dot{m}_k \ , \tag{4.49}$$

we need to multiply the mass flux through the cell face, $\dot{m}_k$, with the value of the transported variable ϕ (which stands for velocity components or scalar variables like temperature, species concentration etc.) at cell-face center. The former is assumed known; in general, it is computed using values from previous outer iteration if the simple Picard-linearization is used. The latter has to be computed by interpolation from cell-center values.

There are many possibilities to choose from when interpolating variables to cell-face center; the simplest one (of second-order accuracy) is the linear interpolation. At the location k', where the line connecting two cell centers on either side of a cell face passes through the face, the linearly interpolated value is:

$$\phi_{k'} = \phi_{N_k}\lambda_k + \phi_C(1 - \lambda_k) \,, \tag{4.50}$$

where

$$\lambda_k = \frac{(\mathbf{r}_k - \mathbf{r}_C)\cdot(\mathbf{r}_{N_k} - \mathbf{r}_C)}{(\mathbf{r}_{N_k} - \mathbf{r}_C)\cdot(\mathbf{r}_{N_k} - \mathbf{r}_C)} \,. \tag{4.51}$$

However, in some cases k' may fall far from cell-face center k; the value of ϕ at cell-face center may be obtained with the second-order accuracy by applying a correction as follows:

$$\phi_k = \phi_{k'} + (\boldsymbol{\nabla}\phi)_{k'}\cdot(\mathbf{r}_k - \mathbf{r}_{k'}) \,. \tag{4.52}$$

The gradient at k' can be obtained by interpolating the gradients from the two cell centers according to Eq. (4.50); these have to be computed anyway, since they are needed to evaluate diffusive fluxes through CV-faces (and often also to compute source terms).

Cell-center gradients can be approximated within second order by using linear shape functions; if we assume linear variation of ϕ between two neighbor cell centers, we may write:

$$\phi_{N_k} - \phi_C = (\boldsymbol{\nabla}\phi)_C\cdot(\mathbf{r}_{N_k} - \mathbf{r}_C) \,. \tag{4.53}$$

We can write as many such equations as there are neighbors for the cell around node C; however, we need to compute only three derivatives $\partial\phi/\partial x_i$. With the help of least-squares methods, the derivatives can be explicitly computed for arbitrary CV shapes [11].

An alternative method of computing gradient at the center of an arbitrary CV is based on a midpoint-rule approximation and Gauss-theorem:

$$\int_V \boldsymbol{\nabla}\phi \, \mathrm{d}V = \int_S \phi\mathbf{n} \, \mathrm{d}S \quad \Rightarrow \quad (\boldsymbol{\nabla}\phi)_C \approx \frac{\sum_k \phi_k \mathbf{n}_k S_k}{\Delta V} \,. \tag{4.54}$$

This method is attractive because it uses cell-face values which have to be computed anyway for convective fluxes (although one does not have to use the same approximation in convective and diffusive fluxes).

Diffusive fluxes require an approximation of the derivative in the direction normal to cell face:

$$\int_{S_k} \Gamma \boldsymbol{\nabla}\phi \cdot \mathbf{n}\, \mathrm{d}S \approx (\Gamma \boldsymbol{\nabla}\phi \cdot \mathbf{n})_k S_k = \Gamma_k \left(\frac{\partial \phi}{\partial n}\right)_k S_k\ , \tag{4.55}$$

where Γ is the diffusion coefficient (viscosity, thermal conductivity). One possibility to compute this quantity would be to interpolate cell-center gradients to cell-face center like in Eq. (4.50), but this may lead to oscillations in the solution since this approximation is insensitive to wiggles with smallest wavelength that can be represented on the grid. This problem can be solved by introducing a third-order diffusion term [11], resulting in the following approximation of the cell-face gradient:

$$(\boldsymbol{\nabla}\phi)_k \cdot \mathbf{n}_k \approx \frac{\phi_{N_k} - \phi_C}{|\mathbf{r}_{N_k} - \mathbf{r}_C|} - \overline{(\boldsymbol{\nabla}\phi)_k}^{\text{old}} \cdot \left(\frac{\mathbf{r}_{N_k} - \mathbf{r}_C}{|\mathbf{r}_{N_k} - \mathbf{r}_C|} - \mathbf{n}_k\right). \tag{4.56}$$

The first term on the right-hand side (which is expressed through unknown values at neighbor nodes and thus contributes to the coefficient matrix when implicit time-integration methods are used) represents a central-difference approximation of the derivative in the direction of a straight line connecting nodes C and N_k (see Fig. 4.2); the underlined term corrects the error due to the fact that we need the derivative in the direction of the cell-face normal n. The first part of the correction is an approximation of the derivative just described, but computed using interpolated cell-center gradients; in the case of a smooth profile of ϕ, the two terms will cancel out in the solution. The remaining term represents the desired derivative in the normal direction. The correction becomes zero if the direction of line connecting C and N_k coincides with the direction of cell-face normal, since then the two unit vectors in brackets are the same and the first term on the right-hand side does not need to be corrected.

Another possibility to compute the derivative in the direction of cell-face normal is depicted in Fig. 4.3. One can define two auxiliary nodes, C' and N_k', which lie on the normal n, and express the derivative through a simple central difference as follows:

$$\left(\frac{\partial \phi}{\partial n}\right)_k \approx \frac{\phi_{N_k'} - \phi_{C'}}{|\mathbf{r}_{N_k'} - \mathbf{r}_{C'}|} = \frac{\phi_{N_k'} - \phi_{C'}}{(\mathbf{r}_{N_k} - \mathbf{r}_C) \cdot \mathbf{n}}\ . \tag{4.57}$$

The variable values at auxiliary nodes can be computed as:

$$\phi_{C'} \approx \phi_C + (\boldsymbol{\nabla}\phi)_C \cdot (\mathbf{r}_{C'} - \mathbf{r}_C)\ ,$$

$$\phi_{N_k'} \approx \phi_{N_k} + (\boldsymbol{\nabla}\phi)_{N_k} \cdot (\mathbf{r}_{N_k'} - \mathbf{r}_{N_k})\ . \tag{4.58}$$

The normal derivative can thus be approximated as:

$$\left(\frac{\partial \phi}{\partial n}\right)_k \approx \frac{\phi_{N_k} - \phi_C}{(\mathbf{r}_{N_k} - \mathbf{r}_C) \cdot \mathbf{n}} + \frac{(\boldsymbol{\nabla}\phi)_{N_k} \cdot (\mathbf{r}_{N_k'} - \mathbf{r}_{N_k}) - (\boldsymbol{\nabla}\phi)_C \cdot (\mathbf{r}_{C'} - \mathbf{r}_C)}{(\mathbf{r}_{N_k} - \mathbf{r}_C) \cdot \mathbf{n}}\ . \tag{4.59}$$

The last term on the right-hand side can be treated as a deferred correction, while the first term contributes to the coefficient matrix.

Surface integrals over faces which coincide with solution domain boundaries may require special approximations (e.g. one-sided differences). Rearrangement of approximations for all terms in the equation for one CV leads to an algebraic equation in which the unknown variables at CV center and at centers of nearest-neighbor CVs feature; for the solution domain as a whole, a system of algebraic equations is obtained that can be solved iteratively.

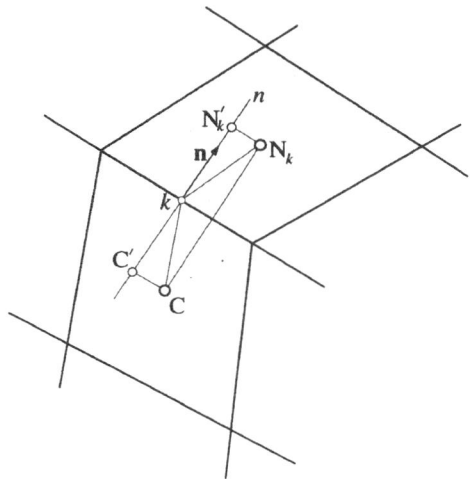

Figure 4.3 On the computation of derivatives in the direction normal to cell face

4.7 Pressure-Correction Algorithm for FV-Methods

We describe here the implementation of the pressure-correction algorithm for all flow speeds into the FV-method described above. It is assumed here that an implicit time-integration method is employed; the necessary modifications for explicit methods are straightforward.

Momentum equations are solved first to yield u_i^{m*} :

$$A_C^{u_i} u_{i,C}^{m*} + \sum_k A_k^{u_i} u_{i,k}^{m*} = Q_{i,C} . \tag{4.60}$$

The source term $Q_{i,C}$ contains the discretized pressure gradient term:

$$Q_{i,C} = Q_{i,C}^* + Q_{i,C}^p = Q_{i,C}^* - \left(\frac{\delta p}{\delta x_i} \right)_C \Delta V . \tag{4.61}$$

If the pressure term is approximated in a conservative way (as a sum of surface forces), the mean pressure gradient over the CV can be expressed – using Gauss-theorem – as:

$$Q_{i,C}^p = - \int_S p\, \mathbf{i} \cdot \mathbf{n}\, dS = - \int_V \frac{\partial p}{\partial x_i}\, dV \quad \Rightarrow \quad \left(\frac{\partial p}{\partial x_i} \right)_C \approx - \frac{Q_{i,C}^p}{\Delta V} . \tag{4.62}$$

Thus, in the analysis that follows we shall assume that the pressure term is expressed as in Eq. (4.61).

SIMPLE-type methods for a colocated variable arrangement (which is the most widely used approach in the case of unstructured grids) require a special interpolation technique to compute cell-face velocities needed to assemble mass fluxes, since a simple linear interpolation leads to oscillatory solutions. The following correction of the interpolated velocity has been proposed by Rhie and Chow [14]:

$$v_{n,k}^{m*} = \overline{(v_n^{m*})}_k - \Delta V_k \overline{\left(\frac{1}{A_{\mathrm{C}}^{v_n}}\right)}_k \left[\left(\frac{\delta p}{\delta n}\right)_k - \overline{\overline{\left(\frac{\delta p}{\delta n}\right)}}_k\right]^{m-1} , \tag{4.63}$$

where the superscript m denotes outer iterations within the new time step, which converge towards the solution at time t_{n+1} .

Only the normal velocity component contributes to the mass flux through a cell face; it depends on the pressure gradient in the normal direction. Since the interpolated velocity implicitly implies interpolation of pressure gradients from cell centers, their contribution is removed and replaced by the gradient computed at cell face, which is described by the terms in square brackets. Here overbar denotes linear interpolation while the double overbar denotes arithmetic averaging. Thus, if the pressure derivative in normal direction at the cell face is computed using central differences, the terms in square brackets will cancel out for a linear or quadratic pressure distribution, since the central difference is second-order accurate at the midpoint between the two nodes.

The correction term is proportional to h^2 and the third derivative of pressure, with h being mesh spacing; it is thus a consistent approximation which tends to zero as the grid is refined and as the pressure distribution becomes smooth. The role of the correction term is to detect pressure oscillations and lead to a pressure correction which smoothes them out. This approach is equivalent to adding a fourth derivative of pressure to the differential equation.

The normal derivative of pressure can thus be approximated using Eq. (4.59); the last term on the right-hand side reflects both the non-orthogonality of the grid and the distance between k and k', see Fig. 4.3. This complicates the pressure-correction equation to be derived below; fortunately, in most applications one can ignore the non-orthogonality and simply use the derivative along the line which connects the two nodes (which is given by the first term on the right-hand side) instead of the derivative in the normal direction. Otherwise, one has to use a kind of predictor-corrector procedure: first neglect the second term on the right-hand side of Eq. (4.59) and take it into account explicitly in the second step.

Note that the usual discretization of momentum equations leads to equal coefficients A_{C} for all three Cartesian velocity components and further below we shall denote it as A_{C}^v .

With the simplified approach, the cell-face velocity at the m th outer iteration is approximated as:

$$v_{n,k}^{m*} = \overline{(v_n^{m*})}_k - \frac{\Delta V_k}{(\mathbf{r}_{\mathrm{N}_k} - \mathbf{r}_{\mathrm{C}}) \cdot \mathbf{n}} \overline{\left(\frac{1}{A_{\mathrm{C}}^v}\right)}_k \left[(p_{\mathrm{N}_k} - p_{\mathrm{C}}) - \overline{\overline{(\boldsymbol{\nabla} p)}}_k \cdot (\mathbf{r}_{\mathrm{N}_k} - \mathbf{r}_{\mathrm{C}})\right]^{m-1} . \tag{4.64}$$

Mass fluxes are computed using $v_{n,k}^{m*}$ and density from the previous outer iteration,

$$\dot{m}_k^{m*} = (\rho^{m-1} v_n^{m*} S)_k . \tag{4.65}$$

They in general do not satisfy the continuity equation; if – for the sake of simplicity – we assume that a fully-implicit Euler scheme is used for time integration, the discretized continuity equation reads:

$$\frac{(\rho^{m-1} - \rho^n)_{\mathrm{C}} \Delta V}{\Delta t} + \sum_k \dot{m}_k^{m*} = Q_{\mathrm{m}}^* . \tag{4.66}$$

The mass imbalance must be eliminated by a correction method. Both velocity *and* density can – and should – be corrected if the flow is considered compressible. The corrected mass fluxes at the m th outer iteration can thus be defined as:

$$\dot{m}_k^m = \dot{m}_k^{m*} + \dot{m}_k' = [(\rho^{m-1} + \rho')(v_n^{m*} + v_n')S]_k \ . \tag{4.67}$$

The mass-flux correction is thus:

$$\dot{m}_k' = (\rho^{m-1}v_n'S)_k + (\rho'v_n^{m*}S)_k + \underline{(\rho'v_n'S)_k} \ . \tag{4.68}$$

The underscored term can be neglected since it involves a product of corrections and thus tends to zero faster than other terms; otherwise it can be taken into account using a predictor-corrector approach, as mentioned above for the simplification that neglects grid non-orthogonality.

The SIMPLE-approximation for velocity correction is (see Eq. (4.18)) proportional to the gradient of pressure correction:

$$(\rho^{m-1}v_n'S)_k = -(\rho^{m-1}S\,\Delta V)_k \overline{\left(\frac{1}{A_C^v}\right)}_k \left(\frac{\delta p'}{\delta n}\right)_k \ . \tag{4.69}$$

If the temperature is, for one outer iteration, frozen, we can express the density correction as a function of pressure correction using an equation of state, see Eq. (4.45):

$$\rho_k' \approx p_k' \left(\frac{\partial\rho}{\partial p}\right)_T^{m-1} = C_{\rho,k}\, p_k' \ . \tag{4.70}$$

The coefficient C_ρ can, for an ideal gas, be determined as follows (see Eq. (4.46)):

$$C_\rho = \frac{1}{RT^{m-1}} \ . \tag{4.71}$$

For non-ideal gases, the derivative in C_ρ may need to be computed numerically. As discussed earlier, the converged solution is independent of this coefficient because all corrections are then zero, but it can influence the *convergence rate* of the method.

The second term in the mass-flux correction, Eq. (4.68) is thus:

$$(v_n^{m*}\rho'S)_k = \left(\frac{C_\rho \dot{m}^{m*}}{\rho^{m-1}}\right)_k p_k' \ . \tag{4.72}$$

With the third term neglected and with pressure gradient approximated as in Eq. (4.64), the complete mass flux correction reads:

$$\dot{m}_k' = -(\rho^{m-1}S\,\Delta V)_k \overline{\left(\frac{1}{A_C^v}\right)}_k \frac{p_{N_k}' - p_C'}{(\mathbf{r}_{N_k} - \mathbf{r}_C)\cdot\mathbf{n}} + \left(\frac{C_\rho \dot{m}^{m*}}{\rho^{m-1}}\right)_k p_k' \tag{4.73}$$

The second-order central-difference approximation for the pressure derivative is the usual choice; however, approximations of higher order are also possible (see Lilek and Perić [9] for an example of a fourth-order FV-method for Cartesian 2D grids).

The value of p' at the cell-face center also needs to be approximated. Any interpolation scheme used for convective fluxes can be applied. Linear interpolation is the usual choice, although first and second-order upwind approximations as well as quadratic or cubic interpolants or a blend of various schemes can also be used. Since higher-order methods usually lead to oscillations near shocks, either blending with first-order upwind scheme or more sophisticated TVD or ENO schemes need to be used; see other chapters for a description of some suitable schemes.

The corrected mass fluxes are required to satisfy the continuity equation:

$$\frac{(\rho^m - \rho^n)_{\mathrm{C}}\Delta V}{\Delta t} + \sum_k \dot{m}_k^m = 0 \,. \tag{4.74}$$

By subtracting Eq. (4.66) from Eq. (4.74), we obtain the following equation:

$$\frac{\rho'_{\mathrm{C}}\Delta V}{\Delta t} + \sum_k \dot{m}'_k + Q_{\mathrm{m}}^* = 0 \,. \tag{4.75}$$

By replacing ρ' and $\dot{m}'$ with expressions given by Eqs. (4.70) and (4.73), we obtain an equation in which the pressure correction is the only unknown; it can be arranged in the usual form:

$$A_{\mathrm{C}}p'_{\mathrm{C}} + \sum_k A_{\mathrm{N}_k}p'_k = -Q_{\mathrm{m}}^* \,. \tag{4.76}$$

The coefficients depend on the approximations used.

As noted earlier, the pressure-correction equation for compressible flows resembles a discretized convection-diffusion equation, as can be seen from Eq. (4.73): the first term on the right-hand side is akin to diffusive flux, cf. Eq. (4.55), while the second term resembles the convective flux of Eq. (4.49). An important feature of the above pressure-correction equation is that it adjusts itself automatically to the local flow features and is thus applicable to flows which locally have high Mach numbers while elsewhere they may be virtually incompressible (e.g. flows around valves and through small openings connecting large regions of low-speed flow).

Compressible flows can be computed using only velocity correction when constructing the pressure-correction equation (like in purely incompressible flows) up to a Mach number of the order of 0.5; in that case, density is frozen during one outer iteration but is updated after the new temperature and pressure, T^m and p^m, are computed. Similarly, if the Mach number is low but density varies due to temperature and/or concentration variation, the same approach can be used. In the case of unsteady flows, sequential solution is usually efficient enough since changes from one time step to the next are not too large; for steady flows, faster convergence could be achieved by a coupled solution method for velocities and temperature, or by using a multigrid acceleration of outer iterations (see [4] for details of such multigrid schemes).

4.8 Implementation of Boundary Conditions

The following boundary conditions are often encountered in engineering applications:

- *Solid walls* where no-slip (zero relative velocity) conditions, and prescribed temperature or heat flux apply.

- *Inlet* where velocity and temperature are prescribed. The same applies to other transported variables (concentration, turbulence quantities etc.).

- *Outlet* where usually zero gradient of all quantities in flow direction is prescribed.

- *Prescribed static pressure*; this is usually the case at an outflow boundary, where the flow subsonic.

- *Prescribed total pressure and temperature*; this is usually the case at inflow boundaries, when the local velocity, pressure and temperature are not known, but the conditions in a reservoir further upstream where fluid is at rest are known. In this case flow direction has also to be specified.

- *Non-reflective free-stream conditions*; this is usually the case at lateral boundaries in a supersonic flow, where shocks should not be reflected.

- *Symmetry plane* where zero gradient normal to boundary applies for all scalar quantities and the velocity component parallel to it, while the normal velocity component is equal to zero.

In FV-methods, boundary conditions are implemented through surface integrals over cell faces that lie in the boundary surface, i.e. through convective and diffusive fluxes. If variable values are prescribed, than convective fluxes can be directly computed; the computed value enters the source term of the equation for the near-boundary CV. In order to compute the diffusive flux through a boundary cell face, we need to compute the gradient at the boundary; one-sided approximations have to be used, which involve the known boundary values and one or more values from near-wall cell-centers.

For an isentropic flow of an ideal gas, the *total pressure* at a boundary is defined as:

$$p_t = p_b \left(1 + \frac{\gamma - 1}{2} \mathrm{Ma_b}\right)^{\frac{\gamma}{\gamma - 1}} , \tag{4.77}$$

where p_b is the local pressure at boundary and the ratio of specific heats at constant pressure and constant volume is $\gamma = c_p/c_v = 1.4$. The flow direction is defined by a unit vector $\mathbf{i_b}$:

$$\mathbf{v_b} = \mathrm{Ma_b} v_s \mathbf{i_b} , \tag{4.78}$$

where $v_s = \sqrt{\gamma R T_b}$ is the local speed of sound, T_b is the temperature at boundary, and Ma $_b$ is the Mach number at the boundary. The total temperature is given by:

$$T_t = T_b \left(1 + \frac{\gamma - 1}{2} \mathrm{Ma_b}\right) . \tag{4.79}$$

The velocities and temperature are computed at boundary cell faces from the above equations; pressure is obtained by extrapolation from interior. Since the local value of the Mach number is not known a priori, the boundary values of velocity components and temperature (and also of pressure) change during outer iterations until a converged steady solution is obtained. For faster convergence, boundary velocities should be corrected within the SIMPLE-step.

When the *static pressure* is prescribed at an outflow, velocity must be extrapolated from upstream nodes; it is computed at cell-face center using similar expression as for internal cell faces, cf. Eq. (4.63), only here interpolation is replaced by extrapolation:

$$v_{n,b}^{m*} = \overline{(v_n^{m*})}_b - \Delta V_b \overline{\left(\frac{1}{A_C^v}\right)}_b \left[\left(\frac{\delta p}{\delta n}\right)_b - \overline{\left(\frac{\delta p}{\delta n}\right)}_b\right] . \tag{4.80}$$

In a SIMPLE-step, $p' = 0$ at the boundary, but the velocity has to be corrected there:

$$v_{n,b}' = -\Delta V_b \overline{\left(\frac{1}{A_C^v}\right)}_b \left(\frac{\delta p'}{\delta n}\right)_b . \tag{4.81}$$

In the mass-flux correction, $\rho_b' = \rho_C'$ can be assumed; if the outlet boundary is far enough downstream, all quantities that need to be extrapolated can be simply taken equal to those at the near-boundary cell center. One-sided differences have to be used to approximate the pressure and pressure-correction gradient at boundary.

At a *non-reflective free-stream boundary*, oblique shocks must cross the boundary without reflection. This can be achieved by computing velocity at the boundary as follows:

$$\mathbf{v}_b = \mathbf{v}_{b,t} + \mathbf{v}_{b,n} , \tag{4.82}$$

where the velocity component tangential to free-stream boundary, $\mathbf{v}_{b,t}$, is obtained by extrapolation from interior, while the velocity component normal to boundary is obtained from:

$$\mathbf{v}_{b,n} = -f(p_b)\mathbf{n}_n , \tag{4.83}$$

where $\mathbf{n}_n$ is the outward pointing unit vector parallel to $\mathbf{v}_b$. The function $f(p_b)$ is computed either from the simple wave theory for an ideal gas in case of compression waves, or from the Prandtl-Meyer function for expansion waves.

In the following section some examples of application of the above FV-method will be given.

4.9   Examples of Application

In order to demonstrate the features of the method described above, we have chosen two representative examples. Since the aim is only to demonstrate the performance of the method and not to seek accurate solutions for particular problems, no detailed analysis of discretization errors will be made; the grids are locally refined a priori in order to produce reasonably accurate solutions typical of industrial applications. However, we do emphasize the importance of error estimation and grid quality control; more details on these issues are available in [4, 15] and numerous other publications.

The first test case deals with flows around a NACA0012 airfoil. Since the method described above is designed to work at all flow speeds, we shall present solutions for laminar flow at low Reynolds number (Re = 10, based on free-stream velocity and chord length), turbulent flow at Re = 3.6×10^6 and a free-stream Mach number around 0.16, transonic flow with free-stream Mach number around 0.8, and a supersonic flow with free-stream Mach number around 1.8. In the first two cases, we chose incoming flow to be at an angle of attack of $25°$, while for the last two cases the angle of attack was $1°$. In all cases the solution domain extended 10 chord lengths ahead, above, and below, and 15 chord lengths behind the foil. The base grid is shown in Fig. 4.4; it has been locally refined for some cases, as will be pointed out later.

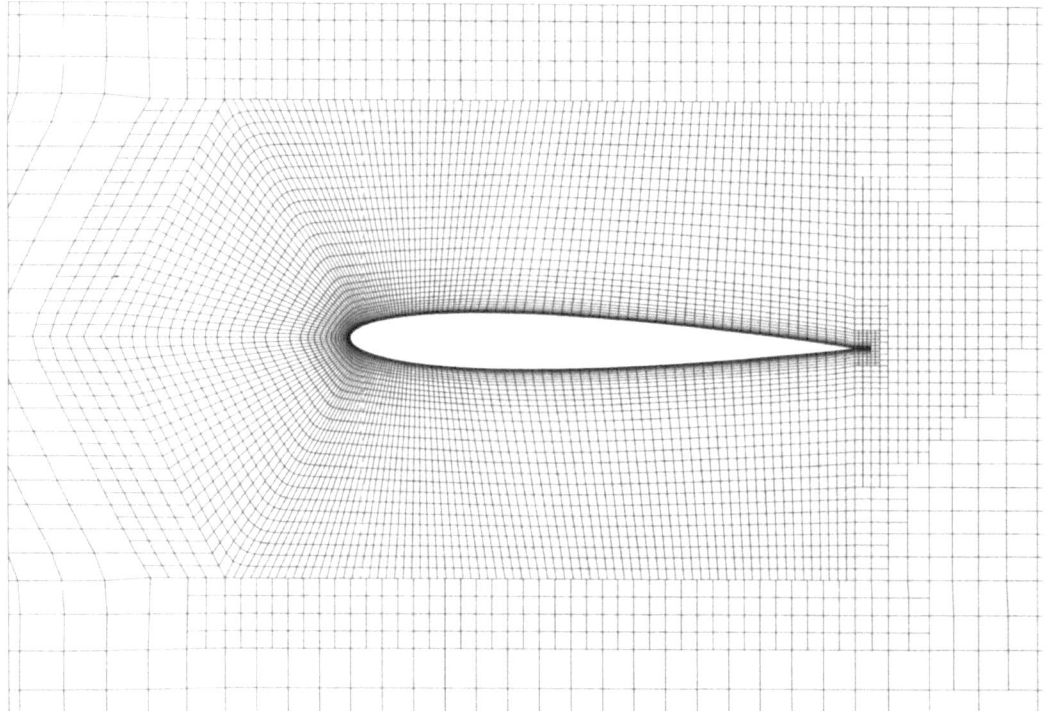

Figure 4.4 Base grid used for computations of flows around a NACA0012 aerofoil

The airfoil was 1 m long in all computations. For turbulent and compressible flows, fluid properties of air at 293 K have been used; for laminar flow, inlet velocity was 1 m/s, density was set to 1 kg/m^3, and viscosity was set to 0.1 Pa s. In the case of compressible flows, viscous effects were neglected, i.e. Euler-equations were solved.

4.9.1 Subsonic Laminar Flow Around Airfoil

In this case fluid velocity was prescribed at all boundaries except for the right-hand side, where outlet conditions (extrapolation in flow direction) were used. Due to low Reynolds number, the flow is elliptic in nature and the rate of convergence is relatively low. This type of flow is specially suitable for multigrid-acceleration, which was not used here in order to demonstrate the performance of the basic method. No other local refinements than those seen in Fig. 4.4 were used; the grid had 10808 CVs.

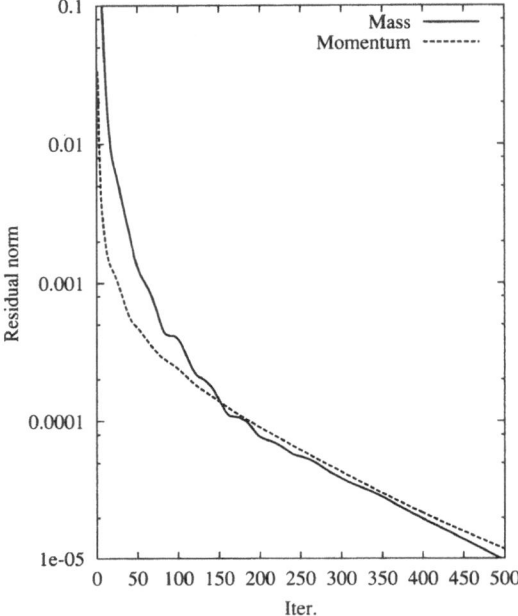

Figure 4.5 Convergence of outer iterations when computing laminar flow around a NACA0012 airfoil at 25 ° angle of attack and a Reynolds number based on chord length around 10

Figure 4.5 shows how the residual norm (normalized sum of absolute values of residuals at all CVs, computed at the begining of each outer iteration after the coefficient matrix and source term have been updated) for mass and momentum conservation equations vary from iteration to iteration. The velocity field is initiated by an approximate potential flow; initially, residual levels are reducing quickly, but after a while a constant reduction rate is achieved. We know from experience that the iteration error is initially reducing somewhat slower than residuals. However, once the residuals attain a constant convergence rate, the rate at which the error is reduced becomes the same as the rate at which the residuals are reduced. Monitoring of velocities and pressure at some CVs also suggests that at the end, no value was changing at the three most significant digits, indicating iteration errors lower than 0.1%.

Figure 4.6 shows pressure and velocity vectors around the foil. Velocity varies gradually near wall and there is no flow separation.

4.9.2   Subsonic Turbulent Flow Around Airfoil

Turbulent flow at high Reynolds number and high angle of attack is difficult to predict correctly. When the grid shown in Fig. 4.4 is used an the k - ϵ turbulence model with wall functions is employed, only a small separation region at the rear of the foil is predicted

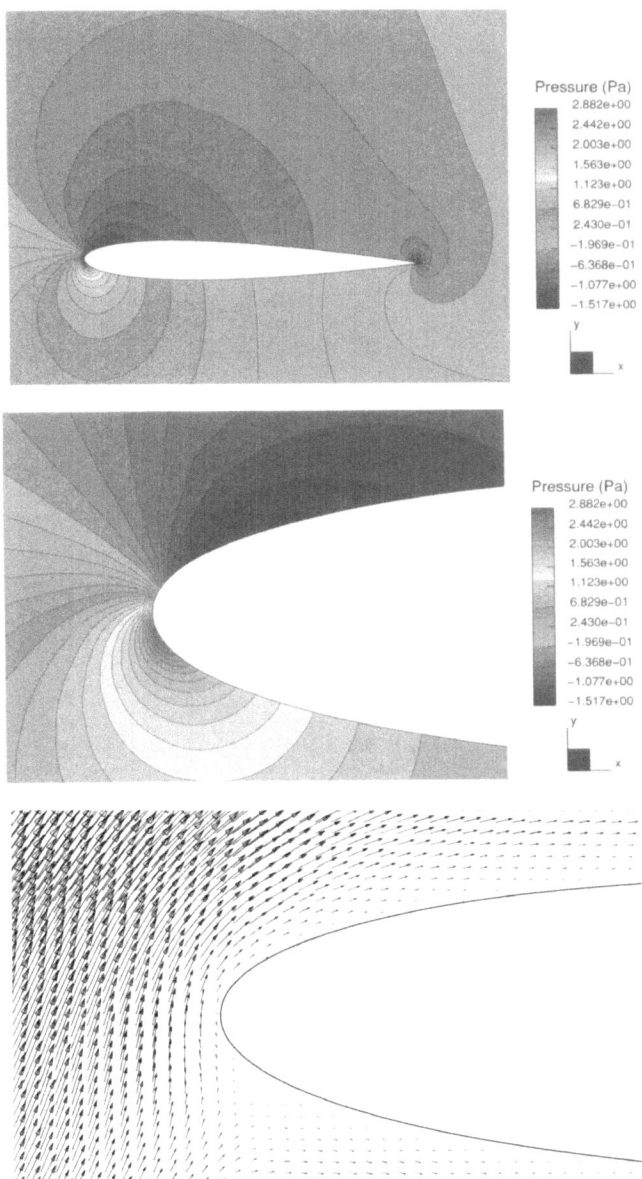

Figure 4.6 Pressure distribution around the foil (upper), detail of pressure distribution (middle), and velocity vectors (lower) around the nose of the NACA0012 airfoil at 25 ° angle of attack and a Reynolds number based on chord length around 10

on suction side. On the other hand, when the grid is refined in the wall-normal direction so that the first cell-center away from wall is at the normalized distance $n^+ \approx 1$ and a low-Reynolds-number turbulence model is used (here the so called SST-version of the k-ω model after Menter [10]), a large recirculation region is predicted with unsteady vortex shedding, so that a transient computation has to be performed. Figure 4.7 shows distribution of pressure, turbulent kinetic energy, and eddy viscosity around the foil at one time instant.

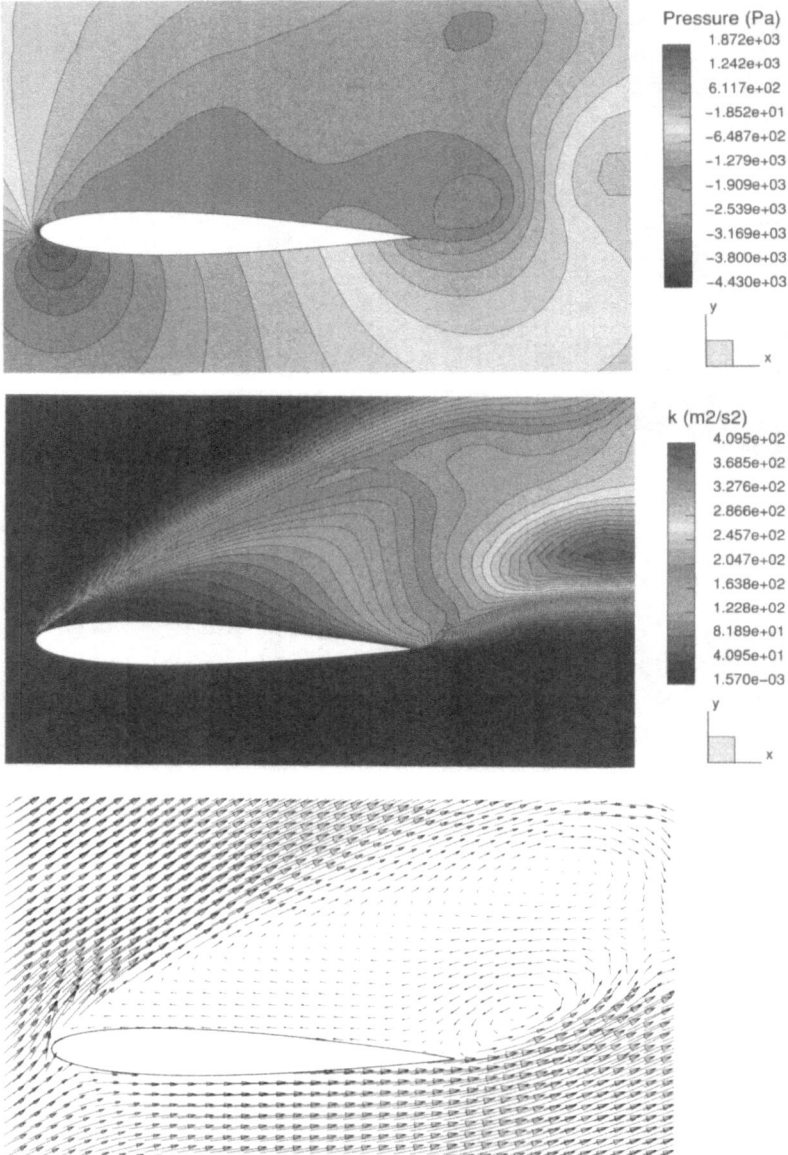

Figure 4.7 Distribution of pressure (upper), turbulent kinetic energy (middle), and velocity (lower) around the NACA0012 airfoil at 25 $^\circ$ angle of attack and a Reynolds number based on chord length around 3.6×10^6

Here modeling errors play an important role, in addition to iteration and discretization errors. Since the aim of this exercise is only to demonstrate the performance of the numerical method, we shall not attempt here to quantify these errors; it is worth noting that the above-mentioned k-ω model has in the past shown superior performance in predicting flows around airfoils compared to other two-equation models.

For the steady flow obtained using the k-ϵ turbulence model and wall functions, around 200 outer iterations were needed to reduce the residual levels by four orders of magnitude. The grid used for the low-Re model had 12050 CVs and very thin cells near wall (with aspect ratios up to 200). The effective viscosity varied within the solution domain between 1.8×10^{-5} Pa s (viscosity of air) and 0.84 Pa s (maximum eddy viscosity in the recirculation region; the contours of eddy viscosity are similar to those of turbulent kinetic energy, see Fig. 4.7). Unsteady simulation was performed using time step of 0.001 s; typically three outer iterations per time step were needed, and it took 42 minutes to compute 1000 time steps on a laptop computer with a Pentium III, 1 GHz processor. During the simulation time of 1 s a fluid particle traveling at the mean speed of 55 m/s moves over a distance corresponding to 55 chord lengths.

4.9.3 Transonic Flow Around Airfoil

Transonic flow was computed for the free-stream Mach number around 0.8, with the following boundary conditions: specified velocity and temperature at inlet, specified atmospheric pressure at outlet, and symmetry conditions at lateral boundaries (the foil in Fig. 4.4 is at an angle of $1°$, the undisturbed flow is parallel to the lateral boundaries of solution domain which are 10 chord lengths away from the foil). Under these conditions a shock is created at the suction side of the foil. Euler equations were solved and convective fluxes were discretized using 95% of linearly interpolated convected variable and 5% of the value at the upstream node (a blend of central and upwind differencing) for momentum and energy equations; in the pressure-correction equation, pure linear interpolation was used.

Figure 4.8 shows the results obtained on the base grid with one refinement in x-direction in the vicinity of the shock. For a better resolution of the shock, one would need to refine the grid further. Shock is also clearly seen in the pressure distribution along airfoil wall. Due to insufficient damping of oscillations at shock caused by linear interpolation, one can see some over- and undershoots.

Computation in this case requires either time marching, stronger under-relaxation, or a good estimate of the solution from a coarser grid. Since only one grid was used here, simulation was performed in the unsteady mode until a steady state was reached.

4.9.4 Supersonic Flow Around Airfoil

For the computation of supersonic flow around airfoil, the following boundary conditions were applied: specified velocity and temperature at inlet, non-reflecting conditions at lateral boundaries, and supersonic outlet at the downstream boundary. The foil was inclined by $1°$ to the free stream, like in the transonic case.

Figure 4.9 shows the convergence of outer iterations in this case. The number of iterations required to reach a solution which is accurate on more than three significant digits is lower than in the case of subsonic laminar flow; this is due the the hyperbolic nature of the equations. The computation took only 137 s on a laptop computer with a Pentium III, 1 GHz processor.

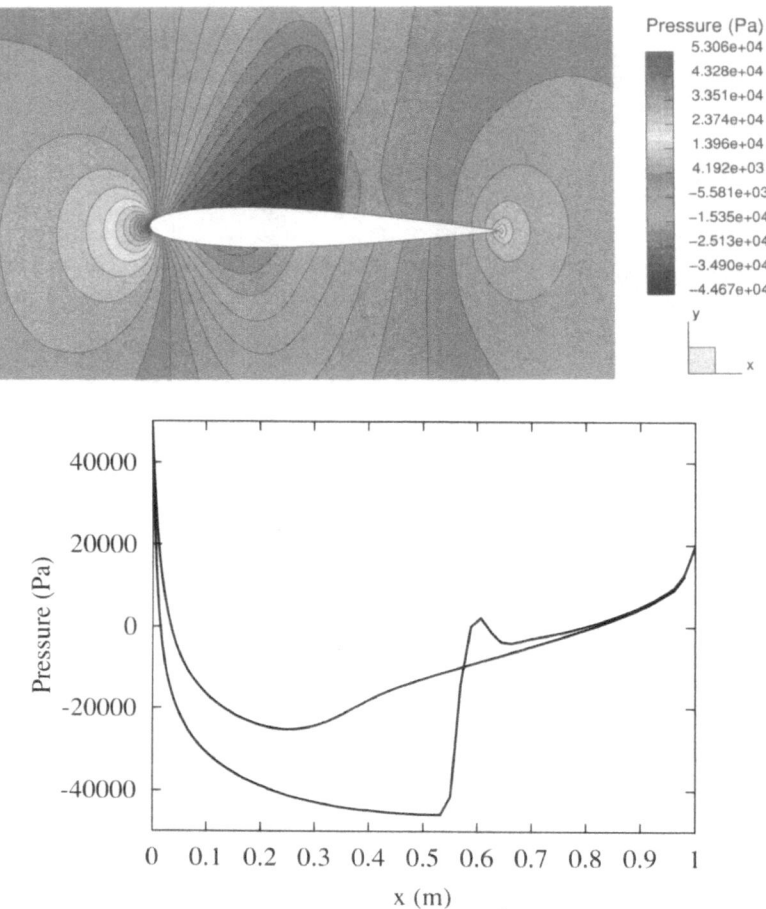

Figure 4.8 Distribution of pressure around airfoil (upper), and along suction and pressure wall (lower) in a transonic flow around a NACA0012 airfoil at 1 ° angle of attack

Figure 4.10 shows the predicted distribution of Mach number and temperature around the foil. A shock forms ahead of airfoil due to its blunt nose; the grid has been once locally refined around the shock after its location has been predicted on the base grid. At the trailing edge another two shocks form, so that in the wake the magnitude and direction of velocity vectors matches that of the free stream.

Temperature distribution resembles closely the Mach-number contours; due to a high Mach number, temperature varies in a wide range – between 252 and 410 K.

This example shows that the pressure-correction method is quite efficient for high Mach number flows; as already indicated, further increase in efficiency is possible by using multigrid methods.

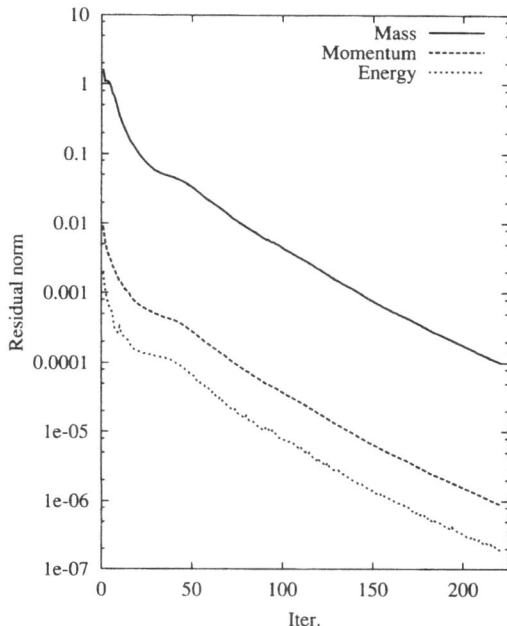

Figure 4.9 Convergence of outer iterations when computing supersonic flow around a NACA0012 airfoil at 1 ° angle of attack

4.9.5 Flow Through an Orifice

The last example demonstrates the self-adaptivity of the method to the local flow structure. We simulate the flow that develops after a sudden opening of a small slot in a wall that separates pressurized air from the atmosphere. Figure 4.11 shows a sequence of pressure distributions at different times after the start of simulation. At the slot entry, stagnation conditions corresponding to 1 bar over-pressure and 293 K temperature were specified; the lower boundary is the symmetry plane, while at the far right and top boundaries, atmospheric pressure and temperature were set. The time step was 5×10^{-6} s and around 7 outer iterations were needed per time step to reduce all residual norms by three orders of magnitude. Since a time-accurate simulations is attempted, a second-order, fully implicit time-integration scheme is used (it assumes a quadratic profile of variables in time and integrates from $t_{n+1} - \Delta t/2$ to $t_{n+1} + \Delta t/2$).

The isobars in Fig. 4.11 shows that a pressure wave, initiated by the sudden slot opening, travels at a speed of sound forming a cylindrical front (in a 2D, plane simulation). After sudden expansion, a strong vortex develops which travels downstream with about 1/5th of the speed of sound.

Figure 4.12 shows distribution of Mach number at three time instants, while Fig. 4.13 shows the velocity vectors (interpolated to a uniform presentation grid) at the corresponding time instants. Some wiggles can be seen ahead of the frontal shock, due to the fact that – like in the previous examples with compressible flow – 95% of central and 5% of upwind differences were used when approximating convective fluxes in momentum and energy equations. Here Navier-Stokes equations were solved so that viscous terms are

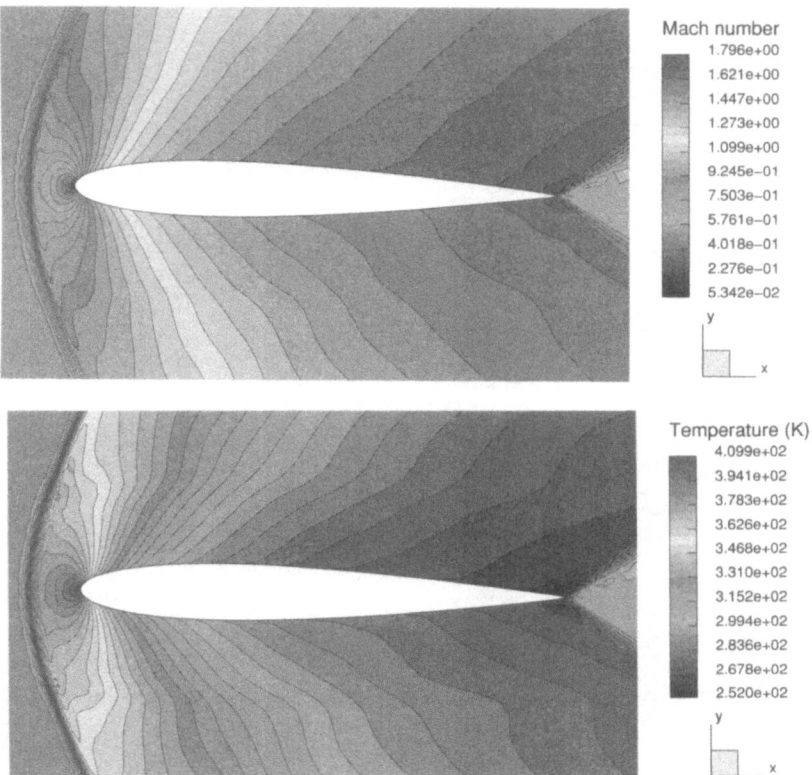

Figure 4.10 Distribution of Mach number (upper), and temperature (lower) in a supersonic flow around a NACA0012 airfoil at 1 ° angle of attack

also taken into account. This figure shows that the Mach number varies from virtually zero (nearly incompressible flow, initially even stagnant fluid at larger distance from the slot) to almost 2; complicated flow phenomena including shocks appear locally and have to be captured in a transient simulation.

4.10 Conclusions

We have shown in the preceding sections how a family of pressure-correction methods can be developed for computing fluid flows at all speeds. Although here a segregated solution method and FV-discretization on unstructured grids have been used, the pressure-correction methods can be extended to other solution strategies and discretization schemes.

The attractive feature of the pressure-correction method for all flow speeds is that it adapts itself to the local flow structure and is equally efficient for incompressible and supersonic flows. The segregated solution strategy can easily be adapted to more complex flows including sophisticated turbulence models (like the Reynolds-stress model, in which

– in 3D – seven additional equations are solved), chemical reactions and multi-phase phenomena.

The results of example computations demonstrated the robustness and reasonable efficiency of the method. The FV method using unstructured grids with cell-wise local refinement can be adapted to allow error-guided dynamic refinement and coarsening of the grid; here only a-priori local refinement has been used. Local refinement offers the possibility to achieve high accuracy while using simple approximations and thus a simple computer code.

Acknowledgments

All computations have been performed using computer code **Comet** of ICCM Institute of Computational Continuum Mechanics GmbH.

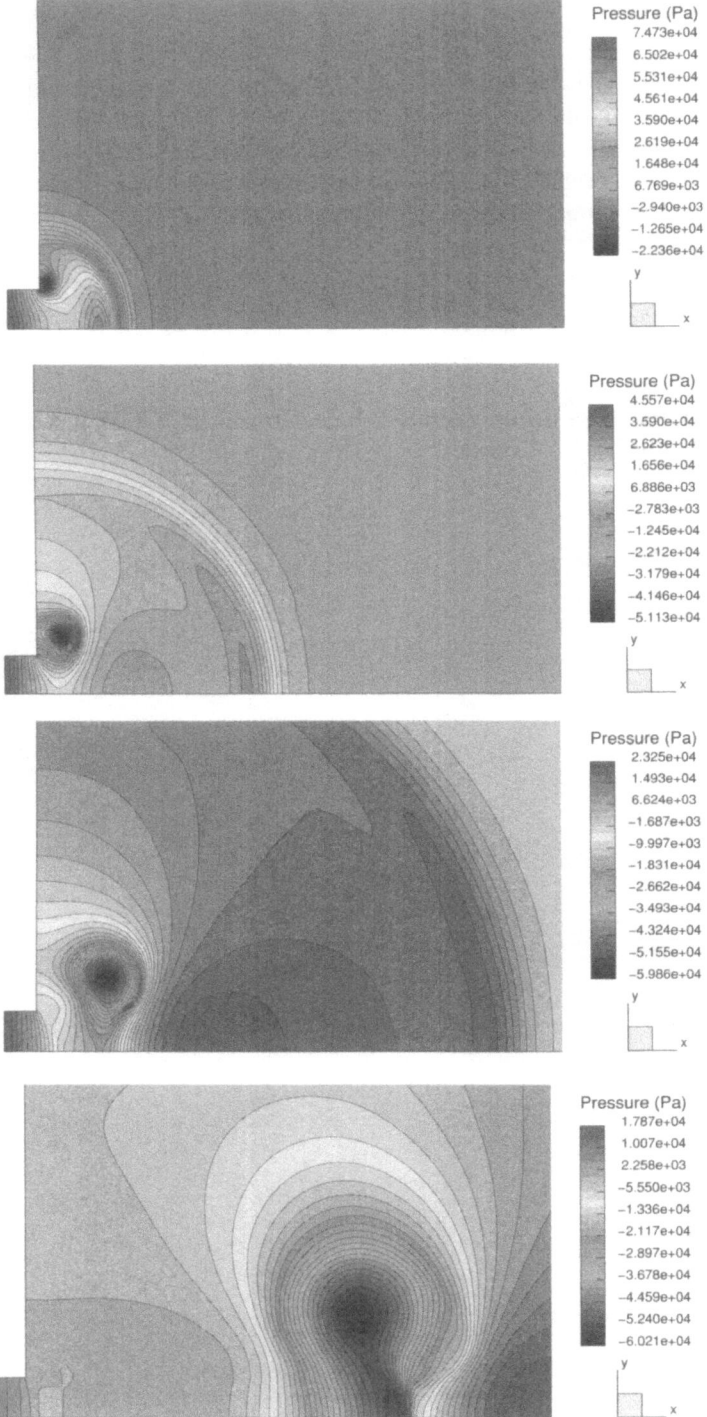

Figure 4.11 Computed distribution of pressure outside the slot after 20, 50, 100, and 300 time steps (from top to bottom, respectively) since opening

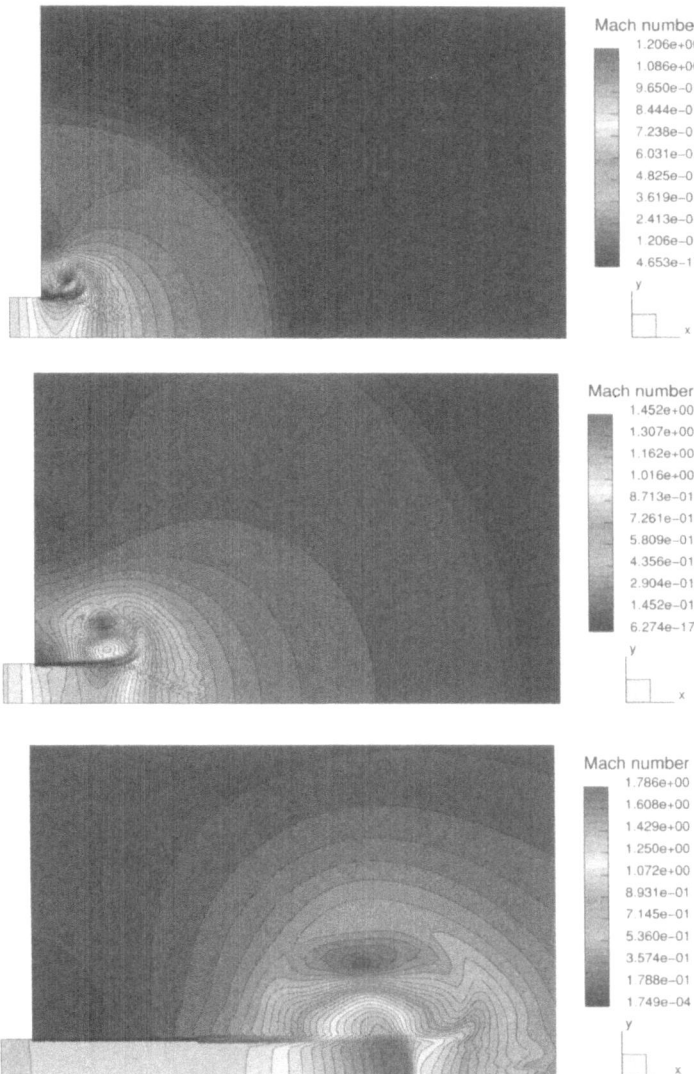

Figure 4.12 Computed distribution of Mach number outside the slot after 50, 100, and 300 time steps (from top to bottom, respectively) since opening

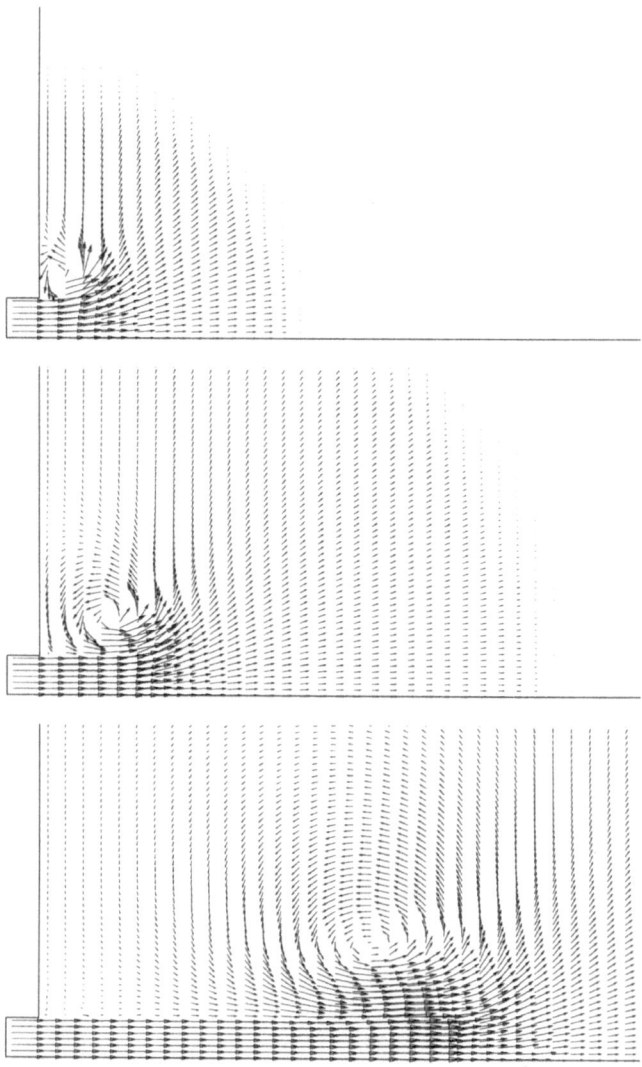

Figure 4.13 Computed distribution of velocity vectors outside the slot after 50, 100, and 300 time steps (from top to bottom, respectively) since opening

Bibliography

[1] D.A. Anderson, J.C. Tannehill, and R.H. Pletcher. *Computational fluid mechanics and heat transfer.* Hemisphere, New York, 1984.

[2] I. Demirdžić, Ž. Lilek, and M. Perić. A colocated finite volume method for predicting flows at all speeds. *Int. J. Numer. Methods Fluids*, 16:1029–1050, 1993.

[3] I. Demirdžić and S. Muzaferija. Numerical method for coupled fluid flow, heat transfer and stress analysis using unstructured moving meshes with cells of arbitrary topology. *Comput. Methods Appl. Mech. Eng.*, 125:235–255, 1995.

[4] J.H. Ferziger and M. Perić. *Computational methods for fluid dynamics.* Springer, Berlin, 2nd edition, 1997.

[5] C.A.J. Fletcher. *Computational techniques for fluid dynamics.* Springer, Berlin, 1991.

[6] C. Hirsch. *Numerical computation of internal and external flows.* Wiley, New York, 1991.

[7] R.I. Issa. Solution of implicitly discretized fluid flow equations by operator-splitting. *J. Comput. Physics*, 62:40–65, 1986.

[8] R.I. Issa and F.C. Lockwood. On the prediction of two-dimensional supersonic viscous interaction near walls. *AIAA J.*, 15:182–188, 1977.

[9] Ž. Lilek and M. Perić. A fourth-order finite volume method with colocated variable arrangement. *Computers and Fluids*, 24:239–252, 1995.

[10] F.R. Menter. Two-equation eddy-viscosity turbulence models for engineering applications. *AIAA J.*, 32:1598–1605, 1994.

[11] S. Muzaferija. *Adaptive finite volume method for flow predictions using unstructured meshes and multigrid approach.* PhD Thesis, University of London, 1994.

[12] S.V. Patankar. *Numerical heat transfer and fluid flow.* McGraw-Hill, New York, 1980.

[13] S.V. Patankar and D.B. Spalding. A calculation procedure for heat, mass and momentum transfer in three-dimensional parabolic flows. *Int. J. Heat Mass Transfer*, 15:1787–1806, 1972.

[14] C.M. Rhie and W.L. Chow. A numerical study of the turbulent flow past an isolated airfoil with trailing edge separation. *AIAA J.*, 21:1525–1532, 1983.

[15] P.J Roache. Perspective: a method for uniform reporting of grid refinement studies. *ASME J. Fluids Engrg.*, 116:405–413, 1994.

[16] J.P. Van Doormal and G.D. Raithby. Enhancements of the simple method for predicting incompressible fluid flows. *Numer. Heat Transfer*, 7:147–163, 1984.

5 Computational Fluid Dynamics and Aeroacoustics for Low Mach Number Flow

5.1 Introduction

The main difficulty in the calculation of sound generated by fluid flow at low Mach numbers is the occurrence of quite different scales. While the fluid flow may be affected by small fluid structures containing large energy, such as small vortices in a turbulent flow, the acoustic waves are phenomena of low energy with long wavelengths that may travel over long distances. These different scales and different physical behaviors of fluid flow and sound propagation lead to a difficult task to construct numerical methods for their approximation. The difficulties associated with low Mach number flow are surveyed ,e.g.in the paper of Crighton [8].

Classical approaches in aeroacoustics are mainly based on analytical solutions of the linear wave equations which are valid in the far field. Often a transformation into the frequency domain is performed. The sound generation is approximated by source terms for these acoustic equations. This procedure is called acoustic analogy. The continuous increase of the power of modern computers in connection with the improvement of the numerical methods enables to go more and more into the direction of direct simulation of fluid flow and acoustics in the time domain. This may reduce the gap between the classical aeroacoustic which is valid in the far field and the sound generation and propagation near the flow field. This is the aim of computational aeroacoustic (CAA). Hence, the field of CAA may be defined to be the domain between computational fluid dynamics (CFD) and classical acoustics. This is an active topic of research embossed by many industrial applications. Due to the multi-scale nature of these phenomena there are a lot of uncertainties and open problems. Numerical errors may generate sound that totally falsifies the physical situation. Even in theoretical aeroacoustics there are still a number of open questions as, e.g., listed by Federochenko [12]. We will consider in this paper some methods and problems associated with this connection of CAA and CFD in the low Mach number regime of fluid flow.

The equations of compressible fluid flow, usually called the compressible Navier-Stokes equations, read as

$$\rho_t + \nabla \cdot (\rho v) = 0, \tag{5.1}$$

$$(\rho v)_t + \nabla((\rho v) \circ v) + \nabla p = \nabla \tau, \tag{5.2}$$

$$e_t + \nabla \cdot (v(e + p)) = \nabla \cdot (\tau v) + \nabla \cdot q, \tag{5.3}$$

where ρ, v, p, e, τ and q denote the density, velocity, pressure, total energy per unit volume, the viscous stress tensor and the heat flux, respectively. These equations represent the conservation of mass, momentum and energy. They are closed by an equation of state $p = p(\rho, \epsilon)$ with the specific internal energy ϵ and the energy relation

$$e = \rho \epsilon + \frac{1}{2} \rho v^2. \tag{5.4}$$

In the following, fluid flows at low Mach numbers $M = v/c$ are considered. Here, c denotes the sound velocity. This means that pressure waves move much faster than the fluid flow. Hence, a fast pressure equalization takes place and the local pressure fluctuations become small. Changes in the fluid velocity can not longer generate strong pressure gradients and thus change the density: The fluid flow becomes incompressible.

Incompressible fluid flow is mathematically modelled by the incompressible Navier-Stokes equations. The assumption of constant density is a common way to obtain the incompressible equations directly from the compressible continuity and momentum equations (5.1) and (5.2). By this, the momentum equation (5.2) may be rewritten as the velocity equation via dividing by density as

$$v_t + (v \cdot \nabla)v + \frac{1}{\rho}\nabla p = \frac{1}{\rho}\nabla \tau, \tag{5.5}$$

while the mass conservation (5.1) leads to

$$\nabla \cdot v = 0. \tag{5.6}$$

Hence, (5.6) is often called continuity equation. The pressure p and the velocity v are determined by the incompressible Navier-Stokes equations (5.5) and (5.6) without respect of any equation of state.

Using the energy relation (5.4) to split the total energy into its internal and kinetic part and using the other conservation equations (5.1), (5.2) together with the equation of state, the energy equation (5.3) may be rewritten as a pressure equation. For the equation of state of a perfect gas

$$p = (\gamma - 1)\,\rho\,\epsilon, \tag{5.7}$$

where γ denotes the adiabatic exponent, the pressure equation reads as

$$p_t + v \cdot \nabla p + \gamma p \nabla \cdot v = (\gamma - 1)\nabla \cdot q. \tag{5.8}$$

As this is a hyperbolic evolution equation for the pressure it is obvious that this equation is a contradiction to the incompressible Navier-Stokes equations (5.5), (5.6). From the latter an elliptic equation for the pressure is obtained by applying the divergence to the velocity equation (5.5) and using the divergence-free condition (5.6) for the velocity. Hence, the role of the pressure in the incompressible limit is more subtle than that indicated by the "constant density" derivation above. In the incompressible limit the pressure looses its thermodynamic meaning and becomes a pure hydrodynamic variable. This behavior is explained in Section 5.3 after introducing an appropriate non-dimensionalisation of the compressible equations in the next Section 5.2. An asymptotic expansion in powers of a global Mach number M is used as considered by Klainerman and Majda [20] to get insight into the $M = 0$ limit and to resolve the discrepancy with respect to the pressure as outlined above.

Based on these considerations, Section 5.4 reviews numerical methods applied to low Mach number fluid flow. The multiple pressure variables method as proposed in [22], [41] and [33] is described. Applied to the compressible Navier-Stokes equations in primitive variables different pressure correction methods are constructed. They may be considered as extensions to the low Mach number regime of incompressible SIMPLE-type schemes proposed by Spalding and Patankar[36]. The other approach to construct low Mach number schemes is to precondition compressible schemes. This is shortly outlined and referred to the literature.

The numerical simulation of problems in aeroacoustics may be divided into two main topics. One is the calculation of the generation of sound in the fluid flow, the other is the propagation of the sound away from the fluid flow region to the far field. The generation as well as the propagation of the sound waves are in principle mathematically described by the same equations, the compressible Navier-Stokes equations. But, for small Mach numbers quite different scales occur. The energy in the fluid flow is much bigger, small scale fluid structures as small vortices may generate sound, while the sound itself is a phenomena with long wavelengths and small amplitudes. If the fluid flow is nearly incompressible and not influenced by sound waves, the numerical simulation of the flow and the sound may be separated. The calculation of acoustic wave propagation is based on wave equations with source terms determined by the fluid flow. In Section 5.5 the acoustic analogy of Sir Lighthill [24] is presented. We continue the asymptotic considerations for small Mach numbers and describe the Janzen-Rayleigh expansion [35], giving the compressible corrections of the incompressible equations. A perturbation analysis around the incompressible solution leads to a system of linear wave equations for the sound propagation in the far field with appropriate source terms for the sound production in the flow field. Numerical results are shown for the sound generation of two rotating vortices.

In all the approaches described in Section 5.5, no special attention is paid to the different length scales in fluid flow and acoustic wave propagation. This is made up leeway in Section 5.6. The asymptotic considerations described in the previous sections may be considered as inner expansions valid in the region of the fluid flow. With appropriate matching conditions they may be matched to outer expansions as proposed by Ting and Miksis [46] and Slimon, Soteriou and Davis [44]. We adopt in this section a multi-scale asymptotic analysis with two space and one time scale as proposed by Klein [21]. This leads by multi-scale averaging techniques to wave equations for long wavelength acoustic waves that may describe thermo-acoustic phenomena at low Mach numbers. Numerical results are shown for a few test problems.

The numerical methods in CAA (Computational Aeroacoustics) and CFD (Computational Fluid Dynamics) are quite different. This is caused by the different properties of the physical phenomena and the different scales. While in CFD the numerical methods have to resolve small scale structures with strong gradients in a robust way, in CAA long wavelength acoustics have to be resolved without falsification of the wave properties such as dispersion and dissipation. The classes of methods used in CAA are addressed and shortly reviewed in Section 5.7. They are compared with high order CFD methods. In more detail we describe a rather new approach to get high order accurate methods, the so called ADER-schemes as proposed in [47] and [42]. Section 5.8 contains our conclusions.

5.2   Non-Dimensionalisation of the Governing Equations

For moderate and high Mach numbers the non-dimensionalisation of the compressible Navier-Stokes equations is usually performed using the basic reference values: length, density and fluid velocity. All other reference values are determined from these via their functional relationship. The left hand side of the compressible Navier-Stokes equations (5.1)-(5.3) then remains unchanged under this non-dimensionalisation replacing the physical variables by the non-dimensionalised ones only. On the right hand side the Reynolds and the Prandtl number appear before the viscous terms and the heat conduction as characteristic numbers, respectively. If the global Mach number as the quotient of the reference values for fluid and sound velocity tends to zero and hence the sound velocity tends to infinity with respect to the fluid velocity, then the choice of different basic reference values for the two different velocities becomes favourable. We do this by using the basic reference values

$$x_{ref}, \rho_{ref}, v_{ref} \text{ and } p_{ref} \qquad (5.9)$$

for length, density, fluid velocity and in addition the pressure, respectively. We emphasize that the number

$$M = \frac{v_{ref}}{c_{ref}}, \qquad (5.10)$$

called the global Mach number characterizes this non-dimensionalisation. Note that the speed of sound is $c = \sqrt{(\gamma p)/\rho}$. Here, we simply take $c_{ref} := \sqrt{p_{ref}/\rho_{ref}}$.

The compressible Navier-Stokes equations for a perfect gas, non-dimensionalised in this way, read in primitive variables as

$$\rho_t + \nabla \cdot (\rho v) = 0, \qquad (5.11)$$

$$v_t + (v \cdot \nabla)v + \frac{1}{\rho M^2}\nabla p = \frac{1}{\rho Re}\nabla \cdot \tau, \qquad (5.12)$$

$$p_t + v \cdot \nabla p + \gamma p \nabla \cdot v = -\frac{\gamma - 1}{Pe}\nabla \cdot q + \frac{M^2(\gamma - 1)}{Re}(\nabla \cdot (\tau v) - v\nabla \cdot \tau). \qquad (5.13)$$

The Mach number M , the Reynolds number Re , and the Peclet number Pe appear in the equations as global dimensionless characteristic quantities being a measure for compressibility, for viscosity, and heat conduction, respectively. The non-dimensionalisation introduces a factor $1/M^2$ in the velocity equations before the pressure gradient. Furthermore, the non-dimensionalised version of the energy relation (5.4) reads as

$$e = \rho\epsilon + \frac{M^2}{2}\rho v^2. \qquad (5.14)$$

For simplicity we often neglect first viscous effects and omit the influence of the heat flux. This results in the Euler equations written in primitive variables as

$$\rho_t + \nabla \cdot (\rho v) = 0, \qquad (5.15)$$

$$v_t + (v \cdot \nabla)v + \frac{1}{\rho M^2}\nabla p = 0, \qquad (5.16)$$

$$p_t + v \cdot \nabla p + \gamma p \nabla \cdot v = 0. \qquad (5.17)$$

The factor $1/M^2$ in (5.12) or (5.16) clearly shows the singular behavior of the incompressible limit. Neglecting the viscous terms, the equations form a hyperbolic system. Their incompressible counterparts contain an elliptic condition: In the divergence-free condition (5.6) no time derivative appears. A Poisson equation for the pressure is obtained combining this condition with the velocity equations. The wave velocities of these non-dimensionalised Euler equations in one space dimension are $v \pm \frac{c}{M}$ and v. Hence, the wave speeds associated with the sound velocity tend to infinity in this limit when the Mach number tends to zero.

The "constant density" derivation of the incompressible Navier-Stokes equations as outlined in the introduction does not explain the role of the pressure in the incompressible limit. The term $\nabla p/M^2$ has to converge to the gradient of the "incompressible" pressure. This pressure then doesn't have the role of a thermodynamic variable anymore. More insight into this limit behavior is obtained by the asymptotic considerations in the next section.

5.3   The Incompressible Limit of a Compressible Fluid Flow

The incompressible limit of a compressible fluid flow is characterized by a velocity of sound that tends to infinity with respect to the velocity of the fluid flow. This means that the pressure waves become very fast, an immediate pressure equalization takes place and the pressure becomes nearly constant. Hence, gradients in the fluid velocity cannot longer generate strong pressure gradients that lead to changes in density: The fluid flow becomes incompressible in the limit.

As the parameter M appears explicitly in the system of partial differential equations (5.15)-(5.17) and tends to zero in the incompressible limit an asymptotic ansatz in powers of the Mach number is a good candidate to get insight into the physical behavior. We assume that the solution of the system can be represented by the following expansion:

$$\rho(x,t) = \rho^{(0)} + M\rho^{(1)} + M^2\rho^{(2)} + O(M^3), \tag{5.18}$$

$$v(x,t) = v^{(0)} + Mv^{(1)} + M^2v^{(2)} + O(M^3), \tag{5.19}$$

$$p(x,t) = p^{(0)} + Mp^{(1)} + M^2p^{(2)} + O(M^3). \tag{5.20}$$

These expansions are substituted into the equations. The unknown expansion functions $\rho^{(i)}$, $v^{(i)}$, $p^{(i)}$ with $i = 0, 1, 2, \ldots$ depend on the space variable x and the time variable t. Collecting all terms multiplied by equal powers of the Mach number M and separately equating these to zero gives a hierarchy of asymptotic equations for the various expansion functions.

The goals of the analysis are to gradually solve the asymptotic equations or to get insight in the behavior of the unknown functions. We investigate especially the functions at leading order. The equations of the incompressible limit is our main interest in this section.

In the velocity equations terms of order M^{-2} and M^{-1} appear and thus we obtain as first conditions

$$\nabla p^{(0)}(x,t) = 0, \qquad \nabla p^{(1)}(x,t) = 0, \tag{5.21}$$

from which it follows that $p^{(0)}$ as well as $p^{(1)}$ depend on time only:

$$p^{(0)} = p^{(0)}(t), \qquad p^{(1)} = p^{(1)}(t). \tag{5.22}$$

Next, we consider the pressure equation of leading order and use (5.22) to obtain the equation

$$p_t^{(0)} + \gamma p^{(0)} \nabla \cdot v^{(0)} = \frac{\gamma - 1}{Pe} \nabla \cdot q^{(0)}. \tag{5.23}$$

This equation is integrated over the whole domain Ω of the fluid flow:

$$p_t^{(0)} \int_\Omega d\Omega + \gamma \, p^{(0)} \int_{\partial\Omega} v^{(0)} \cdot n \, dS = \frac{\gamma - 1}{Pe} \int_{\partial\Omega} q^{(0)} \cdot n \, dS. \tag{5.24}$$

Here, we applied the Gaussian theorem, where n denotes the unit normal, and used (5.22).

Let us first consider the case when heat conduction can be neglected and no compression from the boundary $\partial\Omega$ occurs, i.e.

$$\int_{\partial\Omega} v^{(0)} \cdot n dS = 0, \tag{5.25}$$

often called the incompressibility constraint. No volume flux through the boundary of Ω appears. In this case, the integrated pressure equation (5.24) immediately gives that $p^{(0)}$ is constant in time. Without heat conduction it follows then from (5.23) that the velocity field has zero divergence. Together with the zeroth order continuity and velocity equations we get the basic equations

$$\rho_t^{(0)} + v^{(0)} \cdot \nabla \rho^{(0)} = 0, \tag{5.26}$$

$$v_t^{(0)} + (v^{(0)} \cdot \nabla) v^{(0)} + \frac{1}{\rho^{(0)}} \nabla p^{(2)} = \frac{1}{\rho^{(0)} Re} \nabla \tau^{(0)}, \tag{5.27}$$

$$\nabla \cdot v^{(0)} = 0. \tag{5.28}$$

These are the equations of incompressible fluid flow with variable density. If the density is constant initially, then it is constant for all times. In this case the density equation cancels out.

We shortly discuss the role of the pressure in this singular limit. Two pressure terms $p^{(0)}$ and $p^{(2)}$ survive in the limit $M \to 0$. The leading order pressure $p^{(0)}$ becomes constant in space and time. It satisfies the equation of state for $M = 0$ and may be called the thermodynamic pressure. The second order pressure term $p^{(2)}$ is the agent to balance the inertial forces and to establish the divergence free condition of the velocity. It is called the hydrodynamic pressure. In the equations (5.12) and (5.16) we formally get

$$\frac{1}{M^2} \nabla p \to \nabla p^{(2)} \qquad \text{for} \qquad M \to 0. \tag{5.29}$$

This corresponds quite well to the physical situation. The pressure wave propagation becomes infinite with respect to the fluid velocity and a fast pressure equalization takes place. This means that the pressure in the compressible equations tends to the thermodynamic background pressure $p^{(0)}$ in the limit. Due to the fast pressure waves velocity gradients in the flow cannot longer generate large pressure gradients and no density changes occur. According to (5.14) the total energy equals the internal energy. The velocity fluctuations are very small and do not account to the total energy.

In the case of outer compression and heat conduction the divergence-free constraint (5.28) must be replaced, according to leading order pressure equation (5.23), by

$$\nabla \cdot v^{(0)} = -\frac{1}{\gamma\,p^{(0)}}p_t^{(0)} + \frac{\gamma-1}{Pe\,\gamma\,p^{(0)}}\nabla \cdot q^{(0)}. \tag{5.30}$$

The continuity equation at leading order now reads

$$\rho_t^{(0)} + v^{(0)} \cdot \nabla\rho^{(0)} = -\rho^{(0)}\nabla \cdot v^{(0)}. \tag{5.31}$$

The velocity equation (5.27) remains unchanged.

The system (5.31), (5.27), and (5.30) is closed by the evolution equation for $p^{(0)}$ which is derived from integrated pressure equation (5.24) as

$$p_t^{(0)} = -\frac{\gamma\,p^{(0)}}{|\Omega|}\int_{\partial\Omega} v^{(0)} \cdot n\,dS + \frac{\gamma-1}{Pe\,|\Omega|}\int_{\partial\Omega} q^{(0)} \cdot n\,dS. \tag{5.32}$$

The situation is now as follows. Compression from the boundary, when the incompressibility constraint (5.25) is not valid, leads to a background pressure which is constant in space but a function in time. In this case and further neglecting heat conduction, the divergence of the velocity is constant but non-zero in space and a function in time. The density varies in time as given by (5.31).

We conclude that asymptotic analysis gives insight into the behavior and structure of solutions for fluid equations at small Mach numbers. The incompressible limit of the Euler or compressible Navier-Stokes equations was considered using an expansion in powers of the Mach number. A rigorous mathematical justification of such an analysis under several constraints on the initial data is given by Klainerman and Majda in [20] and Majda and Sethian in [27]. For a barotropic flow $p = p(\rho)$ they show that compressible flow in fact converges to incompressible flow as the Mach number vanishes, if the pressure is uniform initially, except for perturbations of $O(M^2)$ and a single characteristic length scale governs the initial data. Klein [21] extended the formal asymptotic analysis to the compressible equation of a perfect gas with variable entropy and to initial data with long-wavelength pressure perturbations of order $O(M)$ introducing multiple length scales. We will come back to these considerations in Section 5.6.

5.4 Numerical Methods for Low Mach Number Fluid Flow

As shown in the previous section, the relationship between compressible and incompressible equations is quite complex due to the fact that the equations change their type and wave velocities become infinite. All explicit methods as described by Russo and Sonar in Chapter 2 and 3, respectively, run into trouble due to the CFL-stability condition. The time step necessarily tends to zero and the computational efforts increase when the Mach number converges to zero. Several types of methods were designed to handle these aspects. These are extensions of numerical methods either for the incompressible equations or for the compressible equations. In the following we shortly review both approaches.

5.4.1   Compressible Pressure Correction Schemes

The so called pressure correction methods have their origins in the SIMPLE method (Semi-Implicit Method for Pressure Linked Equations) of Patankar and Spalding [36] for the incompressible Navier-Stokes equations. A detailed description is given in the book of Patankar [35]. This method became very common for the numerical simulation of practical problems for incompressible fluid flow. We give first a short survey of the structure of the incompressible SIMPLE-type methods. Then we describe the extension to compressible flow as given by Karki and Patankar [19] for homentropic fluid flow.

Any discretisation of the velocity equation (5.5) of the incompressible equations will give a system of equations for the unknown values of pressure and velocity at every time step. The convective terms introduce some nonlinearity, if they are approximated implicitly in time. If we assume that these terms are linearized or approximated explicitly, then this system of equations is linear and has the form

$$Av^{n+1} + \frac{1}{\rho}\tilde{\nabla}p^{n+1} = f^n, \tag{5.33}$$

where A is the matrix which accounts for all velocity terms at the time level t_{n+1}, the convection as well as the viscous effects, $\tilde{\nabla}p$ is a discrete gradient operator consistent with ∇p, and the right hand side f contains all terms known from the old time level. Several space and time approximations may be considered, finite difference or finite element methods. In their original form, the SIMPLE-type algorithms are applicable to incompressible constant density flow. In the following survey we restrict ourselves to this case.

If an estimation p^0 of the pressure field at the new time level t_{n+1} is available, the corresponding velocity v^0 may be calculated from the discretized velocity equations:

$$Av^0 + \frac{1}{\rho}\tilde{\nabla}p^0 = f. \tag{5.34}$$

A SIMPLE-type scheme is a procedure to iteratively update the velocity and pressure field to obtain the desired solution v^{n+1} and p^{n+1}. The iteration index is denoted by ν, the index for the time step is omitted as long as no misunderstanding can occur. Corrections δv^ν and δp^ν with

$$p^{\nu+1} = p^\nu + \delta p^\nu, \quad v^{\nu+1} = v^\nu + \delta v^\nu \tag{5.35}$$

are introduced. They satisfy the pressure-velocity correction equation

$$A\delta v^\nu + \frac{1}{\rho}\tilde{\nabla}\delta p^\nu = 0, \tag{5.36}$$

which is obtained as the difference of (5.34) evaluated for the different pairs p^ν, v^ν and $p^{\nu+1}$, $v^{\nu+1}$, respectively. This relation is inserted into the discretized divergence-free condition of the velocity field

$$\tilde{\nabla} \cdot (v^\nu + \delta v^\nu) = 0 \qquad (5.37)$$

to obtain the elliptic pressure correction equation

$$\tilde{\nabla} \cdot \left(A^{-1} \frac{1}{\rho} \tilde{\nabla} \delta p^\nu \right) = \tilde{\nabla} \cdot v^\nu. \qquad (5.38)$$

This equation has to be solved to obtain δp^ν.

The main problem is the inversion of the matrix A in (5.36). If the inversion is done exactly, a great amount of computational effort is needed. To circumvent this problem, the pressure-velocity correction relation (5.38) is simplified. The idea is to move cheaply into the proper direction, but not to reach the target in one step. This might be done by decomposing A into the difference $D - B$ where the matrix D contains the diagonal elements of A only and B all the others. Next B is approximated by a matrix which is more easy to convert. In the original SIMPLE algorithm this is done by setting $Bv^{\nu+1} = Bv^\nu$ (see [36]). In this case the matrix B appears on the right hand side of the discrete velocity equation (5.33) and drops out in the velocity-pressure correction relation (5.36) and the calculation of the inverse matrix in (5.38) becomes trivial. A review about other simplifications is given in [43] and in chapter 4 by Peric.

The elliptic problem is completed by specifying boundary conditions. This is a typical problem in a pressure-based method where only boundary values for the velocity are physically relevant. We do not discuss this aspect in the present paper but refer to Gresho and Sani [16]. Any simplification of (5.36) disturbs the consistency of the approximation and an iterative relaxation technique becomes necessary. The result obtained as a solution of (5.38) is used to define a new estimation $p^{\nu+1}$ according to (5.35). The iterative procedure converges to the exact solution as the velocity is calculated by the equation (5.33) without simplification.

This incompressible pressure correction method has been extended to low Mach number compressible fluid flow by Karki and Patankar [19] in the case of a homentropic equation of state. The idea in their approach is to replace the divergence-free condition by the full continuity equation. The density is replaced via the homentropic equation to obtain a pressure correction equation of the state $p = p(\rho)$.

This concept has been extended to general fluid flow with density variations by several authors (see [39], [43] and chapter 4 by Peric) in the following way. Again the elliptic pressure equation is obtained via the continuity equation in which the density is eliminated via the equation of state. In this case, of course, an unknown variable remains, because the pressure is a function of both the density and the internal energy. The usual way is to handle this problem iteratively. In each iteration cycle a pressure correction is calculated at first by the same procedure as in the homentropic case with an internal energy fixed at the old iteration level. After applying the pressure and velocity correction, the transport equation for the internal energy is solved in an explicit way. In case of convergence the solution obtained is a solution of the Navier-Stokes equations. The pressure correction equation for a perfect gas obtained by this procedure has the form

$$\delta p^\nu - \frac{\Delta t\,(\gamma-1)\,\varepsilon^\nu \rho^\nu}{M^2}\tilde{\nabla}\cdot\left(A^{-1}\frac{1}{\rho^\nu}\tilde{\nabla}\delta p^\nu\right) =$$

$$\frac{\varepsilon^\nu}{\varepsilon^n}p^n - p^\nu - (\gamma-1)\,\Delta t\varepsilon^\nu \rho^\nu \tilde{\nabla}\cdot(v^\nu). \tag{5.39}$$

Here, the same notation as in (5.38) is used and additionally ε^ν denotes the internal energy at the iteration level ν. The heat flux is omitted here for simplicity.

The asymptotic analysis also motivates a different procedure. The pressure equation and the velocity equations are used to obtain a pressure correction equation and the density is fixed at the old iteration level. Next we proceed in the same manner as in the incompressible SIMPLE-method. Starting from a pressure estimation p^0, the corresponding velocity v^0 is calculated from the velocity equations of the compressible Navier-Stokes equations by solving

$$Av^0 + \frac{1}{\rho^0 M^2}\tilde{\nabla}p^0 = f, \tag{5.40}$$

where f contains all explicit terms. The pressure correction δp^ν and the velocity correction δv^ν depend upon each other via the relation

$$A\delta v^\nu + \frac{1}{\rho^\nu M^2}\tilde{\nabla}\delta p^\nu = 0. \tag{5.41}$$

A pressure correction equation is then obtained by discretizing the pressure equation. For simplicity we introduce a fully implicit approximation of first order accuracy in time for the divergence term. The convection term is approximated explicitly. Written as an equation for the velocity and pressure correction, it reads as

$$\delta p^\nu + \Delta t\gamma p^\nu \tilde{\nabla}\cdot\delta v^\nu = p^n - p^\nu - \Delta t\gamma p^\nu \tilde{\nabla}\cdot v^\nu - \Delta t(v\cdot\tilde{\nabla}p)^n. \tag{5.42}$$

Replacing the velocity correction via (5.41) leads to the desired pressure correction equation

$$\delta p^\nu - \frac{\Delta t\gamma p^\nu}{M^2}\tilde{\nabla}\cdot\left(A^{-1}\frac{1}{\rho^\nu}\tilde{\nabla}\delta p^\nu\right) = p^n - p^\nu - \Delta t\gamma p^\nu \tilde{\nabla}\cdot v^\nu - \Delta t(v\cdot\tilde{\nabla}p)^n, \tag{5.43}$$

and the compressible SIMPLE-algorithm is defined. The pressure correction δp^ν is obtained by solving this elliptic equation and the pressure $p^{\nu+1}$ by using (5.35). If the required accuracy is reached, then the iteration for the pressure p^{n+1} is stopped, and the velocity v^{n+1} is calculated from (5.33), ρ^{n+1} is obtained from the discrete continuity equation. For the homentropic case, both approaches become identical because the density equation and the pressure equation coincide.

It is expected that for small Mach numbers, especially in the limit $M \to 0$, the comparison with the asymptotic results cause problems. We consider the formal limit of the pressure correction equation in the case $\rho = const$. We multiply (5.39) and (5.43) by the term M^2 and obtain for the limit $M = 0$ formally the elliptic equation

$$\tilde{\nabla}\cdot\left(A^{-1}\frac{1}{\rho^\nu}\tilde{\nabla}\delta p^\nu\right) = 0. \tag{5.44}$$

This means that the pressure correction equation and hence the compressible SIMPLE-method does not converge towards the incompressible pressure correction equation (5.38) and the incompressible SIMPLE-method, respectively. This is due to the fact that the pressure in the compressible equations does not converge towards the incompressible pressure. This fact will be valid for all numerical approximations which do not introduce some sort of pressure decomposition because they determine iteratively the total pressure which becomes constant in space at $M = 0$. The $O(M^2)$ incompressible pressure has to be extracted from this $O(1)$ total pressure. In the following section we consider this issue in more detail and propose the introduction of a numerical pressure decomposition into the SIMPLE-method.

5.4.2 Multiple Pressure Variables (MPV-)Approach

The term $\nabla p / M^2$ occurring in the velocity equation causes the difficulty for $M \to 0$ is. The limit of this term exists as indicated by the asymptotic considerations: $\frac{1}{M^2} \nabla p \to \nabla p^{(2)}$. In the following we introduce a numerical approximation of the pressure decomposition

$$p = p^{(0)} + M^2 p^{(2)} \qquad (5.45)$$

at low Mach numbers avoiding these difficulties. We assume that the Mach number M is small such that $p^{(0)}$ becomes constant in space. In this case the thermodynamic pressure can be obtained as the pressure p averaged over the whole computational domain

$$p^{(0)} = \bar{\bar{p}} := \frac{1}{|V|} \int_V p(x) dx. \qquad (5.46)$$

The pressure $p^{(2)}$ is then given according to the consistency with (5.45) by

$$M^2 p^{(2)} = p - p^{(0)}. \qquad (5.47)$$

Multiple pressure variables are introduced by applying the averaging operator in (5.46) to the discrete values. The basic philosophy of the MPV-approach is indicated by discussing the approximation of the term $\nabla p / M^2$ via the decomposition

$$\frac{1}{M^2} \nabla p = \frac{1}{M^2} \nabla p^{(0)} + \nabla p^{(2)}. \qquad (5.48)$$

By construction, the thermodynamic pressure term is constant in space, so $\nabla p^{(0)} = 0$. The idea in the multiple pressure variables approach is now to replace the approximation of $\nabla p / M^2$ in the velocity equations by an approximation of (5.48) where the singularity is eliminated and which converges towards $\nabla p^{(2)}$.

Because the sound and transport velocity scales spread as M tends to zero, we solve separately the different effects for small Mach numbers using the asymptotic results. This allows us to rescale the problem and also to use different adequate numerical approximations for different physical phenomena. Because we do not know the regime of validity of our asymptotic analysis, all asymptotic considerations are only used to design suitable predictor steps which are performed to precondition the solution of the full equations. We will explain the principle ideas in the following by proposing a compressible SIMPLE-algorithm for small Mach numbers with multiple pressure variables.

In a first step, the pressure p is decomposed into the thermodynamic and the incompressible pressure terms $p^{(0)}$ and $p^{(2)}$ using discrete analogs of (5.45) and (5.47). This first step produces approximate values at the old time level t_n

$$p^{(0),n}, \quad p^{(2),n}. \tag{5.49}$$

We assume now that we have estimates for the change in time of the thermodynamic pressure during the time step $\Delta t = t_{n+1} - t_n$. We denote this update by $dp^{(0)}$ being an approximation of the time derivative multiplied by Δt.

In the second step we proceed in the way proposed in the last section. The time estimate is used as a predictor for the effect of global compression. We define estimates of the pressure variables p and $p^{(2)}$ at the new time level called p^0 and $p^{(2),0}$ as

$$p^0 = p^{(0),n} + dp^{(0)}, \quad p^{(2),0} = 0, \quad \rho^0 = \rho^n. \tag{5.50}$$

The corresponding velocity v^0 is calculated by solving

$$Av^0 + \frac{1}{\rho^\nu}\tilde{\nabla}p^{(2),0} = f^n. \tag{5.51}$$

This equation is obtained from equation (5.40) by performing the pressure decomposition.

Next we define the iteration procedure and introduce a correction of the incompressible pressure term $\delta p^{(2),v}$, the velocity δv^v and the density $\delta \rho^v$:

$$p^{(2),v+1} = p^{(2),v} + \delta p^{(2),v}, \quad v^{v+1} = v^v + \delta v^v, \quad \rho^{v+1} = \rho^v + \delta \rho^v. \tag{5.52}$$

Because an estimate of the thermodynamic pressure is obtained in step 1 and used here as a predictor, we only need a correction of the incompressible pressure. The velocity and pressure correction relation is then obtained in the form

$$A\delta v^v + \frac{1}{\rho^\nu}\tilde{\nabla}\delta p^{(2),v} = 0. \tag{5.53}$$

We note that the difference between this equation and (5.41) is that the term $1/M^2$ disappears, because an $O(M^2)$ correction is considered only.

Again the elliptic equation for the pressure correction is deduced from the pressure equation. Here we also introduce the pressure decomposition to get

$$M^2 p_t^{(2)} + v \cdot \nabla p + \gamma p \nabla \cdot v = -p_t^{(0)}. \tag{5.54}$$

The same type of approximation as in the compressible SIMPLE-method is applied to this equation. The convection term is approximated explicitly, and the third term on the left hand side is approximated fully implicitly, with a linearisation of the pressure term. Inserting the pressure-velocity correction relation (5.53) the elliptic pressure correction relation equation reads as

$$M^2 \delta p^{(2),v} - \Delta t\, \gamma\, p^v \tilde{\nabla} \cdot \left[A^{-1}\frac{1}{\rho^\nu}\tilde{\nabla}\delta p^{(2),v} \right] =$$
$$M^2 \left(p^{(2),n} - p^{(2),v} \right) - dp^{(0)} - \Delta t\, \gamma\, p^v \tilde{\nabla} \cdot v^v - \nabla t\, (v \cdot \nabla p)^n. \tag{5.55}$$

If the pressure correction is obtained by solving this equation, then (5.52) and (5.53) gives the new iterates for pressure and velocity. The new density or the density correction is then obtained by solving the continuity equation. If the iteration is successfully stopped, the total pressure p^{n+1} is defined by adding the value $M^2 p^{(2),n+1}$ obtained by iteration to the right hand side of the first relation in (5.50).

When M tends to zero, all terms multiplied by M^2 vanish. Furthermore, the total pressure p becomes constant in space and ∇p becomes zero. If we have no compression from the boundaries, then $dp^{(0)} = 0$, and (5.55) converges to the corresponding pressure correction equation for the incompressible equations (5.38). The numerical procedure has not to be changed for $M = 0$, because all the terms multiplied by M^2 automatically vanish. The compressible pressure p equals $p^{(0)}$ and the hydrodynamic pressure decouples. It becomes quite obvious from these considerations that it is necessary to introduce some sort of pressure decomposition to obtain the correct limit. We remark that the estimate of the time derivatives of $p^{(0)}$ within this concept are used only as a predictor. The approximation is constructed in such a way that it is consistent with the full equations. This is an important fact, because we do not know how accurate the asymptotic approximations are. However, if the estimates are bad, then the $p^{(2)}$ pressure correction has to capture these errors and the MPV-method will become inefficient.

In these considerations we omitted the heat conduction for simplicity. To consider heat conduction the corresponding terms have to be added to the right hand side of the pressure equation. This is described in detail in [41]. The multiple pressure variables approach has been proposed in [2], see also [33]. A similar procedure introducing such a technique of re-scaling has been proposed by [3] and [11]. A detailed description is given by Peric in Chapter 4 of this book (see also [13]).

To demonstrate the different fields of applications we how in the following a few results of the MPV-approach for typical problems of low Mach number fluid flow. Here, the acoustic waves do not play an important role. The first problem has been proposed as a benchmark problem for the workshop "Modelling and Simulation of Natural Convection Flows with Large Temperature Differences" in January 2000 in Saclay. It is a differentially heated square cavity. The vertical walls have a strong temperature difference, the right one is warm, the left one is cold, while the horizontal walls are thermally insulated. A sketch of this situation with the values used is shown in the Figure 5.1. The big difference in temperature of the two walls and the constant inner state will introduce temperature gradients inside the cavity. Thus the density cannot be considered to be constant. Due to the large difference a Boussinesque approximation is not appropriate. In our calculations, the maximum of the density $\rho_{max} = 2.1$ is reached in the lower right corner at the bottom of the cold wall, while the minimum $\rho_{min} = 0.5$ is located in the upper left corner. The air is heated at the left wall and hence the density decreases, and the air rises along this wall. On the other side at the cold wall, the air in the box is cooled and moves down. Therefore, a clockwise rotating flow occurs. It converges to a steady state. Numerical results are shown in Figure 5.2. The viscosity μ and the thermal conductivity κ were set to a constant value.

In the following we show some results for a flow past a NACA0012 airfoil at the moderate global Mach number $M = 0.4$. The angle of attack is $0°$ and hence, the numerical solution should be symmetric. The viscous terms are neglected in these calculations. The pressure increases at the leading edge to the non-dimensionalised value of 1.13 and the minimum on the lower an upper side is 0.92.

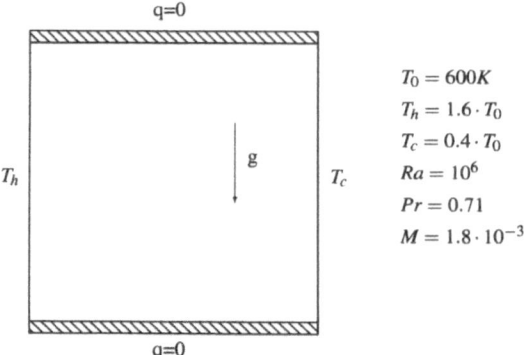

$$T_0 = 600K$$
$$T_h = 1.6 \cdot T_0$$
$$T_c = 0.4 \cdot T_0$$
$$Ra = 10^6$$
$$Pr = 0.71$$
$$M = 1.8 \cdot 10^{-3}$$

Figure 5.1   Square cavity heated at the side faces

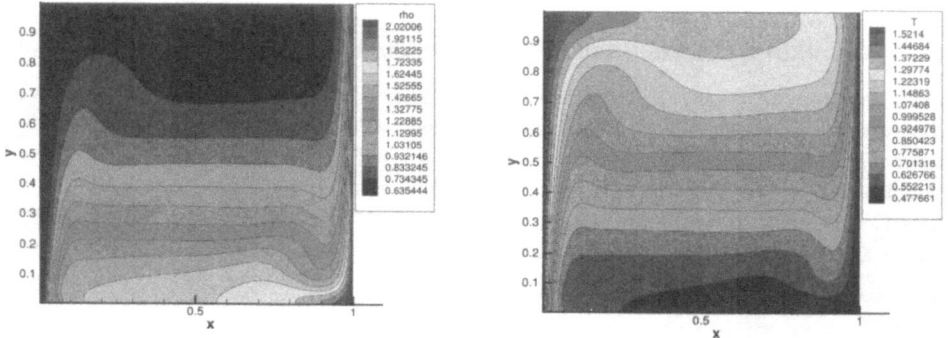

Figure 5.2   Density and temperature distribution at steady state

We did not specify the approximation technique in space discussing the MPV-approach. In the simulations described above we adopted the finite-difference approximation on a staggered grid, well-known from the construction of incompressible pressure correction schemes (see [35]). Here, the scalar quantities, like pressure and density, are calculated in the cell centers, while the velocity components are calculated at the cell faces. The advantage is to have a compact stencil without the problem of the velocity-pressure decoupling. If a collocated grid is used as described in Chapter 4 by Peric, the some stabilisation to avoid the decoupling has to be introduced. The convection terms may be approximated by central differencing on the staggered grid. If the convection terms dominate the flow, then this approximation should be replaced by some upwind-biased differences. Here, we used second order accurate upwind differences where we adopted the extension of the MUSCL-approach of van Leer [51] by Bell et. al. [1], [2]. In the general formulation used above, different time approximation schemes may be possible. The MPV approach may also be used for the extension of incompressible projection schemes [5] to the weakly compressible regime. In this fractional steps approach the convection and diffusion is calculated in a first step, while the sonic term is calculated in a second one (see [1], [50]).

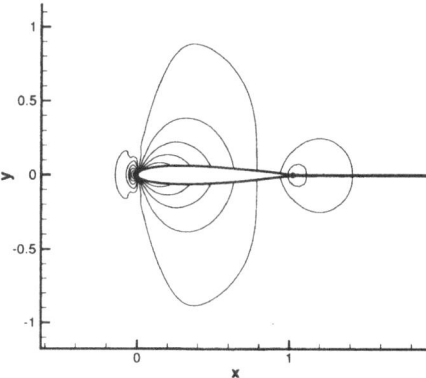

Figure 5.3 Pressure contours of the steady state

5.4.3 Preconditioning

If the standard numerical methods for the compressible Navier-Stokes equations are applied to low Mach number flow, then the methods become inefficient or even produce wrong stationary solutions. This is due to the fact that the difference between the eigenvalues become larger and larger when the Mach number tends to zero. The problem becomes stiff. To handle this a so called preconditioning of the equations may be performed. The principle ideas has its origin in a method for the incompressible equations: Chorin proposed in [6] to replace the $\nabla \cdot v = 0$ equation in the incompressible equation by

$$p_t + c^2 \nabla \cdot v = 0 \tag{5.56}$$

with a constant c. By doing this, a direct pressure velocity coupling has been introduced and the whole system becomes hyperbolic. The infinite propagation rate of the pressure is replaced by the finite constant rate c. The equations are modified by some sort compressibility and Chorin called it the method of artificial compressibility. In the steady state the time derivatives vanish and the modified equations are identical with the stationary Navier-Stokes equations.

For the compressible equations the multiplication of the vector containing the time derivatives by a matrix P^{-1} may introduce a preconditioning of the equations. Using dual time stepping this method of preconditioning may be extended to the simulation of time accurate problems, but needs much computational effort. A survey about preconditioning is given by Turkel, Fiterman, and van Leer [48], Darmofal and van Leer [10] and Weiss and Smith [54]. Guillard and Viozat [17] (see also [52], [34]) proposed schemes where the preconditioning only affects the numerical dissipation term. They used the low Mach number asymptotic results to motivate this approach. This approach is extended taking into account the multiple scale effects by Meister [29].

5.5   Sound Generation and Sound Propagation

Sound generation in the flow field and its propagation are in principle solutions of the same mathematical model, the compressible Navier-Stokes equations. But, due to the quite different scales at low Mach numbers the practical calculation of both phenomena by one numerical scheme turns out to be unsuccessful. Using properties valid for the acoustic waves, simplified mathematical models for sound propagation are obtained from the compressible equations. Various methods for coupling the fluid calculation associated with sound generation and the sound propagation are discussed in this section.

5.5.1   The Equations of Linear Acoustics

In the previous section, we considered the numerical approximation of the fluid flow. This approximation should cover the generation of the sound. The sound propagation away from the fluid flow is, of course, also a solution of the compressible Navier-Stokes equations, but it is a solution with quite different properties. The fluid variables in the far field become nearly constant. The acoustic waves may be considered to be small perturbations of this constant state. Hence, a perturbation analysis around the constant state of a fluid at rest will give equations for the propagation of the acoustic waves valid in the far field. If we assume that the fluid velocity equals zero far away from the fluid flow region, this ansatz reads as

$$\rho = \rho_0 + M^2\rho' \quad , \quad v = Mv' \quad , \quad p = p_0 + M^2p', \tag{5.57}$$

where the primed quantities denote the small fluctuations. These expressions are substituted into the Euler equations. All products of the small acoustic fluctuations and their derivatives are set to zero as usual for a perturbation method.

The resulting evolution equations for the acoustic fluctuations then are given by

$$\rho'_t + \frac{\rho_0}{M}\nabla \cdot v' = 0, \tag{5.58}$$

$$v'_t + \frac{1}{\rho_0 M}\nabla p' = 0, \tag{5.59}$$

$$p'_t + \frac{\gamma p_0}{M}\nabla \cdot v' = 0. \tag{5.60}$$

The combination of the equations (5.59) and (5.60) shows that the propagation of acoustic waves with small amplitudes is isentropic:

$$p' - c_0^2\rho' = \text{const.}. \tag{5.61}$$

The equations may additionally be combined in such a way that the velocity fluctuations are eliminated and a homogeneous wave equation for the pressure as well as for the density is obtained:

$$p'_{tt} - \frac{c_0^2}{M^2}\Delta p' = 0 \quad , \quad \rho'_{tt} - \frac{c_0^2}{M^2}\Delta \rho' = 0, \tag{5.62}$$

usually called the equations of linear acoustics.

5.5.2 Lighthill's Acoustic Analogy

A fundamental idea has been given by Sir Lighthill in the classic paper on the generation of sound by low Mach number eddies [24, 25] and is the starting point of several theories. Sir Lighthill considered the following procedure: The continuity equation is differentiated with respect to time, while the divergence operator is applied to the momentum equations. Then the difference of both equations gives

$$\rho_{tt} - \frac{1}{M^2}\Delta p = \nabla \cdot (\nabla((\rho v) \circ v)) - \nabla \tau. \tag{5.63}$$

The Laplacian of the pressure is shifted to the right hand side and the term $-\frac{c_0^2}{M^2}\Delta\rho$ is added to both sides of the equation to give

$$\rho_{tt} - \frac{c_0^2}{M^2}\Delta\rho = \nabla \cdot (\nabla(\rho v) \circ v) + \frac{1}{M^2}\Delta\left(p - c_0^2\rho\right) + \nabla \cdot \nabla \tau. \tag{5.64}$$

If the density is expressed as the sum of the constant background density and the fluctuations, then the left hand side in equation (5.64) equals the wave equation for linear acoustics (5.62) and the right hand side may be interpreted as the source terms associated with the sound generation. The equation now reads

$$\rho'_{tt} - \left(\frac{c_0}{M}\right)^2\Delta\rho' = \nabla \cdot \nabla T, \tag{5.65}$$

where $T = (\rho v) \circ v + \frac{1}{M^2}(p - c_0^2\rho)I + \tau$ is called the Lighthill tensor.

Up to now only a reformulation of the equations has been performed. On the right hand side as well as on the left hand side the unknown physical variables appear. Having in mind, that for small Mach numbers the acoustic fluctuations are very small in comparison with the flow field, the incompressible instead of the compressible solution may be used to get an appropriate approximation of the Lighthill tensor. If we rewrite the compressible solution as the incompressible limit plus perturbations, then it is obvious that the main part of the product is given by the product of the incompressible variables. If we proceed in that way, then the separation of the calculation of the fluid flow and the sound propagation is established. The simulation of the fluid flow is based on an incompressible flow solver, the calculation of sound propagation is determined via the equations of linear acoustics, and the coupling is introduced by source terms in these acoustic equations. The following basic assumptions are made in this approach: The acoustic waves neither influences the flow field nor the flow influences the sound propagation.

To replace the real problem of sound radiation in a flow field by the problem of sound propagation in a medium at rest with corresponding source terms is called "acoustic analogy". Several different forms of the acoustic analogy have been proposed with different expressions for the sources of sound. There are analogies of the form (5.65) due to Powell [38] and Ribner [40]. Lighthill's acoustic analogy has been extended to fluid flow with solid bodies in the source region by Ffowcs Williams and Hawkings [14].

In classical aeroacoustics the wave equation in the flow field with the approximate source term is solved exactly, e.g. using Green functions. Often the equation is previously transformed into the frequency domain, resulting in a Helmholtz equation. For practical calculations the difficult task is to calculate the source tensor as distributions of monopoles, dipoles and quadropoles from the data of the fluid flow, representing the sound generation in a proper way. Effects as the influence of fluid convection to sound propagation and reflection at solid bodies is are difficult to incorporate in this frame work. Lilley's equation [26] is one such equation that casts the wave operator in a form that take into account the moving medium. Surveys about the "present" classical aeroacoustics can be found in the books of Wagner, Bareiss and Guidati [53] and Goldstein [15].

5.5.3   The Janzen-Rayleigh Expansion

Assuming that the solutions of the compressible Navier-Stokes-equations at small Mach numbers satisfy an asymptotic expansion in powers of the Mach number, we considered in chapter 5.3 the basic equations to find the incompressible limit equations and to get insight into the limit behavior. These considerations showed that the incompressible pressure field has no influence to the leading order density. Compressible effects will occur as higher order corrections in the expansion of the density according to (5.18). The term $p^{(1)}$ in the expansion is constant in space and time, if the boundary conditions do not introduce compression. From the first order pressure perturbation equation immediately follows that the first order velocity term $v^{(1)}$ is divergence free. Hence, this term does not describe compressible corrections to the incompressible limit solution and is dropped together with $\rho^{(1)}$ and $p^{(1)}$.

The second order pressure equation reads as

$$p_t^{(2)} + v^{(0)} \cdot \nabla p^{(2)} + \gamma p^{(0)} \nabla \cdot v^{(2)} = 0 \qquad (5.66)$$

and expresses the compressibility effect of the incompressible pressure leading to a condition for the divergence of the velocity term $v^{(2)}$. Substituting this expression into the second order density equation

$$\rho_t^{(2)} + v^{(0)} \cdot \nabla \rho^{(2)} + \rho^{(0)} \cdot \nabla v^{(2)} = 0, \qquad (5.67)$$

gives

$$\rho_t^{(0)} + v^{(0)} \nabla \rho^{(2)} = \frac{\rho^{(0)}}{\gamma p^{(0)}} \left(p_t^{(2)} + v^{(0)} \cdot \nabla p^{(2)} \right) \qquad (5.68)$$

where the right hand side is the source term responsible for the density changes.

This system may be closed, if all the terms with odd powers of the Mach number are cancelled. The time evolution of the second order velocity term is then given by

$$v_t^{(2)} + \nabla \cdot \left(v^{(0)} \circ v^{(2)} + v^{(2)} \circ v^{(0)} \right) + \frac{1}{\rho^{(0)}} \nabla p^{(4)} = 0. \qquad (5.69)$$

The structure of this system recovers that of the incompressible Navier-Stokes-equations. The time derivative of the hydrodynamic pressure $p^{(2)}$ produces via (5.66) a condition for the divergence of the velocity field. The pressure $p^{(4)}$ plays then the role of a Lagrangian multiplier to meet this constraint. We can proceed in this way to get a series of higher order corrections for the incompressible flow field. This procedure is called the Janzen-Rayleigh expansion (see [49]). It describes the generation of compressible effects inside the flow region as higher order correction of the incompressible solution. The propagation of the acoustic waves into the farfield can not be captured by this expansion. It is valid as inner solution, where the diameter of the inner region is smaller than the acoustic wave length.

5.5.4 Expansion around Incompressible Flow

In Lighthill's acoustic analogy the sound propagation is calculated assuming constant background values for pressure and density and zero velocity as valid near the far field. This is, of course, not a good approximation in the vicinity of the flow, where the flow field will influence the sound propagation, too. In this subsection the point of view is directed to the domain of fluid flow. Instead of using the constant far field values the incompressible solution is taken as the point of expansion. Using the scaling motivated by the asymptotic considerations above we introduce the following perturbation ansatz:

$$\rho = \rho^{(0)} + M^2 \left(\rho^{(2)} + \rho' \right), \tag{5.70}$$

$$v = v^{(0)} + M v', \tag{5.71}$$

$$p = p^{(0)} + M^2 \left(p^{(2)} + p' \right). \tag{5.72}$$

The functions $\rho^{(0)}$, $v^{(0)}$, and $p^{(2)}$ denote again the incompressible solution, while $p^{(0)}$ is the background pressure. The function $\rho^{(2)}$ is introduced as the change of density by the hydrodynamic pressure. According to (5.61) it is given by

$$p^{(2)} - c_0^{(2)} \rho^{(2)} = const., \tag{5.73}$$

where c_0 denotes the background sound velocity.

First we neglect viscosity, heat conduction and compression from the boundary and assume constant density $\rho^{(0)}$. If this ansatz is introduced in the Euler equations (5.15)-(5.17) and the products of the small primed fluctuations as well as their derivatives are dropped, then the following equations are obtained:

$$\rho'_t + v^{(0)} \nabla \rho' + \frac{\rho^{(0)}}{M} \nabla v' = - \left(\rho_t^{(2)} + v^{(0)} \nabla \rho^{(2)} \right), \tag{5.74}$$

$$v'_t + \nabla \cdot \left(v^{(0)} \circ v' + v' \circ v^{(0)} \right) + \frac{1}{M \rho^{(0)}} \nabla p' = 0, \tag{5.75}$$

$$p'_t + v^{(0)} \nabla p' + \frac{\gamma p^{(0)}}{M} \nabla \cdot v' = - \left(p_t^{(2)} + v^{(0)} \nabla p^{(2)} \right). \tag{5.76}$$

The system on the left hand side may be called linearised Euler equations. It is a system of linear wave equations with the wave speeds $v^{(0)} \pm c^{(0)}$ and $v^{(0)}$. The density equation and the pressure equation can be combined to give

$$\left(p' - c_0^2 \rho' \right)_t + v^{(0)} \nabla \left(p' - c_0^2 \rho' \right) = 0 \tag{5.77}$$

showing that the wave propagation is isentropic. If $v^{(0)} \equiv 0$ then the linear acoustic equations are obtained.

We will next consider the modifications, if the viscous terms, heat conduction and compression from the boundary may occur. The incompressible solution $\rho^{(0)}$, $v^{(0)}$, $p^{(2)}$ and $p^{(0)}$ now satisfies (5.31), (5.27), (5.30) and (5.32), while $\rho^{(2)}$ is given by (5.73) again. The corresponding system describing the wave propagation now reads as:

$$\rho'_t + \nabla \cdot \left(v^{(0)} \rho' \right) + \frac{1}{M} \nabla \left(\rho^{(0)} v' \right) = - \left(\rho_t^{(2)} + \nabla \cdot \left(\rho^{(2)} v^{(0)} \right) \right) \tag{5.78}$$

$$\rho^{(0)} v'_t + \nabla \cdot \left(\left(\rho^{(0)} v^{(0)} \right) \circ v' + \left(\rho^{(0)} v' \circ v^{(0)} \right) \right) + \frac{1}{M} \nabla p' = \frac{1}{Re} \nabla \cdot \tau^{(1)} \tag{5.79}$$

$$p'_t + v^{(0)} \nabla p' + \frac{1}{M} \gamma p^{(0)} \nabla v' = \tag{5.80}$$
$$- \frac{\gamma - 1}{Pe} \nabla \cdot q^{(2)} - \left(p_t^{(2)} + v^{(0)} \nabla p^{(2)} + \gamma p^{(2)} \nabla \cdot v^{(0)} + \gamma p' \nabla \cdot v^{(0)} \right)$$

This system is more complicated due to the fact that the terms with $\nabla \cdot v^{(0)}$ do not cancel out and $\rho^{(0)}$ is a function of time and space. The expressions $\tau^{(1)}$ and $q^{(2)}$ denote the $\mathcal{O}(M)$ and $\mathcal{O}(M^2)$ terms of viscosity and heat conduction, respectively. We assumed that $\mathcal{O}(M)$-terms in the expansion of the heat fluxes do not appear.

Without the scaling motivated by the asymptotic considerations, Hardin and Pope have proposed in [18] such an expansion around incompressible flow and called it EIF. In an analogous way the physical quantities are split into incompressible and perturbation parts. This perturbation ansatz is substituted into the compressible equations and the incompressible equations are then subtracted without neglecting the nonlinear terms. In the approach described in this section we neglect products of fluctuations as well as terms of higher order with respect to the Mach number and obtain locally linearized acoustic equations. A similar strategy has been applied by Colella and Pao in [7].

We note that for perturbations in the initial data these perturbation equations have been obtained by Klainerman and Majda [20]. They mathematically justify the asymptotic expansion and the linearized acoustic equations. Slimon, Soteriou, and Davis [44] used the Janzen-Rayleigh expansion and a matching procedure with the outer expansion to get a set of EIF equations.

There may appear some difficulties in practical applications. As source terms for the sound generation on the right hand side in the pressure equation (5.76) the time derivative of the hydrodynamic pressure appears. Considering the incompressible equations we remark that the pressure looses its thermodynamic meaning and plays as balance of force agent the role of a Lagrange multiplier. Often boundary conditions are prescribed to the velocity field only and for the numerical calculation some sort of artificial pressure boundary conditions have to be found. In the case of Neumann boundary conditions, any constant may be added to the pressure without changing the equations. Hence, it may be difficult to get confidential values for the pressure time derivative in numerical simulations. At every time step the pressure establish the divergence-free condition for the velocity without the need of a continuous time development. In the next subsection we show results for a problem whose exact incompressible solution is known and thus this problem is circumvented.

5.5.5   Numerical Example: Co-Rotating Vortex Pair

A typical test case for the validation of a computational aeroacoustic code is the so-called rotating quadrupole. It is generated by a pair of co-rotating vortices of strength $\kappa = \Gamma/(2\pi)$ each, where Γ is the circulation. They are placed at a distance of $2r_0$ and thus each vortex induces a velocity $q = \Gamma/(4\pi r_0)$ on the other. This causes the vortices to rotate around their common midpoint.

For this setting, the exact solution of the potential theory for the incompressible and inviscid flow field as well as the solution of the acoustic far field equations by matched asymptotic expansions can be determined analytically [31]. Therefore, this example is chosen for algorithm validation in 2D. We used the perturbation ansatz around incom-

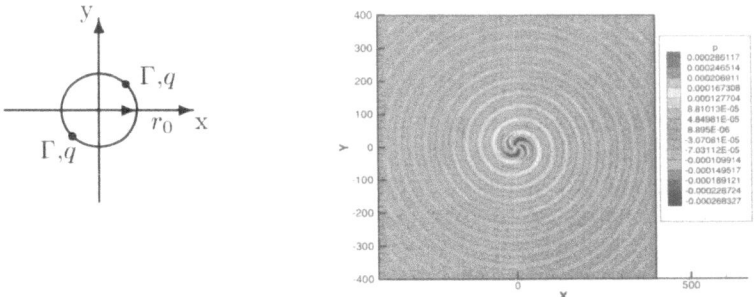

Figure 5.4   Initial setting and contour plot of the acoustic pressure

pressible flow in (5.5.4) to couple the flow field to the acoustic field via embedded domains. With the analytic solution of the incompressible and inviscid flow field the acoustic source terms in equations (5.75) and (5.76) are determined. The propagation of the sound waves is done numerically by solving the linear wave equations (5.75), (5.76), and (5.76). The numerical results are compared to the analytical acoustic solution. We calculated the case with a vortex distance of $2r_0 = 2.0$ and a "rotating" Mach number of $M = 0.095$ which leads to a rotation period of $T = 2\pi$. The computational domain is set to be $[-400, 400] \times [-400, 400]$ and is discretized by 320×320 grid points, i.e. with grid spacing of $\Delta x = 2.5$. The numerical computations were performed with the ADER-scheme of second and fourth order in space and time. We will discuss these schemes and compare them with other appropriate numerical schemes for the wave propagation problem in Section 5.7. Figure 5.4 shows the contour plot of the acoustic pressure. The acoustic source is located in the center and the acoustic waves spiral out of the origin.

Figures 5.5 and 5.6 compare the results of the numerical simulation to the analytical solution. Two points on the x-axes have been chosen at distance $x = 150$ and $x = 300$ from the origin. At these points, the acoustic pressure is plotted versus time. The end time $t \approx 44$ corresponds to about 7 rotational periods. Figure 5.5 shows the results of the second order ADER-scheme. Since the analytic solution is periodic in time and the source terms are calculated from the analytic solution, the amplitude remains constant, no damping can be seen over the time. With growing distance from the acoustic source, i.e. travelling from point $x = 150$ to $x = 300$, the difference between the amplitudes of the numerical wave and the analytical one increases. This is due to the numerical dissipation. Also, a phase error can be seen which grows with the distance, i.e. with the time travelled.

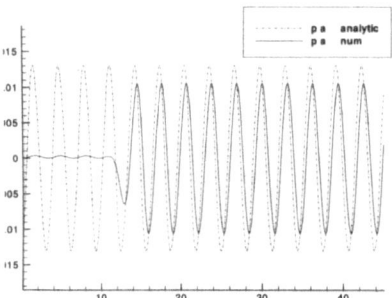 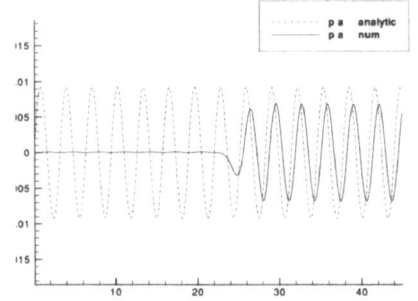

Figure 5.5 Time development of the acoustic pressure at the fixed point (150,0) and (300,0). Comparison of the exact and numerical solution obtained with a 2nd order scheme.

Using the fourth order scheme for the calculation of the sound waves (figure 5.6), nearly no phase error occurs. The quotient of analytical amplitude to numerical amplitude remains constant with the distance. It seems that only the physical decrease of the amplitude of the acoustic wave can be observed. The calculations show, that it is necessary for computational aeroacoustics to use high order schemes. We will come back to this point in Section 5.7.

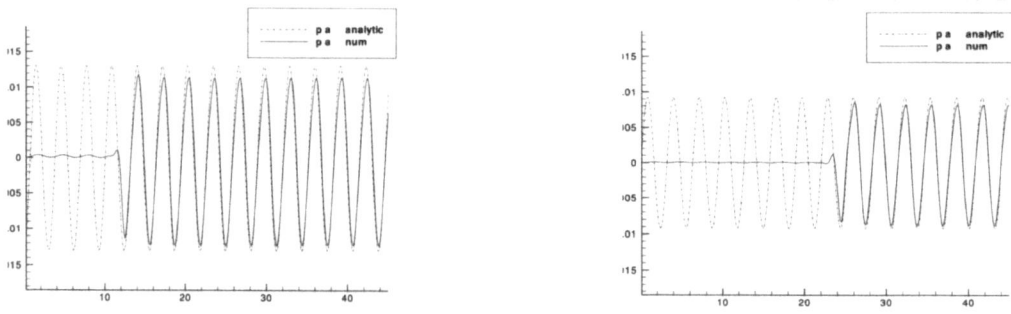

Figure 5.6 Time development of the acoustic pressure at the fixed point (150,0) and (300,0). Comparison of the exact and numerical solution obtained with a 4th order scheme.

5.6   Multiple Scale Considerations

Up to now, the multi-scale behavior has been considered in the formulation of the acoustic equations for the fluctuations. Hence, the different scales in the magnitude of the flow and acoustic effects has been taken into account. But there are several space scales as outlined in the last section. For the fluid flow considered here viscous effects and heat conduction can be neglected. Therefore no additional space scale appears, hence three space scales associated with the acoustics are taken into account. These are the scale of the flow x_{ref}, the scale of the acoustic waves $x_{ac} = x_{ref}/M$, and the length scale

x_{dia} of the flow region under consideration. It is obvious that an asymptotic analogy as considered in the previous section can be successfully applied only, if the Mach number is small enough so that the incompressible flow field is locally a good approximation. This especially will be satisfied, if the diameter of the flow region is small with respect to the wavelengths of the acoustic waves. Under the latter condition the acoustic waves would have only a small influence to the flow field. If these conditions were satisfied, Crow [9] called the fluid flow compact. In this section we still consider a compact flow, but the different length scales of the fluid flow and the acoustic waves are taken into account. The basic assumption that the fluid flow is not affected by the acoustics is still assumed to be valid. Furthermore, we assume that the fluid structures generate the sound and, hence, the flow and the acoustics have the same time reference.

5.6.1 Outer Asymptotic Expansion

Because of the same relevant time scale for the fluid flow and the sound generation, the different velocities of the fluid and the sound introduce different space scales. It turns out that the Janzen-Rayleigh inner asymptotic solution is not uniformly valid in the whole region. Ting and Miksis [46] proposed the asymptotic expansion

$$\rho = \rho^{(0)} + M^3\rho^{(3)} + M^4\rho^{(4)} + ..., \tag{5.81}$$
$$v = M^2v^{(2)} + M^3v^{(3)} + ..., \tag{5.82}$$
$$p = p^{(0)} + M^3p^{(3)} + M^4p^{(4)} + ... \tag{5.83}$$

for the outer region. Here, the expansion functions are assumed to be functions of the large scale space variable $\xi = Mx$ and the time variable t:

$$\rho^{(i)} = \rho^{(i)}(Mx,t), \quad v^{(i)} = v^{(i)}(Mx,t), \quad p^{(i)} = p^{(i)}(Mx,t). \tag{5.84}$$

Using matching conditions with the inner solution Ting and Miksis obtained this form of the asymptotic expansion. If this expansion is substituted into the equations, then the set of perturbation equations

$$\rho_t^{(i)} + \frac{\rho^{(0)}}{M}\nabla v^{(i-1)} = 0, \tag{5.85}$$

$$v_t^{(i-1)} + \frac{1}{M\rho^{(0)}}\nabla p^{(i)} = 0, \tag{5.86}$$

$$p'_t^{(i)} + \frac{\gamma p^{(0)}}{M}\nabla \cdot v^{(i-1)} = 0 \tag{5.87}$$

is obtained for $i = 3, 4$. Ting and Miksis matched these equations to those of the inner solution. They used this procedure to obtain a strategy for the numerical simulation. Slimon, Soteriou, and Davis [44] used the additive composition method described by van Dyke [49]. Here, a composite expansion is obtained by summing the outer and inner solutions and subtracting the part that these solutions have in common.

5.6.2   Multiple Scale Asymptotic Expansion

The functional relationship between the physical phenomena on different space scales is assumed to be characterised by different independent variables. For small Mach numbers an asymptotic expansion of the form

$$u = u^{(0)} + M u^{(1)} + M^2 u^{(2)} + \dots \tag{5.88}$$

is considered where u denotes the vector of the primitive variables ρ, v and p. The expansion functions are now assumed to depend on one time and the two space variables η and ξ:

$$u^{(i)} = u^{(i)}(\eta, \xi, t). \tag{5.89}$$

Here, $\eta = x$ is the local variable, associated with the convective phenomena, while the other independent variable $\xi = Mx$ is called the large scale coordinate and is associated with acoustic wave propagation.

The asymptotic expansion (5.88) with (5.89) is inserted into the compressible equations and as usual the terms multiplied by the same powers of the Mach number are collected to obtain a hierarchy of asymptotic limit equations. In the following we only list those perturbation equations which give insight into the limit behavior. These are the leading order continuity equation

$$\rho_t^{(0)} + \nabla_\eta \cdot (\rho v)^{(0)} = 0, \tag{5.90}$$

the leading, first, and second order velocity equations

$$\nabla_\eta p^{(0)} = 0, \tag{5.91}$$

$$\nabla_\eta p^{(1)} + \nabla_\xi p^{(0)} = 0, \tag{5.92}$$

$$v_t^{(0)} + (v^{(0)} \cdot \nabla_\eta) v^{(0)} + \frac{1}{\rho^{(0)}} (\nabla_\eta p^{(2)} + \nabla_\xi p^{(1)}) = \frac{1}{\rho^{(0)} Re} \nabla_\eta \tau^{(0)}, \tag{5.93}$$

and the leading order and first order pressure equation

$$p_t^{(0)} + (v \cdot \nabla_\eta p)^{(0)} + (\gamma p \nabla_\eta \cdot v)^{(0)} = \frac{\gamma - 1}{Pe} \nabla_\eta \cdot q^{(0)}. \tag{5.94}$$

$$p_t^{(1)} + (v \cdot \nabla_\eta p)^{(1)} + (\gamma p \nabla_\eta \cdot v)^{(1)} + (v \cdot \nabla_\xi p)^{(0)} + (\gamma p \nabla_\xi \cdot v)^{(0)}$$
$$= \frac{\gamma - 1}{Pe} \nabla_\xi \cdot q^{(0)} + \frac{\gamma - 1}{Pe} \nabla_\eta \cdot q^{(1)}. \tag{5.95}$$

The expressions ∇_η and ∇_ξ denote the gradients with respect to η and ξ, respectively. Analogously, the divergence operators are defined.

From the leading order velocity equation (5.91) it follows immediately that $p^{(0)}$ does not depend on η. Next, we consider the first order velocity equation (5.92). To this equation an usual technique in multiple scale asymptotics is applied by averaging the whole equation over the small scale (see [4], [32]). This is done by integrating this equation with respect to the small scale variable η. Both terms in the equation can directly be integrated. The first term involving $p^{(0)}$ does not depend on η and gives the volume of the domain of integration times $\nabla_\eta p^{(0)}$, while the other term gives a surface integral of the variable $p^{(1)}$ over the boundary of the domain of integration. If the Mach number M is small enough and $p^{(1)}$ has a sub-linear growth rate when x becomes large, then the domain of integration may be chosen as large that the first term becomes larger than any given bound. Or vice versa, if we divide by the volume, then the boundary integral divided by the volume may become smaller than any given constant. This means, that in the equation (5.92) both terms have to become zero for their own when the Mach number tends to zero and the sum should vanish. Hence, we have:

$$p^{(0)} = p^{(0)}(t),$$
$$p^{(1)} = p^{(1)}(\xi,t). \tag{5.96}$$

Taking this into account equation (5.94) may be integrated with respect to η over the domain Ω to obtain an evolution equation for $p^{(0)}$. It has the same form as in the one space scale expansion and is given by (5.32). The time evolution of $p^{(0)}$ is determined by the surface integrals of the boundary values of the normal component of the fluid velocity and the heat flux. It describes an overall pressure rise due to compression and heat transfer from the boundary. As in the one scale case the velocity $v^{(0)}$ has the divergence

$$\nabla_\eta \cdot v^{(0)} = \frac{1}{|\Omega|} \int_{\partial\Omega} v^{(0)} \cdot \mathrm{n}\, ds - \frac{\gamma-1}{|\Omega|Pe\gamma p^{(0)}} \int_{\partial\Omega} q^{(0)} \cdot \mathrm{n}\, ds + \frac{\gamma-1}{\gamma p^{(0)}Pe}\nabla_\eta \cdot q^{(0)}. \tag{5.97}$$

Without heat transfer, it becomes constant in space, and furthermore without compression from the boundary it becomes zero. The leading order mass conservation equation (5.90) reduces in the divergence free case to pure convection and otherwise to an adiabatic compression of mass elements along particle paths.

This is quite the same as described in Section 5.3 using the asymptotic expansion for one space scale. If we consider the formal limit equations for $M = 0$ in a bounded domain without heat conduction and compression from the boundary, we obtain the incompressible equations (5.26)-(5.28).

A new phenomena is associated with the first order pressure term $p^{(1)}$. While in the one scale expansion this term turned out to be constant, now the gradient has to be zero only with respect to the small scale variable according to (5.96). This pressure term may carry long wavelength acoustics. We consider this possibility and the circumstances in the following. The large scale average of the leading order velocity equation (5.93) and the first order pressure equation give evolution equations for this large wavelength acoustics:

$$\overline{(\rho v)}_t^{(0)} + \nabla_\xi p^{(1)} = 0, \tag{5.98}$$
$$p_t^{(1)} + \gamma p^{(0)} \nabla_\xi \cdot \overline{v}^{(0)} = \frac{\gamma-1}{Pe}\nabla_\xi \cdot \overline{q}^{(0)}, \tag{5.99}$$

where " $^-$ " denotes the average over the local structures.

Notice that (5.98) contains the average of the momentum $\overline{(\rho v)}^{(0)}$ while (5.99) features $\overline{v}^{(0)}$, so that these equations generally do not represent a closed system. We separate $\rho^{(0)}$ and $v^{(0)}$ into the ξ-scale averages $\bar{\rho}^{(0)}$ and $\bar{v}^{(0)}$ and η-scale fluctuations $\tilde{\rho}^{(0)}$ and $\tilde{v}^{(0)}$. Averaging the leading order mass conservation equation (5.90) over the local structures and using the usual limit arguments shows that the time derivative of $\bar{\rho}^{(0)}$ is zero. Thus, we obtain

$$\overline{(\rho v)}_t^{(0)} = \bar{\rho}^{(0)} \bar{v}_t^{(0)} + \overline{(\tilde{\rho}\tilde{v})}_t^{(0)}. \tag{5.100}$$

Next, we insert this relation into (5.98) to get the equations

$$\bar{v}_t^{(0)} + \frac{1}{\bar{\rho}^{(0)}} \nabla_\xi p^{(1)} = -\frac{1}{\bar{\rho}^{(0)}} \overline{(\tilde{\rho}\tilde{v})}_t^{(0)}, \tag{5.101}$$

$$p_t^{(1)} + \gamma p^{(0)} \nabla_\xi \bar{v}^{(0)} = \frac{\gamma - 1}{Pe} \nabla_\xi \cdot \overline{q}^{(0)}. \tag{5.102}$$

If we assume that we have homentropic basic flow and the leading order density is constant all over the computational domain, then we have $\bar{\rho}^{(0)} = \rho^{(0)}$ and $\tilde{\rho}^{(0)} = 0$. The source terms on the right hand side vanish. The equations (5.101) and (5.102) define nothing else than acoustic wave equations for the large scale quantities $\overline{(v)}^{(0)}$ and $p^{(1)}$ without source term and it can be rewritten in the usual form

$$p_{tt}^{(1)} - \nabla_\xi \cdot \left(c_0^2 \nabla_\xi p^{(1)} \right) = 0 \tag{5.103}$$

with the wave speed given by

$$c_0^2 = \frac{\gamma p^{(0)}}{\rho^{(0)}}. \tag{5.104}$$

Hence, in such a flow no acoustic waves with long wavelength and large amplitudes of the order M will be generated. If large temperature gradients occur and heat conduction cannot be neglected then density variations will occur, because the divergence of the velocity field is non-zero. Hence, the source terms may be non-zero and the acoustic wave equations (5.101) and (5.102) describe the generation and propagation of such thermo-acoustic waves. In a flow with strong temperature gradients those waves may appear with amplitudes of order M, which is much larger than the amplitudes of noise. The formal asymptotic tells that in systems with density variations such as reactive flow or flow with strong heat addition some sort of wave instabilities with a large amplitude may be generated.

Up to now we did not consider the large scale density changes which correspond to the pressure fluctuations $p^{(1)}$. The large scale averaging of the first order density perturbation equations gives

$$\bar{\rho}_t^{(1)} + \nabla_\xi \cdot \left(\overline{\rho^{(0)} v^{(0)}} \right) = 0 \tag{5.105}$$

If $v^{(0)}$ is calculated via the acoustic equations for the large scale effects (5.101), (5.102), the density changes can be calculated directly from the equation above. For homentropic flow the corresponding wave equation of type (5.103) for the density fluctuations $\rho^{(1)}$ is obtained together with

$$p^{(1)} - c_0^2 \rho^{(1)} = const. \tag{5.106}$$

No large scale acoustic waves can be generated.

This formal multiple scale asymptotic has been performed by Klein [21] in the inviscid case using the conservation form of the Euler equations. A mathematical foundation of the multiple scale asymptotic was given by Meister [28].

5.6.3   MPV-Approach for Thermo-Acoustic Applications

If thermo-acoustic waves play a role, then the pressure term $p^{(1)}$ should be introduced in the pressure decomposition for the MPV-scheme as proposed in Section 5.4. To the pressure decomposition (5.45) the term $Mp^{(1)}$ is added. The acoustic pressure $p^{(1)}$ is defined to be

$$p^{(1)} := \frac{1}{M}\left(\bar{p} - p^{(0)}\right), \tag{5.107}$$

where " $\bar{\ }$ " denotes the average over the local structures. The pressure $p^{(2)}$ is then given via consistency arguments by

$$M^2 p^{(2)} = p - p^{(0)} - M p^{(1)}. \tag{5.108}$$

The approximation of the term $\nabla p/M^2$ is now replaced using the decomposition

$$\frac{1}{M^2}\nabla p = \frac{1}{M^2}\nabla p^{(0)} + \frac{1}{M}\nabla p^{(1)} + \nabla p^{(2)}. \tag{5.109}$$

By construction, the thermodynamic pressure term is constant in space, so $\nabla p^{(0)} = 0$. The term $\nabla p^{(1)}/M$ corresponds to the large scale derivative $\nabla_\xi p^{(1)}$ which appears in the thermo-acoustic equation (5.102). According to (5.103) this term leads to a change in time of the momentum average and we can replace this gradient by $-(\overline{\rho v})_t$. The idea is now to replace the approximation of $\nabla p/M^2$ by an approximation to

$$-(\overline{\rho v})_t + \nabla p^{(2)}, \tag{5.110}$$

where the singularity is eliminated and which converges towards $\nabla p^{(2)}$, because the acoustic waves disappear in the $M = 0$ limit.

Because the sound and transport velocity scales spread as M tends to zero, we solve for small Mach numbers the different effects separately using the asymptotic results. This allows us to rescale the problem and also to use different adequate numerical approximations for different physical phenomena. As we do not know the regime of validity of our asymptotic analysis, all asymptotic considerations are only used to design suitable predictor steps which are performed to precondition the solution of the full equations. The pressure correction approach as described in section 5.4.1 is applied in a similar way. The pressure p is decomposed now into the thermodynamic, the acoustic and the incompressible pressure terms $p^{(0)}$, $p^{(1)}$ and $p^{(2)}$ using discrete analogues of (5.46), (5.107) and (5.108). This first step produces approximate values at the old time level t_n

$$p^{(0),n}, \quad p^{(1),n}, \quad p^{(2),n}. \tag{5.111}$$

Estimates for the change in time of the thermodynamic pressure, the acoustic pressure and the averaged momentum during the time step $\Delta t = t_{n+1} - t_n$ are then obtained by solving numerically the corresponding evolution equations as given by the asymptotic analysis. In the second step, we proceed in the way proposed in the last section. The estimates are used as a predictor for the effects of global compression and acoustics and the pressure correction for the hydrodynamic pressure term $p^{(2)}$ is started.

The equations (5.101) and (5.102) involve only the long acoustic wavelength. Numerically, short wave flow components are filtered out prior by the averaging procedures carried out for the pressure decomposition. Therefore, the *space* scale is a factor $1/M$ larger than that of the convective phenomena while the evolution *time* scales match. Nevertheless, we are interested in using explicit schemes avoiding a CFL condition more restrictive than that for the convective terms. This is achieved by making use of the long-wave nature of the data: Without loss of overall accuracy these data can be represented on a much coarser grid with the step sizes multiplied by $1/M$. Using an explicit solver on this coarse grid amounts to a CFL condition involving the averaged flow velocity only because the factor $1/M$ cancels out. Therefore the time step is comparable to that for the small scale convection on the fine grid.

This approach reduces the computational costs in two ways. First, the evaluation is faster since less grid points have to be updated. Secondly, it allows us to stick to an explicit scheme. On the other hand, it requires averaging and interpolation processes to manage the coarse-fine grid communication. These issues are related to those known from the usual multi-grid techniques. But in the situation considered here, the effort can be reduced by making explicit use of the knowledge gained through the asymptotic analysis. Due to the fact that the acoustic variables incorporate large scale contributions only, no additional averaging is necessary. The only averaging necessary is associated with the determination of the acoustic pressure and velocity fields from the total pressure and local flow velocity field, respectively. Straight injection is sufficient for the fine-to-coarse grid restriction, whereas for the coarse-to-fine grid prolongation, straight-forward interpolation techniques with low numerical costs as linear interpolation are sufficient. Errors introduced by the coarse-to-fine interpolation have small wavelength and are therefore quickly captured by the $p^{(2)}$ terms in the corrector step.

In the following, we will show numerical results for some test problems, where the acoustic waves of order M are introduced into the initial data. In the calculations presented below we apply a finite difference method on a staggered grid arrangement which is commonly used for incompressible solvers (see, e.g., [37]). The following calculations are performed with upwind differencing for the convection terms. The usual donor-cell differencing of first order accuracy is extended to second order via the MUSCL-approach of van Leer [51]. This approach has been originally developed for non-staggered grids and can be modified for staggered grids (see [1]). The viscous terms as well as the sonic terms are treated by the usual central differencing.

The first example is a one-dimensional inviscid problem. The initial values are given by

$$
\left.
\begin{aligned}
\rho(x,0) &= 1 + \tfrac{1}{2}\Phi(x)\sin(\tfrac{40\,\pi x}{L}) + M\left(1 + \cos(\tfrac{\pi x}{L})\right) \\
v(x,0) &= \sqrt{\gamma}\left(1 + \cos(\tfrac{\pi x}{L})\right) \\
p(x,0) &= 1 + M\gamma\left(1 + \cos(\tfrac{\pi x}{L})\right)
\end{aligned}
\right\} \quad \text{for } -L \le x \le L \quad (5.112)
$$

with

$$
\Phi(x) = \begin{cases}
0 & \text{for } -\tfrac{1}{L} \le x \le 0 \\
\tfrac{1}{2}\left(1 - \cos\left(\tfrac{5\pi x}{L}\right)\right) & \text{for } 0 \le x \le \tfrac{2L}{5} \\
0 & \text{for } x > \tfrac{2L}{5}
\end{cases} \quad (5.113)
$$

and boundary conditions defined to be periodic. These data generate an acoustic wave of wavelength L propagating to the right. It is overlaid by short wavelength density oscillations which are restricted to a region with length $\frac{2L}{5}$ by the function Φ. For the Mach number we prescribed the value $M = 1/51$ and the length of the computational domain has been chosen in such a way that one complete acoustic wave takes place there: $L = 1/M$.

For this problem we need second order accuracy in space and time, otherwise the solution is strongly dissipated and the waves are damped. In our calculations a fine grid of 1020 grid points is used to resolve the short wavelength density fluctuations which are set in motion by the large wavelength acoustic wave.

The problem is difficult to approximate because short wavelength density oscillations are overlaid to the acoustic wave. The $O(1)$ density fluctuations are set into motion by the long wavelength acoustic pulse. Fig. 5.7 shows the numerical results of the MPV method at time $t = 5.07$. The initial values for density and pressure are plotted by the dashed line and the numerical results by the straight line. The calculation time is chosen in such a way that the acoustic wave has passed the computational domain about 2.5 times and is steepened due to nonlinear effects. This is well-captured in the numerical calculations. The amplitude of the density fluctuations is reduced by some numerical dissipation in the convection step. Again, the acoustic equations have been solved explicitly on the coarse grid. Calculations with an explicit Euler-solver using a second order Godunov-type scheme produced quite similar results, but needed 4086 time steps in contrast to 186 with the MPV-approach.

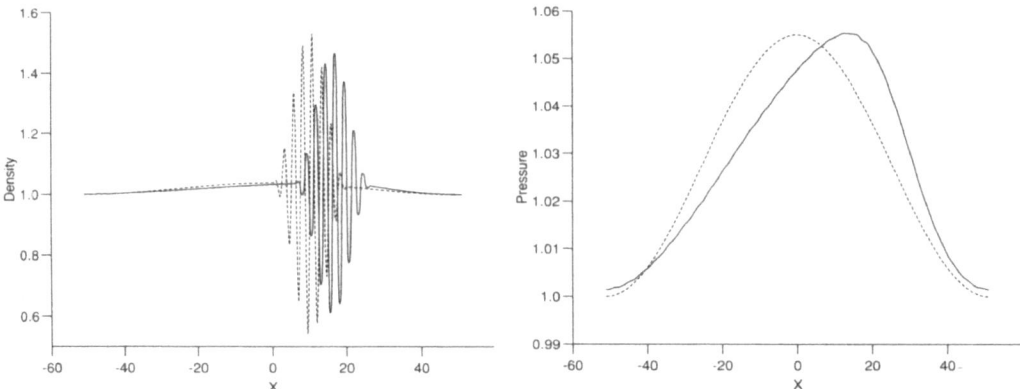

Figure 5.7 Numerical results of density and pressure for the MPV method at times $t = 5.07$ (——), initial data (- - -).

The other example is a two-dimensional inviscid problem at $M = 1/20$. It shows the interaction of a large scale acoustic wave with the small scale flow structures, leading to vorticity generation. In a double-periodic domain with

$$-L \leq x \leq L = \frac{1}{M} \quad ; \quad M = \frac{1}{20},$$
$$0 \leq y \leq L_y = \frac{2L}{5}$$

discretized by 400×80 grid cells, initial data are prescribed as follows

$$\begin{aligned}
\rho(x,y,0) &= & 1 & \quad + \quad 0.2\,M(1+\cos(\tfrac{\pi x}{L})) & +\Phi(y), \\
p(x,y,0) &= & 1 & \quad + \quad M\gamma(1+\cos(\tfrac{\pi x}{L})), \\
u(x,y,0) &= & \sqrt{\gamma}(1+\cos(\tfrac{\pi x}{L})), \\
v(x,y,0) &= & 0.
\end{aligned}$$

With $\rho_y = 0.8/L_y$, the function $\Phi(y)$ is defined by

$$\Phi(y) = \begin{cases} \rho_y y & \text{for} \quad 0 \;\leq\; y \;\leq\; \tfrac{1}{2}L_y, \\ \rho_y(y-\tfrac{1}{2}L_y) - 0.4 & \text{for} \quad \tfrac{1}{2}L_y \;<\; y \;\leq\; L_y. \end{cases}$$

These initial data represent a saw-tooth like density layering in the y-direction. In the x-direction, a right-running acoustic pulse with large wavelength runs over it, setting the density layering into motion. The different density leads to a different speed under the influence of the same acoustic pressure wave. This results in the occurrence of a Kelvin-Helmholtz instability, because different parts of the fluid move at a different rate relative to another. A shear layer with sinusoidal shape is generated and set into motion by the acoustic wave.

Figure 5.8 shows the density at several times. The first plot shows the initial values, the second one the sinusoidal shape at time t=6.0. The initially horizontal interface starts rolling up into large vortical structures. The Kelvin-Helmholtz instability says that in the absence of viscosity small wavelength perturbations grow much faster. This effect can clearly be seen in our calculations at the next times 9.0, 11.0 and 14.0. The smallest structures which can be resolved have a thickness of a few grid zone lengths. Smaller structures are damped out by numerical viscosity. Figure 5.9 shows the generation of vorticity at the same times. In the initially vortex-free fluid flow, a thin region of vorticity is created along the density interface. As the interface starts rolling up, counterrotating vortices are formed. These computations demonstrate, that the large scale long wavelength acoustic pressure fluctuations feed energy to the small scale structures leading to the creation of vortices on the small space scale.

Again, the acoustic predictor was calculated on a coarse grid with space increment $\Delta\xi = \frac{1}{M}\Delta x$. Restriction from fine to coarse grid used "straight injection", i.e. the pure take over of values at the common points of coarse and fine grid. No additional smoothing is necessary, since $p^{(1)}, \bar{\mathbf{U}}$ and $\bar{\rho}$ themselves are obtained by averaging procedures from the full variables $p, \mathbf{U}$ and ρ. The acoustic equations (5.102), (5.103) are solved explicitly with the convection time step, since the Mach number cancels out of the CFL-condition. To complete the acoustical step, the coarse grid values have to be interpolated back to the fine grid. We also performed calculations without calculating the temporal evolution of the acoustic pressure according to the acoustic equations. In this case, the pressure correction must also capture the long wavelength acoustics. More iteration cycles have to be done which leads to an increase of computational time of about 6%.

5.6.4   Multiple Scale Considerations of Noise Generation

We assume that the basic incompressible flow is homentropic i.e. we neglect heat conduction or outer compression. In this case, the long wave length acoustic perturbation of order one cannot be generated by the flow. Hence, the proper multiscale asymptotic expansion should be:

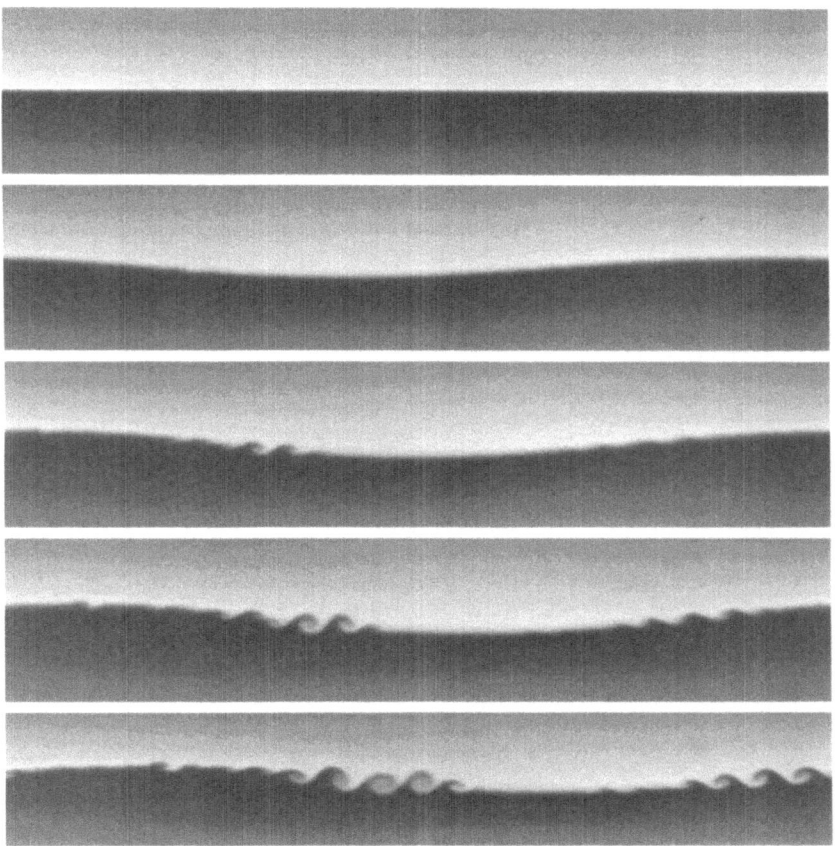

Figure 5.8 Density contours on a 400x80 grid at times t = 0.0, 6.0, 9.0, 11.0 and 14.0.

$$\rho = \rho^{(0)} + M^2\rho^{(2)} + M^3\rho^{(3)} + ..., \tag{5.114}$$

$$v = v^{(0)} + Mv^{(1)} + M^2v^{(2)} + ..., \tag{5.115}$$

$$p = p^{(0)} + M^2p^{(2)} + M^3p^{(3)} + ... \tag{5.116}$$

Now the list of perturbation equations we need to close the system of compressible long wave length corrections is given by:

The 2nd order density equation

$$\rho_t^{(2)} + \nabla_\eta \cdot \left(\rho^{(0)}v^{(2)} + \rho^{(2)}v^{(0)} \right) + \nabla_\xi \cdot \left(\rho^{(0)}v^{(1)} \right) = 0, \tag{5.117}$$

the 1st order velocity equation

$$\rho^{(0)}v_t^{(1)} + \rho^{(0)}\nabla_\eta \left(v^{(1)} \circ v^{(0)} + v^{(0)} \circ v^{(1)} \right) + \nabla_\xi \rho^{(3)} = -\nabla_\xi ((\rho v) \circ v)^{(0)} - \nabla_\xi p^{(2)}, \tag{5.118}$$

and the 2nd order pressure equation

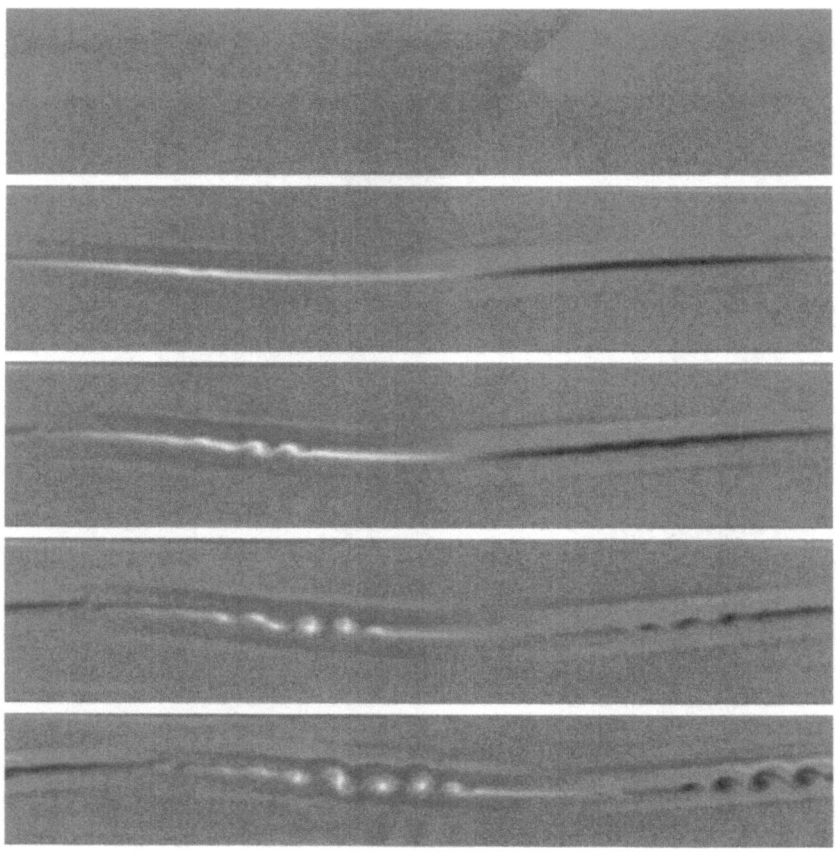

Figure 5.9 Vorticity plot on a 400x80 grid at t = 0.0, 6.0, 9.0, 11.0 and 14.0.

$$p_t^{(2)} + v^{(0)} \cdot \nabla_\eta p^{(2)} + \gamma p^{(0)} \nabla_\eta \cdot v^{(2)} + \gamma p^{(0)} \nabla_\xi \cdot v^{(1)} = 0. \tag{5.119}$$

To get the behavior of the large scale fluctuations we apply the averaging over the small scales to these three equations as done in subsection 5.6.2. The following set of equations is obtained:

$$\overline{\rho}_t^{(2)} + \rho^{(0)} \nabla_\xi \cdot \overline{v}^{(1)} \;=\; 0, \tag{5.120}$$

$$\overline{v}_t^{(1)} + \frac{1}{\rho^{(0)}} \nabla_\xi \overline{p}^{(2)} \;=\; -\nabla_\xi \, (\overline{v \circ v})^{(0)}, \tag{5.121}$$

$$\overline{p}_t^{(2)} + \gamma p^{(0)} \nabla_\xi \cdot \overline{v}^{(1)} \;=\; 0, \tag{5.122}$$

where "$-$" denotes the average. We make the Ansatz

$$p^{(2)} = p_{inc}^{(2)} + p' \quad, \quad \rho^{(2)} = \rho_{inc}^{(2)} + \rho', \tag{5.123}$$

where "$'$" denotes the acoustic fluctuations and insert it into these equations to get

$$\overline{\rho}'_t + \rho^{(0)} \nabla_\xi \cdot \overline{v}' = -(\overline{\rho}^{(2)}_{inc})_t, \tag{5.124}$$

$$\overline{v}'_t + \frac{1}{\rho^{(0)}} \nabla_\xi \overline{p}' = -\nabla_\xi \, (\overline{v \circ v})^{(0)} - \frac{1}{\rho^{(0)}} \nabla_\xi \overline{p}^{(2)}_{inc}, \tag{5.125}$$

$$\overline{p}'_t + \gamma p^{(0)} \nabla_\xi \cdot \overline{v}' = -(\overline{p}^{(2)}_{inc})_t. \tag{5.126}$$

If we assume that the hydrodynamic pressure $p^{(2)}_{inc}$ does not contain a large scale component, then it turns out that the source of noise coincides with the first basic term of the Lighthill tensor, averaged over the small scales. If the diameter of the domain of the fluid flow generating the noise is larger than the wave length of the acoustic waves, then the time derivatives of the averaged hydrodynamic pressure should play a role,too.

This approach should be favourable in the regime when the Mach number is small enough that the acoustic length scale is much larger than the length scale of the fluid convection. The acoustic propagation is captured on a coarse grid and an averaging procedure is necessary to determine the acoustic source terms on the coarse grid. Several wave length of the acoustic waves may fit into the domain of fluid flow, while compressible effects are not important for the fluid flow. The influence of fluid convection to the acoustic waves may be taken into account using the next higher order corrections.

We note that for these basic considerations we always consider a computational domain without interior obstacles. At interior obstacles the time derivatives of $(\overline{p}^{(2)}_{inc})_t$ and $(\overline{\rho}^{(2)}_{inc})_t$ should play the main role of sound source.

5.7 Numerical Aeroacoustics

The numerical approximation of acoustic wave propagation is a difficult task. If we assume that noise of a special bandwidth around 1kHz should be captured, then we have to resolve waves with wavelength of about 0.3 meters. Using a standard finite difference approximation of second order accuracy more than 20 grid points per wavelength are needed to capture the wave propagation over a distance of a couple of wavelengths. For a computational domain of ten wavelengths 200 grid cells in each space direction are necessary. This results for a three-dimensional calculation in 8 millions of grid points for a computational aeroacoustic domain of three meters. This shows the difficulties in computational aeroacoustics in time domain. To reduce the computational effort or to extent the computational domain very efficient numerical methods must be applied on high performance computers. The main criterion for their design and construction is to resolve the basic properties of wave propagation with a small number of grid points per wavelength and at relatively low costs in computer time. Damping and dispersion of propagating waves should be kept as small as possible. Typical numerical approximations as used in CFD are not favorable for an efficient resolution of acoustic waves over long distances. They possess a lot of inherent numerical dissipation to capture strong local gradients and nonlinear convection terms in a robust way. In this section we give a short survey over high resolution schemes in CAA and CFD. Furthermore a comparison of several methods is presented.

To capture an acoustic wave with a small number of grid points it is essential to apply high order schemes. Besides the dissipation of the numerical scheme, the dispersion of the algorithm plays an important role, too. To get a good wave like behavior of the discrete waves, Tam et al. [45] developed a so called DRP (dispersion relation preserving)-scheme. It is based on a standard finite difference (FD) scheme, but the coefficients are not completely determined from Taylor expansion. Using the same stencil but reducing the order of accuracy, one degree of freedom is obtained. This is used to optimize the spectral properties of the numerical numerical scheme. Tam considered the error integral $E(\lambda)$ with $[-\eta,\eta]$ denoting the range of wave numbers of interest

$$E(\lambda) = \int_{-\eta}^{\eta} (\alpha\Delta x - \alpha_{num}\Delta x)^2 d(\alpha\Delta x) \stackrel{!}{=} min, \qquad (5.127)$$

where α and α_{num} denotes the exact and the numerical wave number, respectively. An additional condition beside accuracy to determine the coefficients is then obtained by the requirement that $E(\lambda)$ is minimized. This technique may be applied to space as well as to time finite difference approximations.

Another class of schemes are the so called compact schemes. Here, the finite difference quotients contain beside the values of the physical variables at the different grid points also values of the derivatives. The formula becomes more complicated and results in linear system of equations, which is typically of three-diagonal form and which has to be solved at every time step. The main advantage is that the stencil of the difference scheme is small. A review of compact schemes is given by Lele [23].

The ADER approach by that formally **A**rbitrary high order schemes can be constructed utilising the solution of Riemann Problems for the advection of the higher order **DER**ivatives. It is a finite volume scheme for solving hyperbolic conservation laws [30, 42, 47] and may be designed to have uniform high order accuracy in time and space for smooth solutions. The principle ideas of the ADER approach are shortly reviewed for the one dimensional linear advection equation

$$u_t + a\,u_x = 0 \quad \Longleftrightarrow \quad u_t + F(u)_x = 0 \quad \text{with} \quad F(u) = a \cdot u. \qquad (5.128)$$

A general finite volume scheme solves the discrete, integral formulation

$$u_i^{n+1} = u_i^n - \frac{\Delta t}{\Delta x}\left(\hat{F}(u)_{i+\frac{1}{2}} - \hat{F}(u)_{i-\frac{1}{2}}\right) \qquad (5.129)$$

with

$$\hat{F}(u)_{i\pm\frac{1}{2}} = \frac{1}{\Delta t}\int_{t^n}^{t^{n+1}} F(u_{i\pm\frac{1}{2}})dt \qquad (5.130)$$

$$= a\frac{1}{\Delta t}\int_{t^n}^{t^{n+1}} u_{i\pm\frac{1}{2}}dt \qquad (5.131)$$

$$= a\,u_{i\pm\frac{1}{2}}^{ADER}. \qquad (5.132)$$

For the calculation of the state $u_{i\pm\frac{1}{2}}^{ADER}$ a Taylor series expansion in time is performed at the cell interface:

$$u(t+\tau) = u(t) + \sum_{l=1}^{\infty} \frac{(\tau)^l}{l!}\frac{\partial^l u(t)}{\partial t^l}. \qquad (5.133)$$

Using the Lax-Wendroff procedure the time derivatives are replaced by spatial derivatives according to

$$u_t = -a\, u_x \tag{5.134}$$

$$\frac{\partial^l u}{\partial t^l} = (-a)^l \frac{\partial^l u}{\partial x^l} \tag{5.135}$$

$$= (-a)^l u_x^{(l)}. \tag{5.136}$$

valid for any $l \in \mathbb{N}$. Thus equation (5.133) can be written as

$$u(t+\tau) = u(t) + \sum_{l=1}^{\infty} \frac{(-a\,\tau)^l}{l!} u_x^{(l)}. \tag{5.137}$$

The next step is to calculate the integral mean value:

$$\frac{1}{\Delta t} \int_{t^n}^{t^{n+1}} u_{i+\frac{1}{2}}(t^n+\tau) d\tau = u_{i+\frac{1}{2}}^{(0)n} + \frac{1}{\Delta t} \int_{t^n}^{t^{n+1}} \sum_{l=1}^{\infty} \frac{(-a\,\tau)^l}{l!} u_x^{(l)n} d\tau \tag{5.138}$$

$$= u_{i+\frac{1}{2}}^{(0)n} + \sum_{l=1}^{\infty} \frac{(-a\,\Delta t)^l}{(l+1)!} (u_{i+\frac{1}{2}}^n)_x^{(l)}. \tag{5.139}$$

If the infinite sum is replaced by the sum up to $(k-1)$ one gets a scheme of order k in time and space. It to calculate the spatial derivatives. These derivatives are calculated using Lagrangian interpolation polynomials. For uneven order schemes these reconstruction polynomials can be based on two different stencils. One obtains the final solution of the derivative by solving Riemann problems at the cell interfaces. A crucial point in the ADER approach is, that Riemann problems for the physical variables as well as their derivatives are solved. But the Riemann solver is always the same because a spatial derivative w of u satisfies the same transport equation:

$$w_t = -a\, w_x. \tag{5.140}$$

The ADER scheme can be implemented in the following way:

- Calculate the reconstruction of all derivatives at the cell interface using the Lagrangian Interpolation formula

- Solve the Riemann problems for all derivatives at the cell interfaces

- Calculate the approximate integral mean value $u_{i-\frac{1}{2}}^{ADER}$ according to (5.139)

- Calculate the fluxes at the cell interfaces and update the solution to the new time level

For CAA calculations one normally uses centred schemes which are then of even order. In the following sections we will compare the ADER scheme with some standard CAA schemes. The even order ADER schemes have some special properties which make the schemes much faster. As mentioned before the even order reconstruction polynomials do not have an upwind direction and thus we do not have to solve a Riemann problem.

The ENO/WENO reconstruction technique is described in detail by Sonar in the Chapter 3 of this book. By this reconstruction values at the grid interfaces are calculate from the approximations of the integral mean values. This is done by Newton interpolation in such a way that the stencil is moved to the direction where the function is smoother. The point values then are used in the flux calculation of the finite volume scheme. The time discretisation is performed by Runge-Kutta schemes. For any further details about the ENO- and WENO- approach see the Chapter 3 and the references cited there.

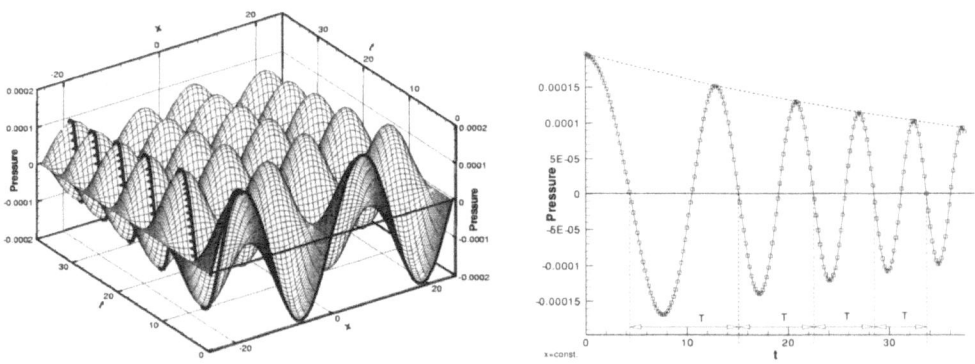

Figure 5.10   Space-time diagram of the standing wave and cross-section at $x = const$

In the following, numerical results of the different schemes mentioned above are compared for two test problems. The first test case is a single standing wave with periodic boundaries. The exact solution is given by $P(x,t) = P_{max} \cos(\omega t) \sin(kx)$ with the exact propagation speed by $C = \frac{\omega}{k}$. The Figure 5.10 shows on the left the typical space-time evolution of the wave approximated by some numerical method. On the right the numerical values are plotted over time for a fixed point in space. The numerical solution shows some damping of the amplitude as well as phase errors. The numerical propagation speed may be determined as $C_{num} = \frac{1}{Tk}$. Considering the amplitude error and the phase error separately, the influence of spatial discretisation and temporal integration may be investigated. The errors in amplitude and phase depend on several discretisation parameters. One is the number grid points per wavelength used in the numerical simulation, i. e., the spatial resolution. The errors will increase, of course, if the grid is coarsened. The minimal number of grid points to represent the wave is five or four.

In the following, numerical results are shown where we simulated the test problem with different numerical methods and varying the number of grid points per wavelength (PPW). The distance of propagation is kept fixed to the value 100 times the wavelength. The amplitude and phase error are determined and plotted in a diagram with the errors as functions of grid points per wavelength. The time steps for the numerical simulation of every method have been chosen maximal under the individual stability restriction of the schemes.

The first Figure 5.11 gives some sort of overview of the numerical results. Relatively large amplitude errors as well as phase errors are produced by the ENO-schemes and WENO-schemes. Of course, they are not designed for linear wave propagation problems. Their strength is the robust numerical approximation of strong gradients and nonlinear convection terms in connection with a robust flux calculation of a FV-scheme. In our

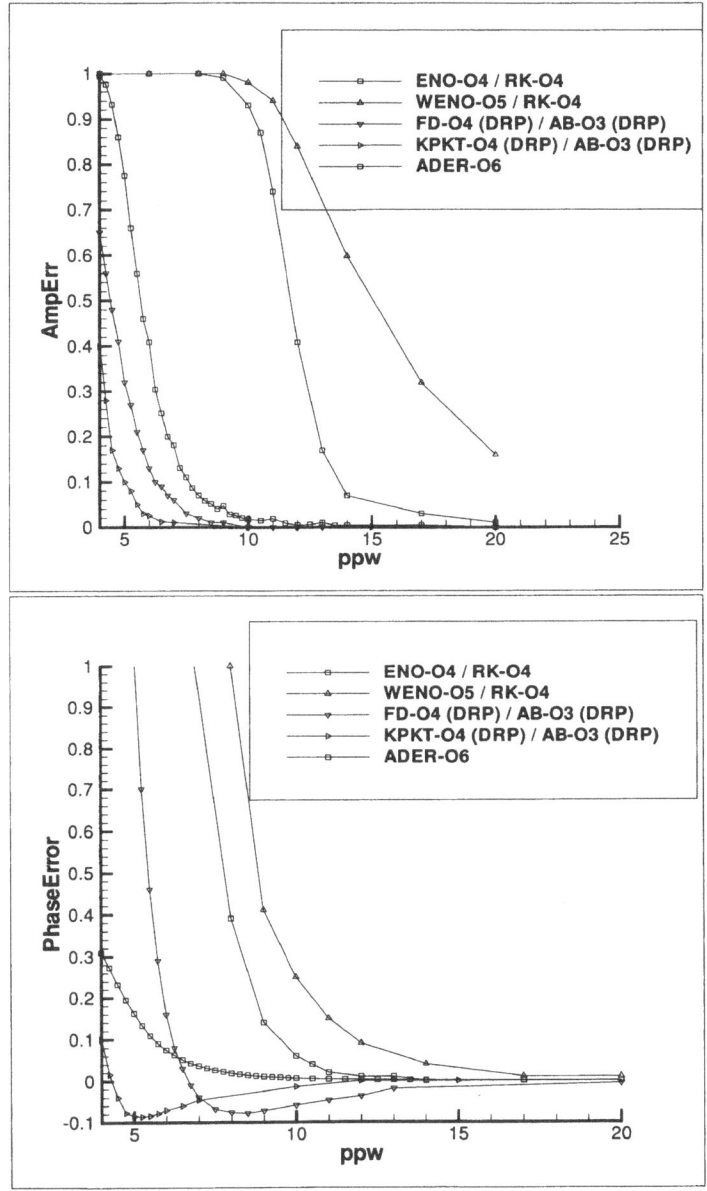

Figure 5.11 Comparison of amplitude error and phase error for ADER schemes, ENO schemes, DRP optimised FD schemes and Compact FD schemes.

calculation the time approximation is done by the Runge-Kutta scheme and the flux calculation by the usual upwind approximation. The interesting fact is the direct comparison of high resolution schemes from CFD with typical aeroacoustic solvers. To preserve the amplitude of the wave over this long distance the ENO-scheme of fourth order accuracy needs about 20 grid points per wavelength. The dispersion errors are almost zero with these discretisation parameters. The compact and the finite difference schemes give for

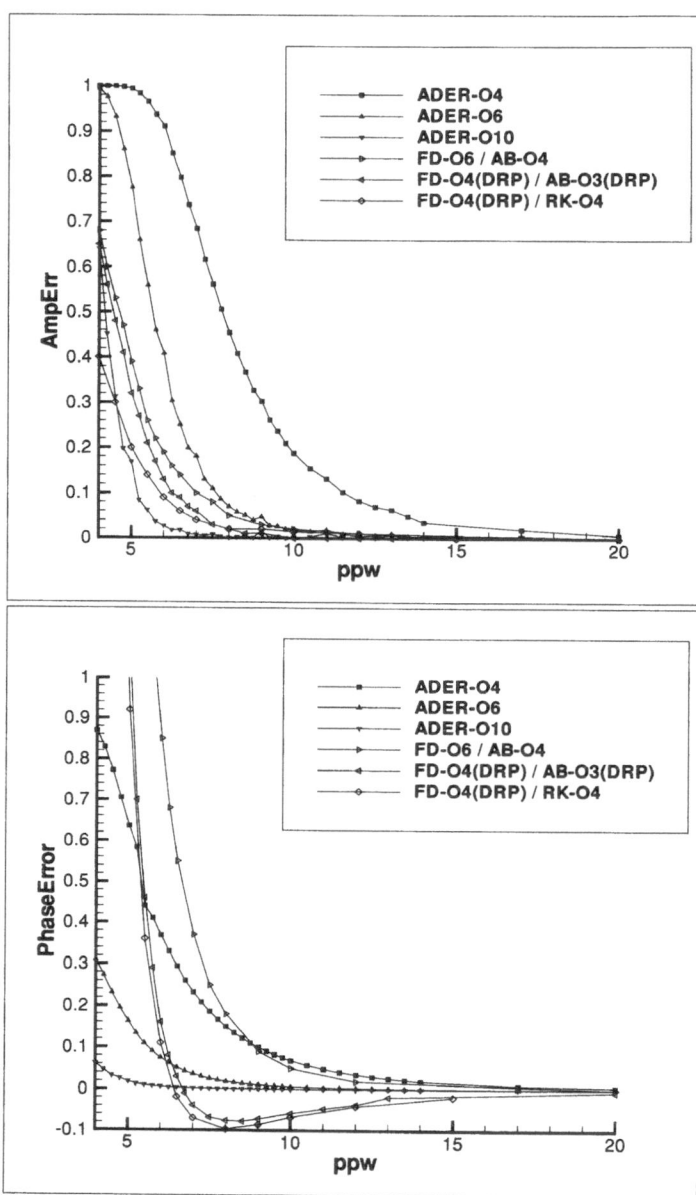

Figure 5.12 Comparison of amplitude error and phase error for ADER-schemes, standard FD-schemes and DRP optimised FD-schemes.

this problem much better results. To avoid any amplitude error, here at about 10 grid points per wavelength are necessary. The results of the ADER-scheme according to the dissipation errors lies in between the CFD and the CAA schemes. Looking at the dispersion errors the picture changes a little bit. Here, the ADER-scheme shows a smaller error with respect to the other methods.

We will go into more detail in the following comparisons. We compare the numerical

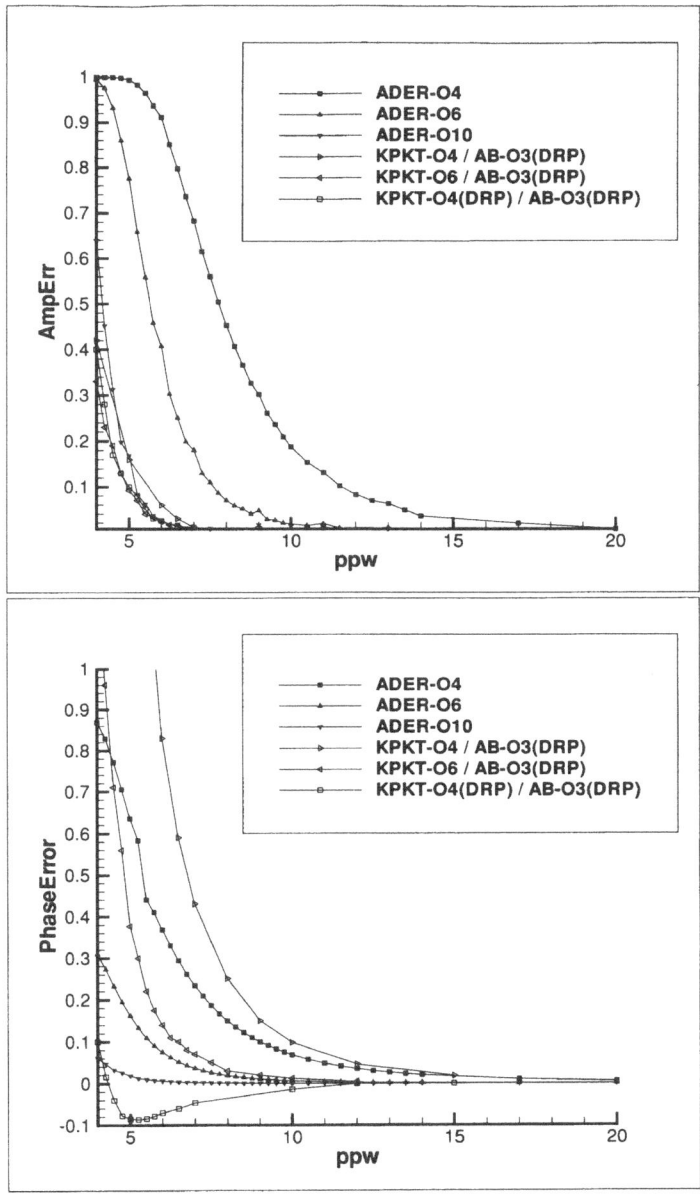

Figure 5.13 Comparison of amplitude error and phase error for ADER schemes, standard Compact FD schemes and DRP optimised Compact FD schemes.

results of the different classes of schemes in every figure with those of the ADER-schemes. In the first Figure 5.12 the amplitude error and the phase error versus the number of points per wavelength is shown for different combinations of central FD-approximations in space with Adams-Bashforth (AB) and Runge-Kutta (RK) approximations in time. The basic scheme is the standard central FD 6th order (FD-O6) with AB 4th order (AB-O4) in time. Using the same 7-point stencil in space, but optimizing the scheme in the DRP-

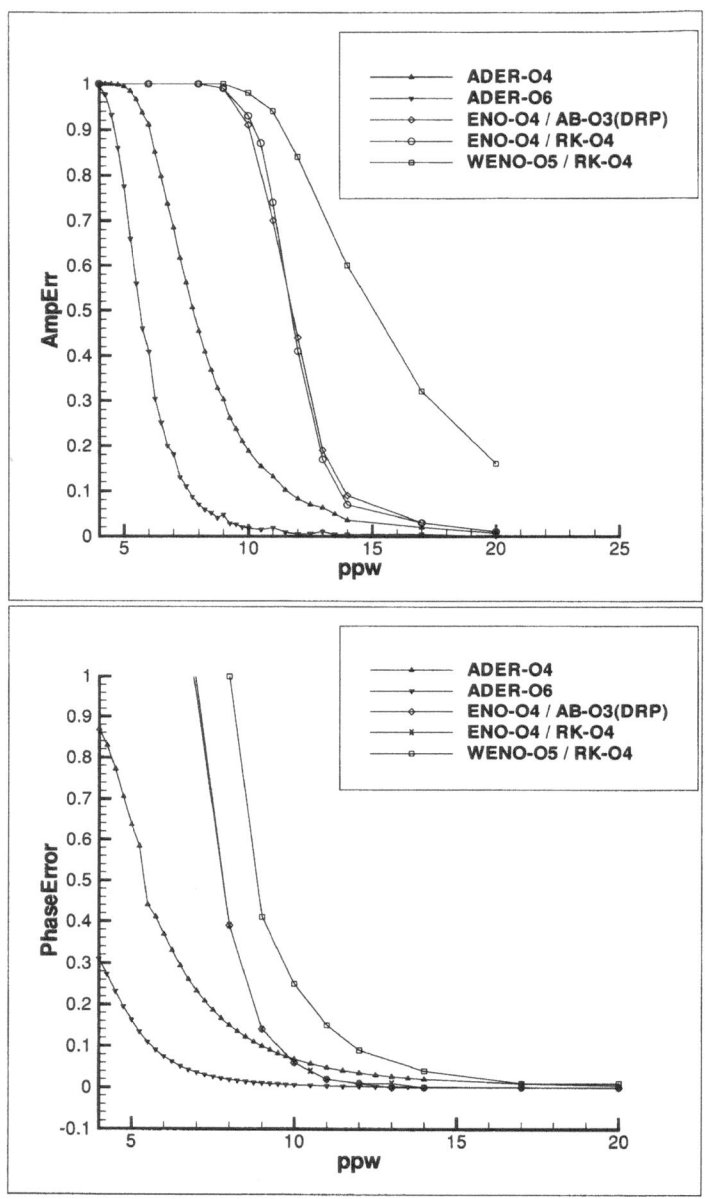

Figure 5.14 Comparison of amplitude error and phase error for ADER schemes and ENO/WENO schemes

sense, the scheme is denoted by FD-O4(DRP). In the same manner, the corresponding time approximation may be optimized using the DRP-technique, e.g., DRP-optimized AB-O3(DRP) scheme is obtained from AB-O4. The upper part of the Figure 5.12 with the dissipation error shows that for the same time integration scheme they nearly coincide, i.e. the amplitude error does not depend strongly on the spatial discretisation. On the other hand, the lower plot in Figure 5.12 shows some independence of the phase error from the time integration scheme. The DRP-optimized schemes show positive as well as negative phase errors, i.e., waves that propagate faster as well as slower than the exact waves. The discrete waves of the non-optimized schemes propagate only slower than the exact ones. The absolute value as well as the gradient of the phase error is smaller for the DRP-optimized schemes than for the non-optimized methods. The ADER-schemes possess a stronger numerical dissipation in general. We have to use a 10th order accurate method to get better results than the other 6th order or DRP-optimized schemes. In this case, we have according to the construction of the ADER-schemes the 10th order accuracy in space and time. The situation for the dispersion looks different. It turns out that even the 4th order ADER-scheme behaves very well and seems to be better than some of the FD-schemes. For the 6th and the 10th order schemes the dispersion errors become very small. This low dispersion behavior seems to be a quite interesting property of this class of schemes.

Additionally, compact FD-schemes have been considered for comparison (Figure 5.13). Those schemes approximate the derivative at point x_i not only from the values of the neighboring points, but also from the derivatives there. Thus, they contain spatially some implicitness which of course means more computational costs. Nevertheless, the number of points needed to resolve the acoustic wave can be reduced. Moreover, the absolute value of the phase error as well as the gradient of the curves are reduced compared to the standard schemes. Therefore, they are worth to be considered for aeroacoustic applications. Figure 5.12, 5.13 also show the results for the new ADER schemes. The ADER-O6 has very good properties concerning the phase error. Over a wide range of ppw the phase error as well as the gradient of the phase error is zero. The amplitude error is not as good as the error of the standard FD and the Compact FD schemes. One interesting point is, that the amplitude error starts to grow where the phase error starts to get significant. This means that the waves which are transported with a wrong wave speed are automatically damped. For demonstration purpose we have included results of an ADER-O10 calculation.

Figure 5.14 gives a comparison of different ENO/WENO schemes to the ADER scheme. The ENO/WENO schemes are high order schemes for flow calculations. In general, they show much bigger errors in both amplitude and phase than the FD schemes. Particularly within the short-wave area a relative error of 1 in the amplitude means that the wave is completely damped out after the regarded distance. The reason is that ENO and WENO schemes run into trouble, if there is more than one strong gradient within the possible stencil. This is the case, if the wavelength and the length of the stencil are approximately the same and therefore several of the steep gradients - with respect to the coarse grid - of the sine wave are situated within the interval. The adaptation of the stencil makes no longer sense, since there are no smoother areas anymore. Therefore a clipping value was introduced which forces the stencil to stay centered up to a certain point.

The test problem of the standing wave with a special wavelength tells of course not the whole truth. We choose another problem, where different wavelength plays a role. This is a Gaussian pulse that is transported over a long distance ($100 \cdot x_{range}$). A Gaussian pulse is a mixture of waves with different frequencies depending on the half-width of the pulse. In a narrow pulse the high frequencies are dominant, while in a wide pulse the maximal portion is at low frequencies. For this test case one cannot distinguish between dissipation and dispersion. It can be expected that a sharp pulse approximated on a coarse grid may produce big errors.

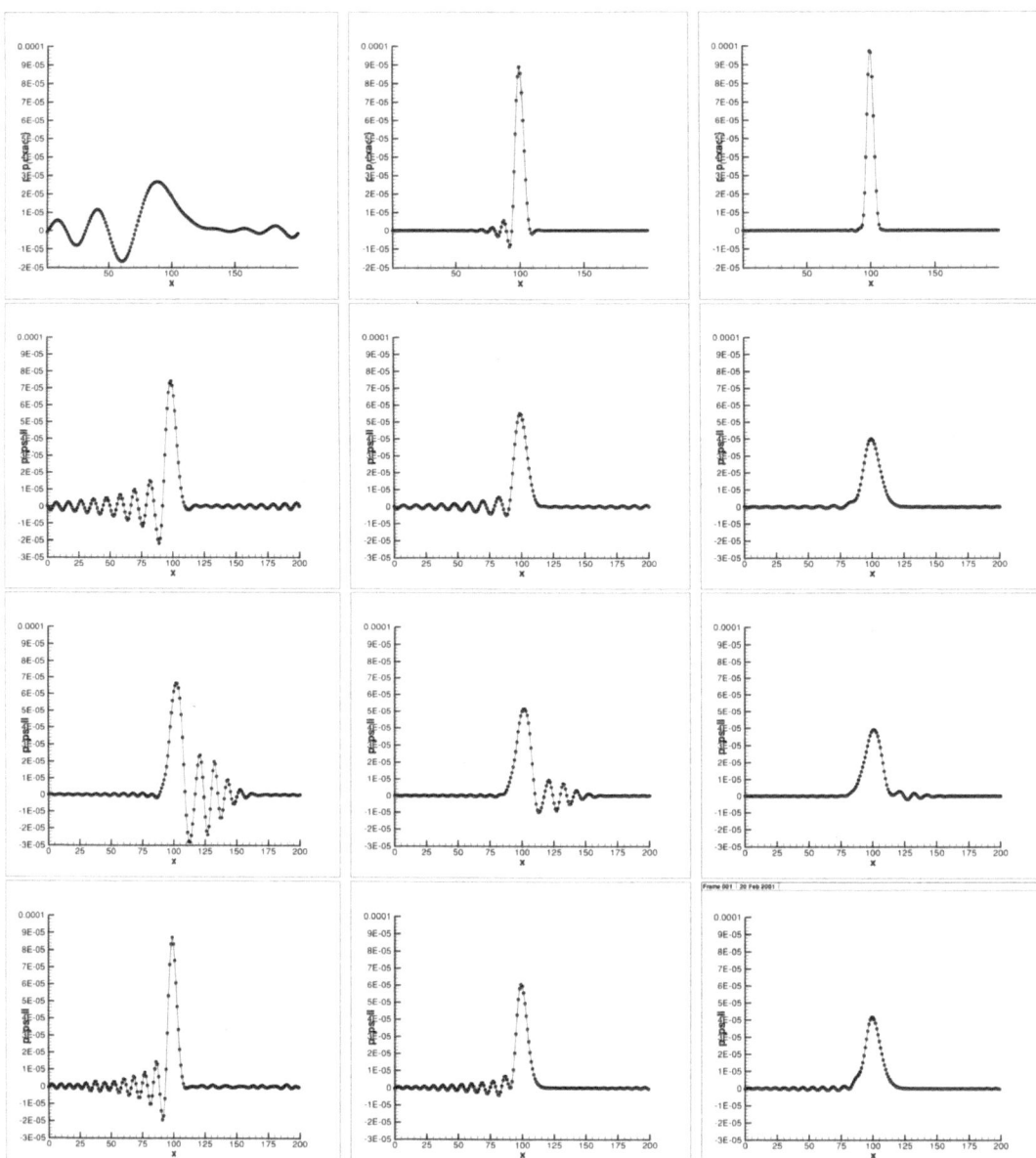

Figure 5.15   Top to bottom: Row 1: ADER O2, O6, O10, row 2: Standard FD scheme $visc =$ 0,10,25 and Adams-Bashforth O4, row 3: DRP scheme $visc = 0,10,25$ and Adams-Bashforth O3 DRP , row 4:Compact FD scheme $visc = 0,10,25$ and Adams-Bashforth O3 DRP. Initial condition is a Gaussian pulse with a half width $\sigma = 3 \cdot \Delta x$.

Figure 5.15 shows the result for a Gaussian pulse with the half width $\sigma = 3 \cdot \Delta x$. Periodic boundary conditions are prescribed. The time step size are chosen again for every scheme near the maximum within the range of stability. Note that almost all schemes considered generate some spurious oscillations. In the first row of Figure 5.15 the results of the ADER-schemes are plotted. While the 2nd order accurate scheme

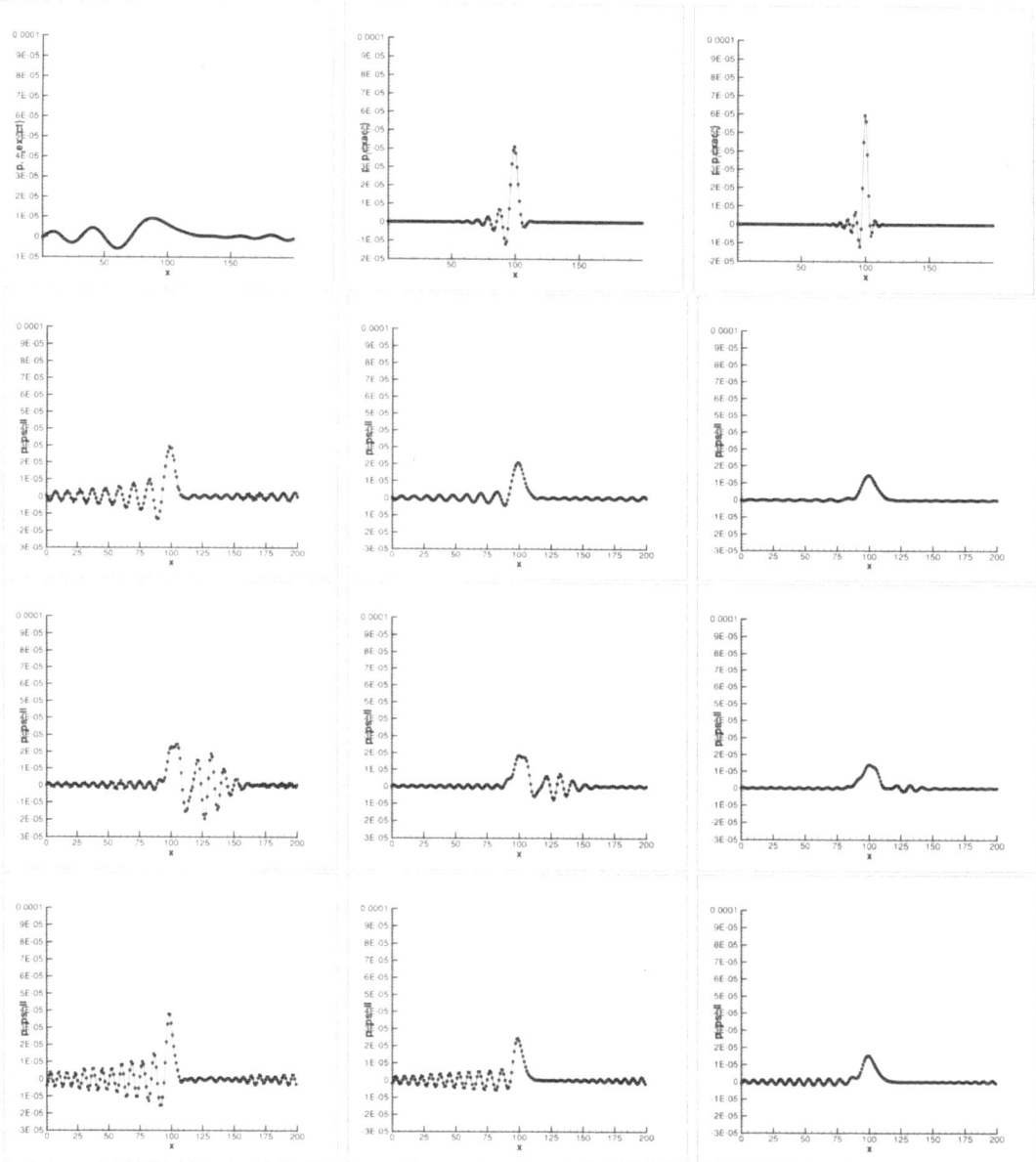

Figure 5.16 Top to bottom: row 1: ADER O2, O6, O10, row 2: Standard FD scheme $visc = 0,10,25$ and Adams-Bashforth 4, row 3: DRP scheme $visc = 0,10,25$ and Adams-Bashforth O3 DRP, row 4:Compact FD scheme $visc = 0,10,25$ and Adams-Bashforth O3 DRP. Initial condition is a Gaussian pulse with a half width $\sigma = 1 \cdot \Delta x$.

(left picture) introduces a lot of dissipation and dispersion so that the solution after that distance is nearly destroyed, the higher order schemes do a much better job. The 6th order scheme shows a very good resolution of the peak with some small spurious oscillations behind. The 10th order scheme gives an almost perfect result. With a very small wiggle behind the pulse it is resolved with the correct height within about 15 grid

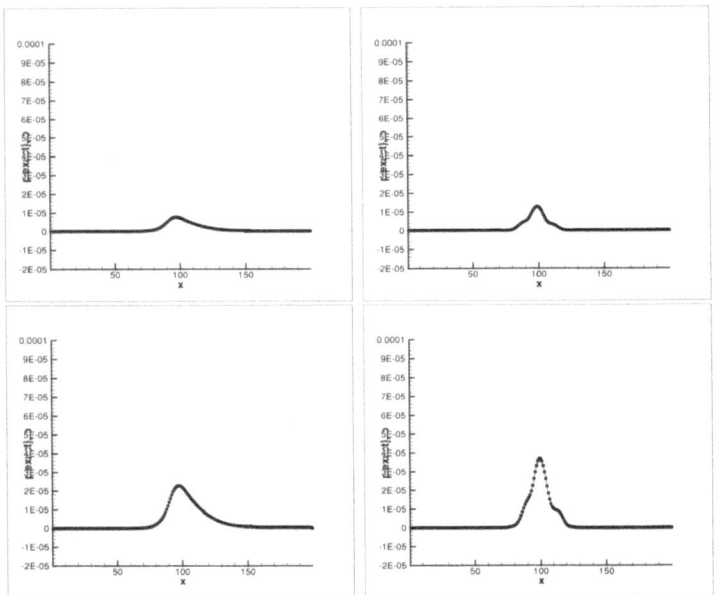

Figure 5.17   Left: ENO O4 and Runge-Kutta O4, Right: WENO O5 and Runge-Kutta O4. Initial condition is a Gaussian pulse with a half width $\sigma = 1 \cdot \Delta x$ for row 1 and with $\sigma = 3 \cdot \Delta x$ for row 2, periodic boundary conditions, distance travelled $x = 100 \cdot x_{range}$, $CFL = 0.9$.

points. The next row shows the results of the standard FD-scheme of 6th order accuracy in space combined with the 4th order Adams-Bashforth time approximation. The results show spurious wiggles behind the pulse. To suppress the wiggles a small amount of linear artificial dissipation is added. The two figures following to the right side show results with different amount of artificial dissipation. The wiggles nearly disappear, but the height of the peak decreases. The third row shows results of DRP-optimized schemes. It clearly can be seen that wiggles appear that are faster corresponding to negative dispersion errors as indicated in Figure 5.12. To reduce these oscillations some artificial viscosity has been introduced again. One can clearly see that this reduces the oscillations but the price is that the amplitude is also damped. In the fourth row results of a compact scheme is plotted. Again the oscillations may be suppressed by adding artificial dissipation.

We conclude for this test problem that the ADER-O6 scheme (figure in the middle at the top row) produces an excellent result concerning spurious oscillations and preservation of the amplitude without any artificial viscosity to be added.

In the Figure 5.16 we show numerical results for the same schemes but now for a Gaussian pulse which has only a half width of $\sigma = 1 \cdot \Delta x$. All other discretisation parameters are the same. This problem is more difficult to capture due to the stronger gradients and the larger number of different wavelength that are present. The results are in principle similar, but the effects become stronger. The spurious oscillations before or behind the pulse become stronger and falsify the physical solution. Adding artificial viscosity damps the spurious oscillations out, but strongly reduces the height of the pulse. There is no free lunch. Once again the the ADER-scheme produces the best results, but now the oscillations generated by the 6th order version of this class of schemes are stronger. Even the 10th order scheme has problems with that test case.

We will compare these numerical methods of CAA with the high order CFD-schemes. Results for an ENO- and a WENO-scheme are shown in the next figure. The stronger numerical dissipation is clearly visible, but there are no overshoots or undershoots in the vicinity of the pressure pulse. If these results are compared with the results of the DRP-schemes or the compact schemes with added artificial viscosity to suppress the spurious oscillations, then it turns out that they are similar. Hence, especially for short distances the propagation of an acoustic wave by a high resolution CFD-method may be performed without the loss of the physical behavior. The results are presented in figure 5.17. On the one hand one can see that the total variation is not increasing for the results of the ENO/WENO schemes. But on the other hand the maxima are strongly damped and the pulse has become much wider.

We did not give any information about the computational effort of the different schemes considered up to now. This is of course an important issue for practical calculations beside the accuracy. The Figure 5.18 shows the computational effort of the schemes for one-dimensional test case considered in this section.

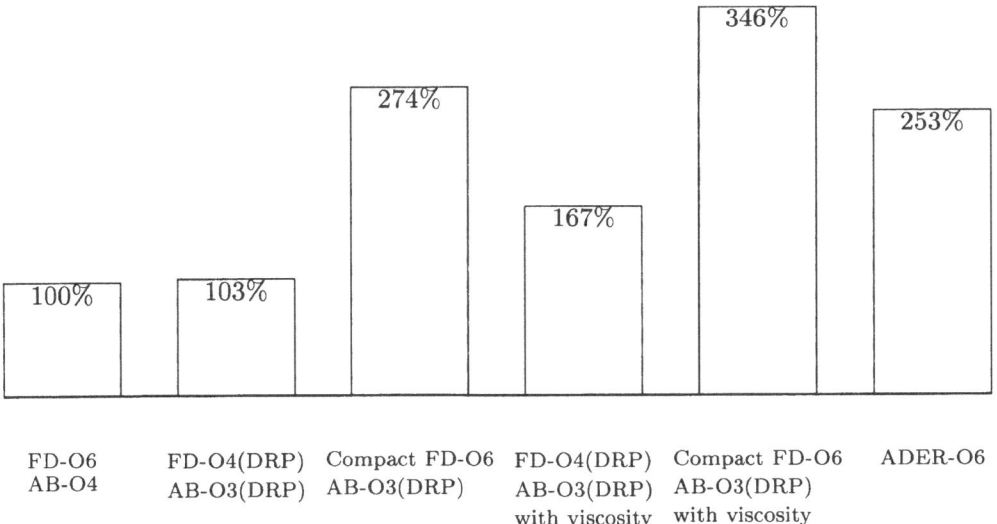

Figure 5.18   Comparison of computational time

The standard FD-O6 scheme with Adams-Bashforth O4 is defined to be the reference with 100%. The DRP scheme which is based on the same stencils needs a little bit more computational time because the restriction on the time step is a little bit stronger. To reduce oscillations we introduced artificial viscosity in the standard FD-schemes and the compact FD-schemes. These artificial viscosity terms increases strongly the computational time. The ADER-O6 scheme is very competitive with respect to the computer times. It is much faster than the compact FD-scheme and only a little bit slower than DRP-schemes with viscosity, while the quality of the solution is much better. Hence, we assume that the class of ADER-schemes may be good candidates for calculating wave propagation in the time domain. In two or three space dimensions the implementation of ADER-schemes become more and more complex and the situation is different to the one-dimensional case. But, two- and three-dimensional ADER-schemes are in development and first results have been obtained. We are optimistic to preserve the good properties of these schemes together with a comparable computational effort.

5.8   Conclusions

The aim in computational aeroacoustics is to close the gap between the calculation of acoustic wave propagation to the far field and the simulation of the fluid flow generating the sound. Hence, the sound generation as well as its propagation near the flow should be captured in the time domain in a similar way as the simulation of the flow is performed. In this chapter we give a survey of different approaches. According to the different interaction of acoustics and flow the strategy in the numerical simulation seems to be as follows.

If the flow is compact in the sense of Crow [9], i.e., the influence of the acoustic waves to the fluid flow can be neglected, then the flow simulation should be based on an incompressible flow simulation. A classical method in aeroacoustics is that the acoustic wave propagation is mathematically modelled by the linear wave equation as valid in the far field flow at rest and an acoustic analogy is used to determine the sound generation as source terms of the linear wave equation. Using a perturbation analysis around the incompressible solution, equations for the acoustic wave propagation may be obtained that seem to be more accurate in the vicinity of the flow. This has been proposed by Hardin and Pope in [45]. We follow this direction in this paper, but scale the different terms in the perturbation ansatz by powers of the Mach number as motivated by a low Mach number asymptotic expansion. For the pressure we propose

$$p = p^{(0)} + M^2 \left(p^{(2)} + p' \right), \tag{5.141}$$

where p' denotes the acoustic pressure fluctuations and $p^{(0)}$ and $p^{(2)}$ denote the background pressure and the hydrodynamic pressure, respectively. In this approach the acoustics and the fluid flow have the same length scale.

If the Mach number becomes very small then the acoustic waves have a much larger length scale than the fluid flow. In this case it seems to be more appropriate to couple sound generation and propagation by matching an inner and an outer asymptotic expansion as proposed by Ting and Miksis [46] and Slimon, Soteriou and Davis [44]. In the formal multiple scale asymptotic analysis, recently introduced by Klein [21], different space scales are considered in the fluid region. In this analysis, evolution equations of thermo-acoustic waves of large amplitudes are discovered. Here, the pressure terms are given by

$$p = p^{(0)} + Mp^{(1)} + M^2p^{(2)} + O(M^3), \tag{5.142}$$

where the acoustic pressure term $p^{(1)}$ only depends on the large scale variable $M\,x$. Within the numerical framework proposed, the asymptotic evolution equations for the large scale components are used in every time step to get an estimate of the time evolution of the long wavelength phenomena. This estimate is used to accelerate the convergence of the pressure correction equation in the multiple pressure variable method. If thermoacoustic effects are absent then a long wavelength component of $p^{(2)}$ may carry the acoustics. The large scale average of the perturbation equations gives linear acoustic equations with source terms. The source terms in the velocity equation consist of the average of the principal part of the Lighthill tensor and in the pressure equation of the average of the time derivative of the hydrodynamic pressure. It is assumed that the first term is the important part in a turbulent flow, while the other source term becomes important at obstacles.

Numerical methods for wave propagation in the time domain should preserve all important wave properties. Typical approximations used in CFD are not favorable for use in CAA. It is shown in this paper, that even high resolution CFD-schemes that are rarely applied in practical flow simulations due to their large amount of computational effort need a lot of grid points to resolve the wave propagation as well as required. They inherently posses too much numerical dissipation. This enables them to capture large local gradients in the flow, which is not necessary in numerical wave propagation. High order finite difference or compact schemes do this job better. By the DRP-technique proposed by Tam [45] the formal order of accuracy is reduced to get a degree of freedom to optimize the approximation of wave properties. Numerical results for wave packages indicate that these methods may generate spurious oscillations which may increase in propagating over long distances. Hence, these oscillations have to be artificially damped out increasing again the numerical dissipation. We think, that the new approach of the ADER-schemes as proposed by Toro [47] is quite interesting in this context. Their computational effort is higher, but their numerical dispersion is significant lower. We did not consider in this paper any finite element method or spectral method.

If the fluid flow is not compact, then interaction between flow and acoustic waves occur and have to be modelled numerically. Here, it seems to be the only general way to base the direct numerical simulation on the compressible equations. According to the problems addressed in our discussion about the numerical approximation of wave propagation this will be a difficult task. We have to assume that the fluid solver, even designed to have a very good accuracy, can capture the propagation of the acoustic fluctuations in an appropriate way over a small distance only. Hence, the strategy must be the following. The computational domain is decomposed into different parts. The solver based on the compressible Navier-Stokes equations is only applied to the very inner region where nonlinear effects play an important role. Into the direction of the far field, next a region is connected where the locally linearized Euler equations are a good mathematical model to resolve the physical situation. The next lower level in the mathematical modelling seems to be the linear acoustics. In this heterogenous domain decomposition the acoustic fluctuations are exchanged at the boundaries only. This strategy, of course, will work only, if appropriate boundary and transmission conditions can be found which do not generate artificial noise that is larger than the physical one.

In this paper we only considered numerical test problems where the noise is generated by the interaction of two vortices. Here, sound of a special frequency is generated. If noise will be calculated as generated by turbulent flow, then the techniques considered can be applied only in the case that a direct numerical simulation (DNS) is used to simulate the flow field. Assuming that the noise generated by large eddies has larger energies than the small scale vortices, a large eddy simulation (LES) of the turbulent flow may also be successful to capture the physical problem well. Both techniques for simulating turbulent flow are still in their beginning for applications to real problems. Hence, the direct simulation of sound from a turbulent flow is still one of the big challenges of high performance computing in the future.

Acknowledgement

This paper is based on the work of a whole group of people. Especially, the author would like to thank R. Alber, R. Fortenbach, M. Ratzel, S. Roller, and T. Schwartzkopff for their support and contribution to this work. The author is also grateful to R. Klein and F. Guidati giving valuable suggestions and discussions. This work was supported by Deutsche Forschungsgemeinschaft (DFG).

Bibliography

[1] J. BELL, P. COLELLA, AND H. GLAZ, *A second-order projection method for incompressible navier-stokes equations*, J. Comput. Phys., 85 (1989), pp. 257–283.

[2] J. B. BELL AND D. L. MARCUS, *A second-order projection method for variable-density flows*, J. Comput. Phys., 101 (1992), pp. 334–348.

[3] H. BIJL AND P. WESSELING, *A unified method for computing incompressible and compressible flows in boundary-fitted Coordinates*, J. Comput Phys., 141 (1998), pp. 153–173.

[4] N. N. BOGOLIUBOV AND Y. A. MITROPOLSKY, *Asymptotic Methods in the Theory of Non-linear Oscillations*, Gordon & Breach Sience Publishers, Inc., 1961.

[5] A. J. CHORIN, *A numerical method for solving incompressible viscous flow problems*, J. Comput. Phys., 2 (1967), pp. 12–26.

[6] ——, *Numerical solution of the navier-stokes equations*, Math. Comp., 22 (1968), pp. 745–762.

[7] P. COLELLA AND K. PAO, *A projection method for low speed flows*, J. Comput. Phys, 149 (1999), pp. 245–269.

[8] D. CRIGHTON, *Computational aeroacoustics for low Mach number flows*, in Computational Aeroacoustics, J. Hardin and M. Hussaini, eds., Springer Verlag, 1993, pp. 50–68.

[9] S. CROW, *Aerodynamic sound emission as a singular pertubation problem*, Studies in applied mathematics, XLIX (1970), pp. 21–43.

[10] D. L. DARMOFAL AND B. VAN LEER, *Local preconditioning: Manipulating mother nature to fool father time*, in Computing the Future II: Advances and Prospects in Computational Aerodynamics, M. Hafez and D. Caughey, eds., John Wiley and Sons, 1998, p. 30.

[11] I. DEMIRDZIC, Z. LILEK, AND M. PERIC, *A collocated finite volume method for predicting flows at all speeds*, Int. J. Numer. Meth. Fluids, 16 (1993), pp. 1029–1042.

[12] A. FEDORCHENKO, *On some fundamental flaws in present aeroacoustic theory*, Journal of Sound and Vibration, 232 (2000), pp. 719–782.

[13] J. FERZIGER AND M. PERIC, *A Computational Method for Fluid Dynamics*, source code available from ftp.springer.de, 1996, Springer Verlag.

[14] J. FFOWCS WILLIAMS AND D. HAWKINGS, *Sound generation by turbulence and surfaces in arbitrary motion*, Phil. Trans. Roy. Soc., 264 (1969).

[15] M. E. GOLDSTEIN, *Aeroacoustics*, McCraw-Hill, New-York, 1976.

[16] P. M. GRESHO AND R. L. SANI, *On pressure boundary conditions for the incompressible navier-stokes equations*, Int. J. Num. Meth. Fluids, 7 (1987), pp. 1111–1145.

[17] H. GUILLARD AND C. VIOZAT, *On the behavior of upwind schemes in the low Mach number limit*, Computers and Fluids, 28 (1999), pp. 63–86.

[18] J. HARDIN AND D. POPE, *An acoustic/viscous splittting technique for computational aeroacoustics*, Theoretical and Computational Fluid Dynamics, 6 (1994), pp. 323–340.

[19] K. C. KARKI AND S. V. PATANKAR, *Pressure based calculation procedure for viscous flows at all speeds in arbitrary configurations*, AIAA Journal, 27 (1989), pp. 1167–1174.

[20] S. KLAINERMAN AND A. MAJDA, *Compressible and incompressible fluids*, Comm. Pure Appl. Math, XXXV (1982), pp. 629–651.

[21] R. KLEIN, *Semi-implicit extension of a godunov-type scheme based on low Mach number asymptotics I: One dimensional flow*, J. Comput. Phys., 121 (1995), pp. 213–237.

[22] R. KLEIN AND C.-D. MUNZ, *The multiple pressure variable (MPV) approach for the numerical approximation of weakly compressible fluid flow*, in Proceedings of "Numerical Modelling in Continuum Mechanics", M. Feistauer, R. Rannacher, and K. Kozel, eds., Charles University Prag, 1995, pp. 123–133.

[23] S. LELE, *Compact finite difference schemes with spectral like resolution*, J. Comput. Phys., 103 (1992), pp. 16–42.

[24] M. LIGHTHILL, *On sound generated aerodynamically - 1. General theory*, Proc. Roy. Soc. London, 211 (1952), pp. 564–587.

[25] ——, *On sound generated aerodynamically - 2. Turbulence as a source of sound*, Proc. Roy. Soc. London, 222 (1954), pp. 1–32.

[26] G. M. LILLEY, *On the noise from jets*, Noise Mechanisms, AGARD-CP-131 (1974), pp. 13.1–13.12.

[27] A. MAJDA AND J. SETHIAN, *The derivation and numerical solution of the equations for zero Mach number combustion*, Combust. Sci. and Tech., 42 (1985), pp. 185–205.

[28] A. MEISTER. Asymptotic Single and Multiple Scale Expansions in the Low Mach Number Limit. *SIAM J. Appl. Math.*, Vol. 60 (1): 256–271, 1999.

[29] A. MEISTER. Analyse und Anwendung Asymptotik-basierter numerischer Verfahren zur Simulation reibungsbehafteter Strömungen in allen Mach-Zahlbereichen, 2001. Habilitationsschrift, Universität Hamburg.

[30] R. MILLINGTON, E. TORO, AND L. NEJAD, *Arbitrary high order methods for conservation laws I: The one dimensional scalar case*, tech. rep., Manchester Metropolitan University, Department of Computing and Mathematics, June 1999.

[31] B. E. MITCHELL, S. K. LELE, AND P. MOIN, *Direct computation of the sound from a compressible co-rotating vortex pair*, J. Fluid Mech., 285 (1995), pp. 181–202.

[32] J. A. MORRISON, *Comparison of the modified method of averaging and the two variable expansion procedure*, SIAM Review, 8 (1966), pp. 66–85.

[33] C. MUNZ, S. ROLLER, K. GERATZ, AND R. KLEIN, *The extension of incompressible flow solvers to the weakly compressible regime, to appear in* Int. J. Numer. Meth. in Fluids, (2001).

[34] H. PAILLÈRE, S. CLERC, C. VIOZAT, I. TOUMI, AND J.-P. MAGNAUD, *Numerical methods for low Mach number thermal-hydraulic flows*, in Computational Fluid Dynamics '98, K. Papailiou, D. Tsahalis, J. Périaux, and D. Knörzer, eds., vol. 2, ECCOMAS, Wiley, 1998, pp. 80–89.

[35] S. PATANKAR, *Numerical Heat Transfer and Fluid Flow*, McGraw Hill, New York, 1980.

[36] S. V. PATANKAR AND D. B. SPALDING, *A calculation procedure for heat, mass, and momentum transfer in three-dimensional parabolic flow*, Int. J. Heat Mass Transfer, 15 (1972), pp. 1787–1806.

[37] R. PEYRET AND T. TAYLOR, *Computational Methods for Fluid Flow*, Springer, Berlin, 1985.

[38] A. POWELL, *Theory of vortex sound*, The journal of the Acoustical Society of America, 36 (1964), pp. 177–195.

[39] C. RHIE, *Pressure based Navier–Stokes solver using the multi grid method*, AIAA J., 27 (1989), pp. 1017–1018.

[40] H. RIBNER, *Aerodynamic sound from fluid dilatations*, Advances in Applied Mechanics, 8 (1964), pp. 103–182.

[41] S. ROLLER AND C.-D. MUNZ, *A low Mach number scheme based on multi-scale asymptotics*, Comput Visual Sci, 3 (2000), pp. 85–91.

[42] T. SCHWARTZKOPFF, C. MUNZ, E. TORO, AND R. MILLINGTON, *Ther ader approach in 2d for a system of linear hyperbolic pdes*, in Proceedings of GAMM workshop on 'Discrete Modelling and Discrete Algorithms in Continuum Mechanics', T. Sonar, ed., Berlin, 2001, Logos Verlag.

[43] W. SHYY, *Elements of pressure-based computational algorithms for complex fluid flow and heat transfer*, Advances in Heat Transfer, 24 (1994), pp. 191 – 275.

[44] S. SLIMON, M. SOTERIOU, AND D. DAVIS, *Development of computational aeroacoustics equations for subsonic flows using a mach number expansion approach*, J. Comput. Phys., 159 (2000), pp. 377–406.

[45] C. K. W. TAM AND J. C. WEBB, *Dispersion-relation-preserving finite difference schemes for computational acoustics*, J. Comput. Phys., 107 (1993), pp. 262–281.

[46] L. TING AND J. MIKSIS, *On vortical flow and sound generation*, Society for Industrial and Applied Mathematics, 50 (1990), pp. 521–536.

[47] E. TORO AND R. MILLINGTON, ADER: *High–order non–oscillatory advection schemes*, in Proceedings of the 8th International Conference on Nonlinear Hyperbolic Problems, February 2000. Preprint.

[48] E. TURKEL, A. FITERMAN, AND B. VAN LEER, *Preconditioning and the limit of the compressible to the incompressible flow equations for finite difference schemes*, in Computing the Future: Advances and Prospects for Computational Aerodynamics, M. Hafez and D. Caughey, eds., John Wiley and Sons, 1994, pp. 215 – 234.

[49] M. VAN DYKE, *Perturbation Methods in Fluid Mechanics*, Annotated Ed., Parabolic Pess, 1975.

[50] J. VAN KAN, *A second-order accurate pressure-correction scheme for viscous incompressible flow*, SIAM J. Sci. Stat. Comput., 7 (1986), pp. 870 – 891.

[51] B. VAN LEER, *Towards the ultimate conservative difference scheme V. A second-order sequel to godunov's method*, Journal of Computational Physics, 32 (1979), pp. 101–136.

[52] C. VIOZAT, *Implicit Upwind Schemes for Low Mach Number Compressible Flows*, Tech. Rep. 3084, Institut National de Recherche en Informatique et en automatique (INRIA), January 1997.

[53] S. WAGNER, R. BAREISS, AND G. GUIDATI, *Wind Turbine Noise*, Springer-Verlag, 1996.

[54] J. WEISS AND W. SMITH, *Preconditioning applied to variable and constant density flows*, AIAA J., 33 (1995), pp. 2050–2057.